Mischievous
Creatures

ALSO BY CATHERINE MCNEUR

Taming Manhattan:
Environmental Battles in the Antebellum City

Mischievous Creatures

The FORGOTTEN SISTERS
Who TRANSFORMED
EARLY AMERICAN
SCIENCE

CATHERINE McNEUR

BASIC BOOKS

New York

Basic Books
Hachette Book Group
1290 Avenue of the Americas, New York, NY 10104
www.basicbooks.com

Printed in the United States of America

First Edition: October 2023

Published by Basic Books, an imprint of Hachette Book Group, Inc. The Basic Books name and logo is a trademark of the Hachette Book Group.

Print book interior design by Jeff Williams

Library of Congress Cataloging-in-Publication Data

Names: McNeur, Catherine, author.

Title: Mischievous creatures : the forgotten sisters who transformed early American science / Catherine McNeur.

Other titles: Forgotten sisters who transformed early American science

Description: First edition. | New York : Basic Books, 2023. | Includes bibliographical references and index.

Identifiers: LCCN 2023010253 | ISBN 9781541674172 (hardcover) | ISBN 9781541674189 (ebook)

Subjects: LCSH: Morris, Margaretta Hare, 1797–1867. | Morris, Elizabeth Carrington, 1795–1865. | Botanists—Pennsylvania—Philadelphia—Biography. | Entomologists—Pennsylvania—Philadelphia—Biography. | Women scientists—Pennsylvania—Philadelphia—Biography. | Women scientists—Pennsylvania—Philadelphia—History—19th Century. | Scientists—United States—History—19th Century. | Germantown (Philadelphia, Pa.)—Biography.

Classification: LCC Q141 .M375 2023 | DDC 509.2/273—dc23/eng/20230530

LC record available at https://lccn.loc.gov/2023010253

ISBNs: 9781541674172 (hardcover), 9781541674189 (ebook)

LSC-C

Printing 1, 2023

For my parents
John and Clare McNeur
and
for my children
Carter and Julia Pellegrini

CONTENTS

A Note on Names

Throughout the book, I often refer to Margaretta Hare Morris and Elizabeth Carrington Morris by their first names, since referring to both as "Morris" would get confusing. I am sensitive to the fact that in the past historians referred to women by their first names and men by their surnames, a habit that made female subjects feel more familiar than professional and authoritative. That is not my intention, though I do hope these remarkable women will become all of the above to you: familiar, professional, and authoritative.

Introduction

SISTER SCIENTISTS

Elizabeth Carrington Morris (1795–1865) and Margaretta Hare Morris (1797–1867) were sisters and scientists. When they sat for these paired portraits, they chose brooches that signaled their shared passion for natural history. *(Special Collections, University of Delaware Library)*

IN THE MIDDLE OF APRIL 1851, WHEN THE RAIN HAD PAUSED long enough to venture outside, Elizabeth Carrington Morris knelt beneath a blossoming pear tree to weed. The ground, thick and wet, was rich with "dockens," as she called them, a European plant related to buckwheat that thrived in neglected spaces. Her garden in Germantown, just outside of Philadelphia, was far from neglected, but Elizabeth had been busy that spring tending to sick relatives, and in that time the dockens had flourished, threatening to overrun her carefully cultivated refuge. Fifty-five years old and a well-established botanist, Elizabeth had put on the rough dress she reserved for gardening, long leather gloves, and black rubber boots, determined to restore order in her flower bed. Some people ate the leaves of certain species like curly dock (*Rumex crispus*), raw in salads or cooked down in stews, but Elizabeth considered them to be an unsightly nuisance. As she dug around the base of the tree, uprooting as much of the spindly plants as she could with her trowel and tossing them in a nearby basket, she caught sight of cicada larvae wriggling around in the upturned soil. While other gardeners might have leapt back, squeamish at uncovering this glistening "bevy of Locusts," Elizabeth sprang to her feet in delight, calling out for her sister to grab their specimen jars.[1]

For years, Elizabeth had been immersed in all things cicada thanks to her younger sister, Margaretta Hare Morris, an entomologist who specialized in agricultural pests. The two women lived together, hiked together, and debated new scientific theories together. They spent long hours in their garden puzzling over the insects and diseases that troubled the plants Elizabeth cultivated. Entomology and botany intersected quite literally in the sisters' garden. Two years Elizabeth's junior, Margaretta had been investigating seventeen-year cicadas and their impact on orchards since she first discovered them attached to the roots of her suffering fruit trees in 1846. While naturalists had established the creatures' seventeen-year life cycle in the eighteenth century, no one knew for certain how the larvae survived underground for so long, though most presumed their presence was benign. Finding them sucking from the roots of her trees, Margaretta had concluded that the creatures subsisted "unseen, and unsuspected, draining the life blood from our most valued fruit trees and timber."

Some trees could handle the ravenous, if lethargic, grubs, but the Morris sisters' apple and pear trees had withered and failed to produce fruit. Margaretta had spent the last five years sending specimens and reports on her findings to the country's most prominent scientific associations and agricultural journals, while monitoring cicadas under bell jars filled with dirt and severed roots.[2]

In the decade leading up to her cicada discoveries, Margaretta had faced skepticism from her male peers who publicly belittled her in agricultural and scientific journals. Before presenting her latest theory to the scientific community, she knew she would need to be more strategic. Starting in 1846, the sisters invited male scientists to visit their garden and witness Margaretta's cicada discoveries for themselves. While Margaretta and Elizabeth chatted with their guests, their gardener dug up a tangle of roots from a sickly tree to reveal cicadas lined up with their long, sharp proboscises deeply embedded. These reenactments worked. With each visit, Margaretta gained another ally ready to endorse her findings, and as she amassed support, respect for her work in both scientific and agricultural circles grew. By 1850, she had established herself as one of the country's most notable agricultural entomologists and, alongside the astronomer Maria Mitchell, was one of the first two women elected to the American Association for the Advancement of Science (AAAS).[3]

While Margaretta was developing her career by maneuvering through the ranks of male entomologists, her older sister was content working behind the scenes to collect and catalog specimens. The reason Elizabeth was so excited to find cicadas emerging from the weeds in April 1851 was not because Margaretta needed more evidence—if anything, her sister had too much. Instead, Elizabeth was eager to share them with Harvard zoology professor Louis Agassiz, who, she bragged to a friend, had told her "he wanted enough to supply *all* the cabinets in Europe." Margaretta had already given Agassiz specimens for Harvard's collection, but this set would make it possible for him to gift the North American insects to other zoologists. Elizabeth counted, weighed, and bottled three hundred to mail to him, offering several hundred more to her elated chickens. She had been doing this kind of thing with botanists for years, supplying them with rare plant specimens that they could share with others. Gifts like these and other forms of support made Elizabeth indispensable to her peers. They fueled an exchange of knowledge and helped other scientists expand both their collections and their influence. Her behind-the-scenes work was integral for the advancement of the natural sciences and involved significant training and expertise in local environments, but very little credit or acclaim.[4]

The jar of cicadas made its way in a small wooden crate from the Morris sisters' Pennsylvania garden to Agassiz's office in a converted bathhouse in Cambridge, Massachusetts. Agassiz and his student assistants then either wedged the jar onto his cramped shelves filled with preserved fish, snakes, and other creatures floating in alcohol, or repackaged them and sent the specimens across the Atlantic to the private and public natural history cabinets of his friends and acquaintances. The cicadas were passengers on a journey American science was only just beginning. Universities that had previously only offered science classes as part of their medical programs had begun to hire more botanists, chemists, geologists, and zoologists, expanding their offerings and opening scientific schools. For decades, women's seminaries had a rich assortment of science classes, but as it became possible to find jobs in these fields, schools for men followed suit.

In the nineteenth century, as the sciences were gaining a foothold in the United States, men of science were beginning to tell stories about the origin and growth of their fields. Like many other women, Margaretta and Elizabeth Morris, no matter their contributions or accomplishments, were left out of these tales. This book is about restoring these sisters' lives to the history of science while making sense of how power dynamics have impacted environmental knowledge and the stories we tell.

ONCE THE CICADAS EMERGED ON THEIR OWN THROUGHOUT the mid-Atlantic states a few weeks later, Americans of all ages, classes, races, and genders were swept up in the entomological event of the season. The loud drone that became the soundtrack for the summer, the abandoned nymphal shells left clinging to tree bark, and the clumsy insects flying into pedestrians inspired newspaper articles, poems, diary entries, and constant commentary. Curious enthusiasts stood shoulder to shoulder with seasoned naturalists, collecting and trading their specimens.

The emerging cicadas were the latest novelty, but the public's preoccupation with plant names, rock formations, chemical reactions, and constellations had been steadily on the rise throughout the nineteenth century. This was a golden age for popular science. Some thought they would unlock the mysteries and wonders of the natural world, while others believed science would help them get closer to God. For many, the combination was what made science irresistible. Regardless of its precise appeal, there was a collective societal desire to better know the environment. Women and men crowded lecture halls to hear scientists speak as a form of evening entertainment. Magazine and newspaper editors incorporated popular science articles into their issues to keep up with their readers' insatiable appetite for more knowledge. Authors wrote books aimed at mothers looking to structure their children's curiosity about the world. Primary schools, particularly those for girls, advertised their advanced science classes as a way to attract new students. While not everyone could afford a microscope, they could flatten the frond of a lady fern between the pages of a book or collect apple moth caterpillars in a box to watch their transformation.

Margaretta and Elizabeth were equal parts avid consumers of the nineteenth century's popular science culture and qualified experts with the ability to influence its development. Like other engaged members of the public, they took the classes, they attended the lectures, they purchased the books. They were conversant in the Victorian era's floriography, or the "language of flowers," where specific blooms carried hidden messages for recipients, such as oleander symbolizing "caution," and chamomile "strength in adversity." Like many of their contemporaries, they also filled their albums with paintings of butterflies and plants. What made them exceptional, though, was that they were connected to the burgeoning professional circles that allowed them to become "insiders." Margaretta and Elizabeth were trained experts who hobnobbed with other scientific luminaries. They traversed both worlds, and that was critical in this transformational moment when professionalization meant that scientists were becoming increasingly distant from the popular culture that had helped them rise to prominence. The Morris sisters were popular science writers who encouraged others to learn about their environments and perform experiments whenever possible. They not only collected specimens for their own use but also donated their most interesting finds to natural history museums, like Margaretta's taxidermied *Dasyprocta niger*, a South American rodent, and Elizabeth's *Actinia marginata*, a sea anemone she found off the coast of Rhode Island. Elizabeth knew enough about seaweed to differentiate what might be known from what might be new before sending dried specimens to an algae specialist in Dublin. Margaretta went further still, jumping into major entomological debates about threats to crops, whether that involved the wheat flies that exacerbated the Panic of 1837, or even the mysterious source of the potato blight that devasted Ireland while simultaneously threatening potato yields in the United States.[5]

Margaretta worked less common images into her personal album, like this aphid-eaten iris with a triumphant damselfly perched on a bud, having devoured the tiny pests. *(Collection of S. C. Doak)*

By joining contentious public debates and connecting with scientists throughout the country, Margaretta's discoveries became well-known, which was a highly unusual feat. Charles Darwin, a rising naturalist in England, was in the midst of drafting *On the Origin of Species* when he heard that Margaretta witnessed water beetles transporting fish eggs from lake to lake in Pennsylvania. This discovery had the potential to shape his discussion of species distribution in his book, but he first had to determine if he could trust her observations. While Darwin ultimately decided he did not believe the "lady in N. America," modern biologists have confirmed that Margaretta's observations were accurate. Still, Darwin was far from alone in dismissing findings on such prejudicial grounds.[6]

So much of this story is about power and how it shapes our knowledge of the environment. By not trusting someone because of their gender, race, age, or class, we lose crucial information. We erase their contributions and discount their observations. Margaretta's and Elizabeth's respective lives and work illuminate the frustrating hurdles they faced as women, despite the privileges afforded to them by their wealth and race. The sisters developed strategies and methods to counter the distrust, the exclusion, even the attacks. Despite all of that, they have been erased from the historical narrative. They have been forgotten.

By recovering the Morris sisters' story, we have an opportunity to read the history of professional science through their experiences. Scientific culture and practices changed dramatically over the course of the nineteenth century, and even just during their lifetimes. Women scientists and the work they did were integral to that professionalization, but the process of drawing lines around what qualified as science also ultimately relegated them to the periphery. This is why we keep rediscovering hidden figures in the history of science. Over the course of the Morris sisters' lifetimes, they witnessed the lives and deaths of those considered the founding fathers of their fields—many of whom had tutored them. When these men died, their friends and colleagues wrote a first draft of the history of botany and entomology, centering their friends in the narratives while sidelining others. Adding women like the Morris sisters back into the story of American science combats the kind of marginalization that led to their erasure in the first place.

Margaretta and Elizabeth were far from the only women shaping the sciences in early America. Jane Colden (1724–1766), a colonial woman in their grandmothers' generation, learned botany from her father but soon surpassed his skills and would eventually identify and describe 326 plants in the Hudson Valley of New York, several of which were previously unnamed. Colden exchanged data and plant descriptions with botanists in England, becoming an invaluable overseas correspondent. In Margaretta and Elizabeth's own generation, the sisters Emma Hart Willard (1787–1870) and Almira Hart Lincoln Phelps (1793–1884) were advocates for women's education, running seminaries and writing best-selling science textbooks for women and children. Similarly, Sarah Mapps Douglass (1806–1882), a talented botanical illustrator and Black abolitionist, taught science and art to her students at the Free African School for Girls and later the Institute for Colored Youth in Philadelphia. Maria Mitchell (1818–1889)— the astronomer who was elected to the American Association for the Advancement of Science with Margaretta—was internationally famous for having discovered a comet in 1847 and later joined the faculty at Vassar College to inspire and educate young female scientists. The writer and naturalist Mary Treat (1830–1923) published popular science books and articles in major magazines, while also sharing her findings on carnivorous plants with Asa Gray and Charles Darwin. And these are just the women that we hear about. The list would be longer still if it included women working in medicine.[7]

There were also lesser-known women who were working in the sciences at the same time as the Morris sisters and whose contributions only faintly survive. Women like Isabella Batchelder James (1819–1901), a friend of Elizabeth's, who researched the physiological source of plant odors. She corresponded with other American botanists and wrote anonymous articles about American trees, science books, and biographies of naturalists. Another is Sarah Coates Harris (1824–1886), who studied botany and herbalism, exchanged letters with other botanists, and delivered a series of lectures to ladies in Ohio on "Anatomy, Physiology, and Hygiene," while also fighting for women's rights.[8]

Women who pursued work in the sciences in the nineteenth century were rarely aware of their predecessors. Elizabeth Morris, as far as I have found, had never heard of Jane Colden and her botanical work in the colonial era. That loss of continuity, even community, caused a lot of damage. When women and other marginalized people broke barriers by becoming a "first" in their specialty, they had the potential to inspire others who had been excluded to believe that what had seemed impossible was actually possible. Obscuring these stories forced those who followed to reinvent strategies in order to join scientific communities and be taken seriously. It also kept those less emboldened or supported from entering the profession at all.

Thanks to the popular enthusiasm for the sciences and growing educational opportunities, though, scientific women were everywhere: collecting specimens, writing articles, illustrating textbooks, and sponsoring the production of groundbreaking books. They played a central role in popularizing the sciences and making them accessible to others. As men of science tried to establish themselves as trained experts within the scientific cultural zeitgeist, they wrestled with how to balance communication with the public and specialists. Men like Asa Gray, Charles Darwin, and Louis Agassiz embraced the opportunity to write for multiple audiences. Others balked, viewing public outreach as degrading or a needless distraction. Desperate to be taken seriously, they sought to separate themselves from popular science—and from the work women were doing.

William Whewell coined the term "scientist" in 1834 in an essay about a woman with a talent for turning complex scientific phenomena into understandable prose for popular audiences:

the British astronomer and mathematician Mary Somerville. In his review of Somerville's book *On the Connexion of the Physical Sciences*, Whewell proposed the idea that women's brains worked differently than men's. This biological difference, Whewell believed, allowed Somerville to artfully synthesize current scientific findings while men stumbled. He called it "the peculiar illumination" of the female mind. While he suggested that a woman's brain enhanced rather than diminished her ability to understand science on a higher level, he also claimed that they were inherently different from male scientists in a way that precluded their equality, something that would haunt the Morris sisters as they struggled to be taken seriously in the burgeoning American scientific community.[9]

Inspired by the connections Somerville drew between the ever-subdividing disciplines of science, Whewell felt that there needed to be a new term like "scientist" for these people who sought "knowledge of the material world" in order to bring unity to the expanding community. However, not many men were eager to embrace Whewell's inclusive "scientist" to describe themselves and instead clung to "man of science" or "naturalist" among other designations even into the early twentieth century. Whewell's struggle to create an inclusive label and its difficulty gaining traction was indicative of the power hierarchies within the scientific community. Those unwilling to drop "man of science" considered the elite scientific circles to be inherently, commonsensically masculine. Hierarchies are difficult to shake.[10]

While in hindsight there is a clear trajectory toward professionalization in the nineteenth century, it was a muddy and uneven process for those living through it. It began far earlier in the century than we usually assume, starting in the decades before the Civil War and continuing through the early twentieth century. One could argue that it has never fully finished, since scientists continue to rewrite the rules that govern good practice.

Still, throughout the nineteenth century there were markers that showed the gradual formalization of the field, as scientists defined their community rules and training, determined whom to exclude, and began to author their own historical narrative.

Much of the formalization of professional practices happened in the decades after the Morris sisters' deaths in the 1860s, but the women still bore witness to changes in their own statuses even during their lifetimes. Though the renowned Harvard botanist Asa Gray had treated Elizabeth as a valuable collector and friend early in their relationship in the 1840s, that changed quickly as his professional star rose and his dependence on her specimens diminished. In 1859, after Asa Gray published his "Japan Paper" on the distribution of plant species—an essay that would have significance for Charles Darwin's *On the Origin of Species*—he noted in a letter to Elizabeth that while he had promised her a copy, he had to first distribute them to "working botanists." She needed to wait.[11]

Employment was crucial if science was to become a profession. At the start of the century, few people could make a living in the sciences in the United States, and many worked day jobs as bankers, doctors, and lawyers to support scientific work in their spare time. Some, like Margaretta and Elizabeth, relied on the wealth of relatives. Gradually, more job opportunities made it possible for people to devote themselves fully to science in an institutional capacity, whether at universities, in government positions, or in natural history museums. It began slowly, with a professorship in chemistry and natural history at Yale in 1802, followed by growing demand for science writers and teachers in the 1820s, and a market for collectors of specimens in the 1840s. By the 1850s, the United States Patent Office had filled its ranks with men and women, and New York had even hired a state entomologist. After the Civil War, funding for railroad surveys, the expansion of natural history museums, and the opening of agricultural schools thanks to the Morrill Act of 1862 meant budding scientists could convert their practice into a career. Some of these jobs did not pay well, but they inspired the creation of others like them, which in turn considerably lowered the economic barrier of entry into the profession.[12]

Some scientists were wary of the ways money might taint scholarship. Toward the end of the century in the Gilded Age, there were men who balked not only at being called "scientists" but also at being called "professionals." They equated professionalization with commercialization, the enemy of "pure science," and chose to refer to their work instead as an avocation or calling. While these distinctions were meant to rein in ethical issues like geology professors profiting from mining contracts, they had deep roots in a collective disdain for those who either did not come from wealth or were unable to secure a salaried position. Ironically, given that women scientists in this era were generally expected to work for free or close to it, they ought to have been held up as ideal scientific investigators. That was not the case.[13]

The more formalized scientific employment became in the late nineteenth century, the more women were relegated to peripheral jobs. Even after women won access to doctoral programs at universities, they remained in subservient positions performing what was considered "women's work." These jobs included calculators at observatories, illustrators at botanical gardens, and assistants at natural history museums. Home economics departments, which developed at land-grant universities at the turn of the twentieth century, became one of the few places where women scientists could find professorships in numbers elsewhere unseen. Prior to this formalization and expansion of scientific work, though, women like Margaretta Morris and Elizabeth Morris had more leeway to be on something closer to equal footing with their male peers. When it was not profitable work, more women could compete. Professionalization and the turn toward more merit-based job opportunities actually edged women into gender-segregated and lower-paid jobs with fewer chances to be promoted, let alone credited for their work.[14]

Scholarly journals were also rapidly transforming over the course of the nineteenth century, as scientists developed their communities and became architects of their profession. In the 1840s, Margaretta published her findings in multiple kinds of journals, including the *Transactions of the American Philosophical Society* as well as the *Proceedings of the Academy of Natural Sciences*. While peer review would not gain the prestige and power it currently holds until the mid-twentieth century, the Philosophical Society used a committee of members to vet Margaretta's research before deciding to publish it in their *Transactions*. At the same time that she published for an audience of scientists through the journals, she also wrote about her findings in agricultural journals predominantly read by farmers. These journals—continuously referenced into the twentieth century by entomologists and other biologists—served as a cacophonous space to exchange scientific data, with fast-paced publication times and feedback from readers. Not all of what was published was credible, but these were important conversations for botanists, entomologists, and agricultural chemists, as well as farmers. By the turn of the twentieth century, these kinds of scientists would establish more specialized journals that gave them space to speak solely to their peers, such as the American Entomological Society's *Journal of Economic Entomology*, which began publication in 1908. With the new journals, the foundation of scientific exclusion grew stronger with the advent of a new kind of professional gatekeeping.[15]

Professionalization by its very nature involves exclusion. The bylaws of the American Philosophical Society and the Academy of Natural Sciences did not explicitly limit their membership to men, but in practice, the organizations did. In 1841, the Academy of Natural Sciences granted membership to Lucy Say, an illustrator and naturalist in her own right, as a show of gratitude for her donation of specimens that belonged to her deceased husband, the entomologist Thomas Say. And in 1859, they voted to admit Margaretta Hare Morris for her lifetime of entomological work. However, in both cases, the women were given a specific type of membership that spared them from paying dues, but also excluded them from attending meetings, let alone voting. The AAAS began admitting women in 1850 with Margaretta Morris and Maria Mitchell, but when the numbers of women joining spiked in the 1870s and 1880s, the Association decided to create a two-tiered membership with professional scientists designated as "fellows." It had already been hard for women to gain membership to the AAAS. Whereas men became members by default if they were involved in local scientific associations or employed in the sciences, women—who were often excluded from those local groups and jobs—had to be nominated by current members. This new tiered system further formalized a sexual segregation among members. Other scientific organizations adopted similar practices or devised systems in which women faced tougher requirements than men. The unspoken racial exclusion was even starker.[16]

Not coincidentally, it was over the course of the nineteenth century that the definition of the word "amateur" transformed. It had originally been an elevated rank, suggesting that someone was doing the work because they loved it, not because it would earn them money. It implied a certain purity of intention. However, by the middle of the nineteenth century and onward, it gradually took on a negative connotation and meant someone who was unskilled and unqualified, the opposite of an expert. Given how murky the process of professionalization was, the line between amateurs and experts was blurred, and those without formal training continued to participate in organizations and activities. Still, the more derogatory the term became, the more it was tied to the activities of women who faced more obstacles gaining access to the kinds of training and employment that would have classified them as professionals or experts.[17]

The denigration of women in the sciences was intricately caught up in gendered insecurities. Some men worried that botany, in particular, was perceived as a female vocation. They felt the need to prove that men, too, could study flowers and plants. These ideas were still prevalent in 1887 when J. F. A. Adams challenged the idea that botany was suitable for "young ladies and effeminate youths, but not adapted for able-bodied and vigorous-brained young men." Writing in *Science*, Adams argued that botany should be "pre-eminently a manly study," given that the work took place outdoors and required intellectual rigor. That Adams felt this argument had to be made and in one of the leading journals for the profession illustrates the scale of women's involvement in the natural sciences during that era.[18]

Not everyone subscribed to these concerns about the masculinity of the sciences. Margaretta and Elizabeth found many allies among their peers who were willing to use their positions of power to amplify and defend the work the women were doing. Some of these men were those who witnessed the cicada discoveries Margaretta made and publicly endorsed her work. Others, like the botanists Asa Gray and William Darlington, welcomed the contributions women made and saw advantages in working with all botanical enthusiasts. The culture among entomologists was not nearly as welcoming, but Margaretta still sought out friends and allies.

Beyond anxieties about gender, many American scientists felt the need to prove that they had the intelligence and training to keep pace with their more established European counterparts abroad. American naturalists rushed to name North American species before Europeans claimed priority. It was not all competitive, though, and scientific conversations were transatlantic, with letters, books, and scientists themselves traveling back and forth, establishing connections and sharing specimens. In some ways, American scientists mimicked what they saw occurring in Britain and elsewhere—the AAAS was modeled after the British Association for the Advancement of Science, for instance. In other ways, America excelled—European travelers were often surprised by the high level of science being taught in girls' schools in the early nineteenth century.

Still, there were lasting insecurities that stemmed from America's colonial past, and professionalization could serve as a balm for that. It had the potential to clean up the messier attempts at science that might reflect poorly on American work—flowery prose, airy theorizing, and a lack of rigor among them. As was true with the arts, architecture, and literature, many Americans sought to prove that democratic societies could generate creative advances, new knowledge, and innovative technologies. However, if America's leading organizations invited all scientists in as equals—no matter their race, gender, or class—there was fear that Europeans might mock the organizations as inferior. These insecurities were the enemy of equality. They prevented many potential scientists from sharing their knowledge and stifled collective advances.[19]

It was within this shifting scientific landscape in the nineteenth century that marginalized scientists like the Morris sisters found ways to participate and become instrumental in its growth and professionalization. Their specimens, their experiments, their publications, and their engagement in transnational exchanges of data were all part of this transformation. Ironically, the professionalization that they helped to advance would ultimately wrest opportunities from women in the generations that followed. The more legitimate, respectable, and profitable science became, the harder it was for women to participate on equal footing, and the more intentionally they were consigned to obscurity.

THIS BOOK WAS NEVER SUPPOSED TO BE ABOUT THE MORRIS sisters. As an urban environmental historian by training, I had been planning to write about a city tree that Americans love to hate: the *Ailanthus altissima*, or Tree of Heaven. When I was at the New-York Historical Society researching the history of that tree, one of the archivists pointed me in the direction of the William Darlington Papers, since Darlington had once written about trees and other plants he classified as weeds. As I sat with the finding aid, scanning for any possible leads, I noticed that there was an enormous collection of letters from someone named Elizabeth C. Morris. Curious as to who she was, I googled her name and found very little—not even a snippet on a Wikipedia page. A bit more digging led me to some information about Margaretta Morris, and I learned that the two sisters were scientists. Rushing to catch a flight home, I scribbled their names in my notes and wondered why I had never heard of them before.

The next month, sitting in Harvard's Ernst Mayr Library of the Museum of Comparative Zoology, I found myself deep in the papers of Thaddeus William Harris, an entomologist who had written about a caterpillar that had infested urban trees (though not the Tree of Heaven) in the 1850s. There in Harris's papers, I coincidentally came across a series of letters from Margaretta Morris. Her words made me pause. She wrote about the difficulty and loneliness of being a woman in the sciences—how her access to colleges was limited, how she lacked peers to compare notes with. "I have panted for the sympathy of someone who could appreciate my love of the science, and overlook my want of that learned love derived from books that are, generally speaking, out of *woman's* reach," Margaretta wrote. "The book of nature, however, has been widely spread before me, and countless hours of inexpressible happiness I have had in the study, there." After reading that, I decided the Tree of Heaven could wait.[20]

If it was Elizabeth whose letters introduced me to the women, it was Margaretta with her energetic voice and desperate longing to connect with other entomologists who captured my imagination. She could not pass a spiderweb without pausing to investigate which insects were tangled in it. Whoever she sat next to at a dinner party could expect to get an earful about the moth or beetle she was currently studying under the bell jars in the sisters' library. From the handful of letters that were in the Mayr library, I also got the sense that she had been defending herself and her findings against skeptical entomologists who were quick to dismiss her. After describing a dispute with another entomologist about the behavior of a fly, she wrote, "This formidable opposition roused me to renewed exertion." The pushback she received from critics did not silence her; it fueled her determination.[21]

Margaretta's steadfast desire to be heard convinced me to change the focus of my book. She was facing unrelenting criticism and all kinds of obstructions, and I was struck by how determined she was to persevere. Her struggles felt relatable across centuries, and her courage was inspiring. It also brought into sharp relief how so much of our knowledge about the environment has been filtered through power relationships based on gender, class, and race. The privileging of some experts over others has resulted in lost knowledge about species, relationships within ecosystems, even how things have changed over time. I wanted to explore how the Morris sisters personally navigated these issues and what effect these power structures had on their work and legacy.

Margaretta, in many ways, fit the mold of similar pathbreakers by defying expectations and rising above doubters, but Elizabeth did not. While fiercely opinionated, Elizabeth was terrified of criticism and overstepping boundaries. Still, after reading her letters and uncovering dozens of her anonymous articles, I came to find her just as fascinating as her sister. While Margaretta was more exceptional in terms of the public-facing work she did, Elizabeth made choices common among women in the nineteenth century. She bristled at the idea of publishing anything under her own name, and the attacks on her sister likely reinforced that decision. She was far from alone. Publishing anonymously allowed women the chance to have their scientific findings given serious weight. Elizabeth shared her passions on the pages of popular agricultural journals, encouraging her readers to take botany seriously. Whatever hesitations she had about publicity, she never stopped obsessing over ferns and wildflowers, and she collected as much as she could, actively assisting other botanists like Asa Gray and William Darlington for the sake of science, just as she had assisted Louis Agassiz with the cicadas. Elizabeth seems to have been unconcerned with getting credit for the work she did.

Studying two sisters made it possible for me to distinguish their personal choices from the limitations they faced during their lifetimes. Margaretta and Elizabeth show how two different women reacted to the cultural restrictions and freedoms of their generation. They came from the same wealth, they had the same resources and even the same education, and yet they chose different paths while following their passions. Margaretta and Elizabeth made decisions—one forging into public debates, the other embracing anonymity—to develop their careers as they preferred, amid the gendered conventions of the nineteenth century. Margaretta was honored as exceptional during her lifetime because she engaged in the sciences in the same ways that men did—publishing findings under her own name, jumping into contentious entomological debates, proving to male peers again and again that she belonged in their world. Elizabeth's work as an anonymous science writer, illustrator, collector, even a community builder, however, was more in line with the kind of work that would be categorized as "women's work"—popularizing, appreciating, disseminating, and consuming. As those boundaries between gendered work calcified over the nineteenth century, the value attributed to one kind of work over the other felt inherent, even natural.

Mischievous Creatures follows the lives of Margaretta and Elizabeth from their childhood curiosities about the natural world through the ups and downs of their careers as they navigated antebellum American science while it professionalized around them. The chapters weave between the two sisters, and by juxtaposing their different approaches to scientific work, we can come to understand how the division of labor marginalized the very women that science relied upon to maintain its position in the nation's intellectual life. The sisters faced personal and intellectual crises at two pivotal points in their lives and careers, in both young adulthood and middle age. Their agile navigation of changing scientific structures during those moments reveal as much about them and their persistence as it does about nineteenth-century America and the course of professionalization. While Margaretta was driven by the desire to solve entomological puzzles, Elizabeth's community-building skills ultimately made it possible for her sister to succeed. These soft skills, as we might call them today, built a network of relationships with male scientific celebrities whose support would prove crucial to the widespread acceptance of Margaretta's work. When Margaretta made her cicada discoveries, it was likely Elizabeth who strategized about how to rally endorsers. Their lives were forever entangled, just like botany and entomology, and they were stronger for that.

The voices of the Morris sisters were buried deep in the archives. They were part of the major scientific conversations happening in the middle of the nineteenth century, yet they were nearly completely silenced in the stories scientists told about themselves. Even though Margaretta had chosen a more public scientific life than Elizabeth, both sisters barely made it into the historical record. These women were wealthy, white, and well connected to the scientific community, and yet, even with those privileges, they were practically invisible. Theirs are just two of countless silenced voices, but their story offers us a window into the realities that women scientists of the time faced as they navigated professional transformations.[22]

For more than a generation, historians have been hard at work rediscovering forgotten scientists' lost contributions. This has often involved parsing the role that power—in the form of misogyny, empire, wealth, and white supremacy—has played in shaping the sciences. By broadening the scope of the field, historians have also been unearthing stories of science taking place in parlors, gardens, classrooms, and kitchens, which has, in turn, revived many underappreciated, uncredited discoveries and collaborations. Reframing what counts as science and calling into question who gets to define those boundaries similarly expands the narrative beyond the great men celebrated in biographies and honored with awards. Remembering can be an act of resistance.[23]

Not every voice and life is recoverable, though. I was able to learn a lot about Margaretta and Elizabeth thanks to their being related to a wealthy, politically notable family—a fact that meant archivists saved some of their papers over the last two centuries. Other things were certainly lost, such as their daily diaries and scientific notebooks, which they occasionally make reference to but seem to have not survived. Much of their scientific writings only exist because they were tucked in the collections of male scientists. All biographers try to understand the lives of their subjects with incomplete information, but for these overlooked women, the records are even less complete. We are used to assuming that women of the past were quiet, but that perception is built on the fact that fewer sources preserve their voices. From what I have been able to excavate of the Morris sisters' writings, though, these women were opinionated and passionate—anything but quiet.

So many marginalized scientists from this time period have scarce or nonexistent archives, and it can prove frustrating when we want to make sense of not only their accomplishments but also how they confronted obstacles along the way. Where historians have come up empty-handed, fiction writers have filled in archival absences by re-creating the lives, thoughts, victories, and struggles of real and imagined scientists from this era. The popularity of characters like Elizabeth Gilbert's Alma Whittaker, Esi Edugyan's Washington Black, Barbara Kingsolver's Mary Treat, and Tracy Chevalier's Mary Anning speaks to our collective desire to understand what these early scientists were thinking as they made space for themselves and their ambitions in a complicated period of fertile growth and contracting opportunities. These captivating stories remind us of the universal frustration of not being taken seriously because of something beyond our control like our gender or race. They also remind us of the joy of observing something new about our environment, the boundless wonder involved in figuring out how things work, and the specific delight in sharing those discoveries with others who care as much about them as us. That was exactly what drew me to Margaretta and Elizabeth. The Morris sisters struggled, yes, but they also could not get enough. They were hungry for new knowledge. They wanted nothing more than to fundamentally know the natural world. They were persistent.[24]

No matter the adversity Margaretta and Elizabeth faced, they had an enormous advantage of not being alone. Whether digging up cicada larvae, arguing over the best recipes for homemade soap, discussing the latest issues of scientific journals, or caring for each other when sick, Margaretta and Elizabeth had each other. When Margaretta came under attack, it was Elizabeth who encouraged her to stand up for herself and her methods. Alone these sister scientists may never have achieved what they accomplished together.

One

WORLD OF WONDERS

MARGARETTA MORRIS AND ELIZABETH MORRIS WERE HAP-
piest outdoors. They certainly spent considerable time there as
adults—Margaretta galloping on horseback across fields to visit
neighbors or hunched over the edge of a lake, losing track of time
while observing water beetles. Elizabeth, for her part, took off for
the forest with her tin vasculum case thrown over her shoulder
filled with her lunch, tools, and papers to separate whatever specimens she gathered.[1]

As scientists in midlife, they reminisced with peers about what
drew them to their specialties—botany for Elizabeth, entomology
for Margaretta. Like anyone looking backward with hindsight,
they sought roots as if the seeds had been sown early. And all of
these formative memories were outdoors.

Elizabeth, ever the romantic, remembered having her eyes
cast wistfully toward distant worlds far from her childhood home
in Philadelphia. "I used to want to set out on *my* travels, in search
of adventures and flowers," she wrote, "like the heroines of the
fairy tales, for then as now, flowers were my diamonds." As an

adult, she would come to possess bookshelves filled with bound descriptions and lithographs of these exotic plants, an enviable herbarium with pressed specimens, and a carefully tended garden full of gifts from botanists around the world. Those men who sent the seeds, rhizomes, and plants were the ones able to go on the expeditions Elizabeth dreamed of.[2]

Margaretta's origin story remained closer to home, as she recalled "the world of wonders opened to the solitary Child, in a rotten apple, or a decaid log and pear." While butterflies might have paired well with the flowers her sister and many other American girls collected, Margaretta was less interested in those than she was with the insects typically considered pests. Housekeeping manuals and agricultural journals regularly gave tips on how to kill all kinds of insects with arsenic, tobacco, cobalt, and boiling water, but Margaretta preferred to capture and observe the creatures. Nothing about her obsession with her "little friends the insects" embarrassed her as a child or an adult, and she breathlessly filled letters and conversations with friends about her latest findings.[3]

The word "scientist" had not even been coined yet when Madge and Libby—as they were known to their family—were born in 1795 and 1797. However, American "men of science" were busy collecting, drawing, and naming the creatures and plants they discovered. Thomas Jefferson, who won the presidency in 1800 when the girls were toddlers, was not only tinkering with his own experiments and defending the size and vigor of North American mammals to European critics, but also sponsoring the Lewis and Clark expedition that would involve collecting a wide range of plants and animals from across the continent. Meanwhile, Yale University offered Benjamin Silliman the very first American professorship in natural history and chemistry in 1802, beginning a trend that would slowly establish the sciences' home in universities across the United States. In order to educate himself prior to accepting the position, Silliman moved to Philadelphia to attend lectures at the University of Pennsylvania's medical school. Almost immediately, he met the Morris sisters' older cousin, Robert Hare, who maintained a chemistry lab in the basement of their mutual boardinghouse. The young men became fast friends.[4]

For two young children interested in science, Margaretta and Elizabeth could not have been born in a better city. Philadelphia, the temporary capital of the country from 1791 until 1800, had emerged as a center for scientific inquiry. The American Philosophical Society was right in the center of the city amid the red brick row houses and buildings devoted to the government. Men of science met in Philosophical Hall to discuss their latest findings and read reports sent from scientific correspondents scattered around the country and across the Atlantic. Charles Willson Peale had moved his natural history museum, the first in the country, into the same building in 1794. Across town, the botanical gardens of William Bartram had become a tourist and scientific destination. Scientific lectures, whether at the Philosophical Society, Peale's museum, the University of Pennsylvania, or in other spaces, drew students and the general public alike, and were a regular feature on Philadelphians' calendars. In 1804, when the explorer Alexander von Humboldt traveled to South America to collect specimens and develop theories, he made sure to take a detour and sail north to Philadelphia to meet with the American leaders of science and Thomas Jefferson, in particular. In this era before universities were the center of intellectual life, Philadelphia's wealth of institutions made it a hub of scientific activity.[5]

Madge and Libby began their lives about a mile south of Philosophical Hall at the Peckham estate. Their home sat on a hilltop overlooking the Delaware River, with terraced gardens rippling down toward the other country estates of Philadelphia's social elite. Their extended family had deep roots in Pennsylvania's founding. Their grandfathers, great-uncles, and great-grandfathers had been mayors of the city, and their family's names inspired street and

The American Philosophical Society's building that housed Peale's Museum stands behind the trees in William R. Birch's 1799 print. Institutions like these made Philadelphia a center for science in the early republic. *(Library Company of Philadelphia)*

town names throughout Pennsylvania.[6] Their parents, Luke Hudson Morris and Ann Willing Morris, inherited not only recognizable names but also real estate and wealth. Their home was full of expensive mahogany tables, chairs, and chests, paintings in gilt frames, as well as shelves and trunks full of books. Like everyone, Margaretta and Elizabeth were shaped by where they came from. And for two children who would come to find joy closely observing the structure of irises and the habits of caterpillars making a home under apple tree bark, their family's encouragement helped set them on their path. A home where they could embark on childhood adventures in the garden likewise made this intellectual exploration possible.[7]

Located in the relatively rural Southwark neighborhood, Peckham was where the sisters' grandparents, Elizabeth Hudson Morris and Anthony Morris III, had retreated to in 1758. Elizabeth—a strong-minded, independent woman, not unlike her future granddaughters—was a renowned Quaker preacher who had traveled from Philadelphia to England, Ireland, and Scotland on a speaking tour. She was thirty when she returned and married the widowed Anthony, taking his six motherless children under her care. After giving birth to her first son, William, she suffered greatly from a debilitating and "unspeakable poverty of spirit" for nearly seven years. She wrote in her diary that she struggled to "properly attend to the cares of so large a family." She had, for unknown reasons, also been blocked from preaching by the church, which had been both an intellectual outlet and a central part of her identity. Witnessing his wife's distress, Anthony readily agreed to abandon the bustling, overwhelming city for the refuge of their country home. After a few years and some recuperation from a variety of illnesses, including a bout of "nervous fever" that nearly took Anthony "to the brink of the grave," they had their second son, Luke Hudson Morris, in 1760.[8]

While the family suffered greatly from physical and mental illnesses, they did not have to worry about money. Anthony had inherited the family's brewery and, like his wife, a large amount of real estate throughout the city. Margaretta and Elizabeth's grandparents were one of the few families in Philadelphia that owned a "pleasure carriage" pulled by stylish horses. They were also among the small percentage of families who had their homes labeled on the 1752 Scull and Heap map of Philadelphia, which spoke to their wealth and stature in the social hierarchy, if not the grandness of their home itself. They enslaved three people: Pompey, Sabrina, and Eleanor Sneed, a practice that Pennsylvania Quakers had been alternately denouncing and justifying for more than a century. By the time the American Revolution erupted in 1776

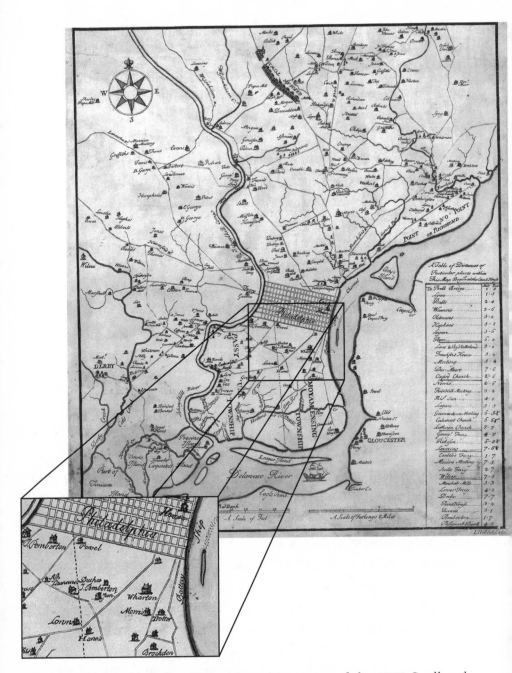

Peckham, labeled as "Morris" on this portion of the 1752 Scull and Heap map, was the home where Margaretta and Elizabeth were born and the site of their early explorations in the garden. (*Library of Congress*)

and Luke was a teenager, his extended Quaker family began granting manumission to the people they enslaved. That fall, it had been announced at the Philadelphia Annual Meeting at the Arch Street Church that members who did not free their enslaved laborers would face disownment by the Society of Friends. The Morris family complied. A few years later in 1780, Pennsylvania would pass the Act for the Gradual Abolition of Slavery, the first of its kind that served as a model for other states. While it stopped the slave trade and blocked the importation of slaves into Pennsylvania, those already enslaved continued to work uncompensated, as would their children, until they were twenty-eight years old.[9]

Peckham was a large, comfortable mansion, which likely included several other buildings to house all who labored there. It was also a working farm with livestock, horses, gardens, and an orchard. As a child, Luke Morris meandered around the property—the same lands that his daughters would fondly remember for their flower and bug explorations—gathering botanical specimens and cultivating a garden. His daughters would recollect how as an adult he had loved deciphering the world of plants.[10]

In the middle of the Revolutionary War, when Luke was just twenty, his father died, followed a few years later by his mother. The many acres of property owned by his parents were divvied up among the siblings and half siblings, and Luke ultimately inherited his childhood home. Luke had stayed out of the war out of respect for his devout, pacifist mother. However, after her death in 1783, he was made captain of the 5th Battalion of the Pennsylvania Militia and then soon after became lieutenant of Delaware County. He also apparently studied law, occasionally tacking "Esquire" to the end of his name.[11]

Meanwhile, less than a mile north in what is now known as Society Hill, Margaretta and Elizabeth's future mother Ann Willing lived with her parents, Charles Willing and Elizabeth

Carrington Willing. Her parents had met and married in Barbados amid the sugar plantations, returning to settle in Philadelphia before Ann, the younger of two daughters, was born in 1767. Like the Morrises, the Willing family was wealthy and well connected, with extensive real estate holdings throughout the city. Ann's grandparents had been among the largest slaveholders and slave traders in Philadelphia earlier in the eighteenth century. The next generation of Willings had mixed feelings about slavery. One aunt—Mary Willing Byrd—ran an enormous plantation in Virginia, while another—Elizabeth Willing Powel—donated money to abolitionists. Ann's parents, though they had likely enslaved people during their time in Barbados, sent their daughter to a progressive girls' school opened by the famed Quaker abolitionist Anthony Benezet.[12]

Gregarious and outgoing, Ann Willing was a significant figure in Philadelphia high society when she met Luke Morris. Nearly a century later, when a series of antiquarians and genealogists wrote the histories of Philadelphia's wealthiest families, they would note Luke Morris's birth and death dates, and an occasional reference to him being a "gentleman." Ann, however, garnered extensive entries. They celebrated her as a "lady of remarkable attainments" and "great mental energy," reflecting her delight in telling stories to relatives who later passed down these tales to the authors.[13]

Ann particularly loved telling stories about her childhood and her proximity to political power. She met many Revolutionary-era celebrities in the homes of her relatives. She delighted in recalling how as a baby, she prattled to George Washington while sitting on his knee at her aunt Elizabeth Willing Powel's townhouse. Her aunt hosted lavish parties and was particularly close with the Washingtons and others, passionately arguing politics with them, while also connecting Ann to this world of powerful politicians. Ann remembered learning about Ben Franklin's famous experiments with electricity from Franklin himself at her grandmother Ann Shippen Willing's home. As an eleven-year-old, she even attended the wedding of Benedict Arnold and her relative Peggy Shippen, though she quickly distanced herself from the infamous character when recounting the event.[14]

Given their similar social circles, it is likely Margaretta and Elizabeth's parents met at a social event in Philadelphia near the end of the American Revolution. However, they most certainly did not meet at the Mischianza, a theatrical, lavish party for British soldiers during the war at Luke Morris's neighbor Thomas Wharton's abandoned estate. Though many of her young, marriageable peers attended, dressed in Turkish costumes, Ann was proud to have refused that traitorous invitation. Regardless of where they met, with both of Luke's parents deceased, he must have felt free to marry Ann, whose family had long ago shed their Quaker faith in favor of Episcopalianism.[15]

Married at Christ Church in 1786, Luke and Ann soon settled into family life at Peckham. Almost immediately, Ann was pregnant with her first daughter, Abigail, who was born the next spring in 1787. Two years later, a baby named Elizabeth was born, though she lived just two months. After that, Ann had five more babies in quick succession, with her namesake, Ann, born in 1790, followed by Thomas in 1792. Elizabeth Carrington Morris was born on July 7, 1795, and Margaretta Hare Morris was born on December 3, 1797, Luke and Ann's fifth and sixth children. Finally, the baby of the family, Susan, was born in 1800. It was a busy fourteen years for this large family, with Ann pregnant or caring for an infant for practically the full stretch. The 1800 census shows that the Morris household was bustling with eighteen residents that included an additional four adults and six children beyond Luke and Ann's six, likely a combination of servants and relatives. Certainly, a wealthy family such as this did not raise their children single-handedly, and at least one of the residents would have been a governess or nurse. Meanwhile, Luke Morris continued to manage real estate for both his and his aunt's properties.[16]

BESIDES THE FAMILY WEALTH THAT WOULD LATER SUPPORT Margaretta and Elizabeth's scientific practice and make it possible to purchase books and scientific journals, paints and pens, microscopes and magnifying glasses, their parents had a passion for education. The family had a history of well-educated women on both sides, so it was not surprising that the two parents took their young daughters' education as seriously as they took their son's. Philadelphia was famous for providing educational opportunities for girls, stemming from the Quaker belief in the intellectual and spiritual equality of women and men. By the time the girls were born, the city was much more diverse religiously, but educational opportunities continued, especially for Philadelphia's elite. While it is unclear which specific schools Margaretta, Elizabeth, and their siblings attended, it does seem that their parents preferred a combination of tutors and formal schools, making their children's education a priority.[17]

Just as Ann and Luke were starting their family and considering how they might educate their children, Benjamin Rush—the famed physician, politician, educator, and one of their family friends—published what he saw as a risky pamphlet arguing that girls should be educated not only for their sake but also for the sake of the young country. Having cofounded the Young Ladies' Academy in Philadelphia, Rush gave a lot of thought to the kind of education he felt girls needed in a republic, believing it should veer away from the superficial and decorative toward more practical skills in mathematics, history, geography, chemistry, and astronomy. Not only was this necessary to run their future households and family businesses, Rush argued, but girls needed a rigorous education so they could better educate the next

generation of American patriots. Worried that this was "so contrary to general prejudice and fashion," Rush dedicated the pamphlet to Margaretta and Elizabeth's great-aunt, Elizabeth Powel,
mentioning that his revolutionary statements had the endorsement of "such a respectable and popular name."[18]

While many wealthy and middling Philadelphia girls continued to be educated by private tutors, the Young Ladies' Academy
that Rush helped found attracted national attention, with Martha
Washington, members of the House of Representatives, as well as
state politicians attending the commencements. Combining the
goals of the Enlightenment with the rhetoric of the Revolution,
girls' education was soaring at the turn of the nineteenth century
and becoming increasingly mainstream, if still primarily accessible
to wealthy white girls. Similar academies began opening around
the city and across the country.[19]

In the midst of all of this in 1794, Philadelphia printer and
bookstore owner Mathew Carey produced 1,500 copies of Mary
Wollstonecraft's *A Vindication of the Rights of Woman*, confident
that he could sell them all. Purchased or borrowed, passed along
to friends and family, read aloud in parlors, Wollstonecraft's
take-no-prisoners message that women ought to be as much a
part of educated civil society as men lent credence to what supporters of women's education were already saying. The Philadelphia educator James Neal was eager to draw connections between
Wollstonecraft's popular book and the work being done at the
Young Ladies' Academy when he wrote *An Essay on the Education
and Genius of the Female Sex* the next year.[20]

Margaretta, Elizabeth, and their sisters benefited from the up-
swell of support for female education during the early republic. What
had been accessible at only the very progressive schools, like the one
their mother attended, was now becoming widespread throughout the city. As young children, the sisters would have learned the

basics of literacy and mathematics, and as was fashionable at the time, they would also have been encouraged to explore and investigate the natural world as they wandered through the gardens and the wheat field, past the pigs, cows, and horses at Peckham. Their father was responsible for that realm of their education at home. He was one of the earliest members of the Philadelphia Society for Promoting Agriculture, an organization intended to spread scientific information about farming while keeping the young United States competitive with English exports. Decades later, Elizabeth would recount to friends that her father had instilled a curiosity about plants in all his children. She considered him "a respectable botanist in the then existing state of the science."[21]

Luke and Ann would have taken their young children to visit Peale's natural history museum up in Philosophical Hall near the State House, close as it was to their great-aunts' and great-uncles' homes. The Morris children were accustomed to visiting the center of the city, crossing the streets paved with river rocks and walking down the brick sidewalks, sometimes nearly tripping over the bricks pushed up by the roots of Lombardy poplar trees that lined many of the city streets. The large family would have to weave through the bustle of pedestrians, passing Absalom Jones's African Episcopal Church of St. Thomas, a center for Black Philadelphian life, on their way to the museum. Philadelphia was growing increasingly diverse, racially, economically, and religiously, and the Morris children would have witnessed that on their many journeys into the city.[22]

With admission set at twenty-five cents, the natural history museum was accessible to most, and Peale had intended it that way. In practice, though, his patrons tended to be white and well-to-do, and many of the people the Morris children passed on the sidewalk before entering the museum probably would not have felt as comfortable inside as they did. Peale wanted children to view the plant and animal specimens from all over North

America and take pride in the natural riches of their continent. With impressive exhibits of preserved birds, amphibians, fish, insects, mammals, fossils, and minerals on display behind glass, it was hard for visitors—Margaretta and Elizabeth included—not to be inspired to start their own collections of similar treasures. Margaretta would have been particularly enchanted with the four thousand insects in cases near the windows that she could examine closely under magnification.[23]

Peale also wanted women as well as men to attend the science lectures he hosted at the museum, explicitly stating as much in the advertisements he posted in newspapers: "It is my wish to behold LADIES among my hearers; for female education cannot be complete without some knowledge of the beautiful and interesting subjects of natural history." This helped to cultivate a culture in which women often outnumbered men in science lectures around town. The Morris children's proximity to these opportunities—this museum, the lectures, the scientists who orbited around Philosophical Hall, even their parents who wanted to ensure they grew up well educated—left an imprint on the girls and their budding interests.[24]

AND THEN, IN AN INSTANT, EVERYTHING CHANGED. IT IS UN-clear exactly how Luke Morris died on March 20, 1802, though by all accounts it was unexpected. He was trained as a lawyer, yet he died without a will. Perhaps he suffered from the same "poverty of spirits" that had afflicted his mother, or perhaps there had been a sudden accident or illness. Whatever the cause of his death, his children and wife—who otherwise seemed interested in preserving family history—said very little in the decades that followed about their father and husband who died at the age of forty-one, when Margaretta and Elizabeth were just four and six. While they refrained from writing about it as adults, this event proved a major turning point in their lives.[25]

Ann, a widow at only thirty-four, had to find a way to keep afloat with her six young children ranging in age from one to fifteen. Even if he had not been a lawyer who would have known better, it was rare for a man as wealthy as Luke Morris to die without a will. This meant the estate went into probate, which put Ann and her children in a vulnerable position, particularly in Pennsylvania where the widow's share of the estate was not protected from creditors. Two weeks after burying her husband, Ann signed a wobbly signature to the court documents that made her the administratrix of Luke's estate. She now had to settle his debts and inventory every item in the house, all the while tending to her household, farm, and children.[26]

Ann turned to family for help as she adjusted to her new circumstances. Her parents had died years ago, and her sister's family was 2,000 miles away in Barbados, but her uncle Thomas Willing, then the president of the Bank of the United States, stepped up as a trusted financial advisor, and her widowed aunts Mary Willing Byrd and Elizabeth Willing Powel swept in to help their struggling niece. Her aunt Mary, who provided her with a modest trust, had experience. She had been widowed a quarter century earlier in 1777 when her husband, Colonel William Byrd, in debt and accused of being a Loyalist, died by suicide. Forced to protect her family's property from both British and American soldiers while paying off her husband's debts, Byrd was herself considered a traitor at times, though she emerged from trial without charges. Ann's other aunt, Elizabeth Powel, had been widowed in 1793 when Samuel Powel died during the yellow fever epidemic. She proved herself an outstanding investor as she took over management of the family's real estate holdings, acquiring properties across the city. Powel would ultimately help her young, widowed niece in the best way she knew how: with real estate.[27]

AT FIRST, IT SEEMED THAT ANN WAS SET ON REMAINING AT
Peckham with her children. In July 1803, a year following her husband's death, she even purchased a small orchard's worth of young
peach and apple trees to plant on the property. As was true for her
daughters, she found solace outside, getting her hands dirty in the
garden. Planting trees and cultivating an orchard hardly seemed a
move that a woman planning to uproot her family would make.[28]

Within a few years, though, Ann changed course and decided
to move her children out of Peckham, away from her extended
family in central Philadelphia, to Germantown, a bustling village
roughly eight miles north of their home in Southwark. Germantown, which was outside the boundaries of Philadelphia at the
time, was in the process of transforming from a summer getaway
for those looking to escape the disease-riddled city into a bustling village with multiple churches, a busy thoroughfare, and a
year-round population. It had rural features, including deep lots
that held farms, but also felt very much like a village, with homes,
shops, and churches lining Main Street. Still, when James Mease
wrote the *Picture of Philadelphia* guidebook in 1811, he warned
his readers that while it featured a healthy and beautiful setting,
Germantown had "little to interest or detain strangers."[29]

Ann initially rented a modest stone house on that street,
where her six children would share two bedrooms. Saddled with
the costs and responsibilities of running Peckham and settling
her late husband's debts, Ann decided this made the most sense
financially. While they lived in the house, Margaretta and Elizabeth's older sister Ann rebelliously joined two teenaged friends
in etching their names in the window with the date 1807, a tidbit
that late-nineteenth-century antiquarians loved to point out in
their histories of Germantown.[30]

Regardless of whether she ever saw her grandniece's van-
dalism, Elizabeth Powel was concerned about how Ann's family

When the Morrises moved to Germantown, they initially rented this house, now known as the Howell House, which could barely fit them all. The home is seen here in 1920 with additions added in the mid-nineteenth century. (*Library Company of Philadelphia*)

would fare in such cramped quarters. Eager to get them better situated, Powel purchased a larger, recently built home down the street on a narrow but extensive plot of land. Powel offered the house to them free of charge, with the hope that it would give the young family space to flourish. By the time Powel had made the offer to Ann, or "Nancy," as she preferred to call her, the decision was all but made. Powel was assertive, telling her niece, "I think you had best remove as soon as possible." Should Ann argue that her current landlady would not let her out of the lease, her aunt was ready with a solution: she would pay off the rest of her rent. Promising her niece that the new house was in good order thanks to the neat and methodical previous homeowners, Powel counseled Ann that a messy house would cause domestic life to spiral out of control. "Punctuality regularity and neatness are the concomitants of Good breeding," Powel added. Perhaps this was a sly critique of her niece's sometimes unwieldy brood. Whatever strings might have been attached, whatever advice and criticism she might have endured, the offer was a lifeline, and Ann took it.[31]

Ann and her children—including ten-year-old Margaretta and twelve-year-old Elizabeth—settled into the two-story stone house with latticed windows that they named "Morris Hall." Their new home sat right on the corner of Main Street (later renamed Germantown Avenue) and High Street. Trellises clung to the stucco walls, enabling vines to climb up and over the façade, connecting their home with the lush gardens around it. The family would have all the benefits of rural life with trees, an elaborate botanical garden, and plenty of fresh air. The Morris family's home was also right on the busiest street, keeping them in close proximity to neighbors, the post office, Germantown Academy, and

By 1808, Ann Morris settled with her children into the house they called "Morris Hall" on Main Street and High Street in Germantown, seen here in 1890. *(Library Company of Philadelphia)*

stores. While the house was far smaller than Peckham had been, it was still sizable. Ann and her children would need to adjust to life on their own, with some occasional hired help.[32]

The beautiful botanical gardens that served as their backyard—where the children wandered through in search of new flowers and creatures—had a storied history that extended back to the settlement of a German religious community, known as the "Mystics of Wissahickon." Dr. Christopher Witt, an English physician, and his companion Daniel Geissler left the pietist community after the death of its leader in 1708 and settled in Germantown. At that time, the property became best known for its garden, which Witt and Geissler cultivated carefully into one of the earliest botanical gardens in the colonies. Witt used the garden for medicinal plants, but he also nurtured native and foreign plants that he collected and traded with the famed botanist Peter Collinson in England. Through their mutual connections with Collinson, Witt and John Bartram, whose botanical garden and nursery in Philadelphia would become far more famous, developed a relationship as well, making extensive visits to each other's gardens. While the property changed hands several times before Ann Morris and her children settled in, the extensive garden secluded behind trees and shrubs would provide an immediate link to the outdoor life that Margaretta and Elizabeth had enjoyed at Peckham.[33]

As Ann and her children settled into Germantown, they quickly became part of the community thanks to Ann's gregarious nature. In 1811, Ann joined with a dozen others who helped to establish Germantown's Episcopal church, St. Luke's, down the street from their new home. While her husband had been a Quaker and was buried at the Arch Street meetinghouse in central Philadelphia, Ann had held tight to her own Episcopalian faith, and in 1813 she had her children baptized privately by William White, the presiding bishop of the Episcopal Church and a close friend of Elizabeth Powel's. Margaretta and Elizabeth would never be as devoted to religious life as their mother, but they went along with these changes. All of these steps—the move out of her husband's childhood home, the baptism of her teenaged children, the rooting of herself in a new community—were signs that Ann was establishing herself and rebuilding life on her own terms. Her children and her grandchildren would long admire her pluck and resilience. They also internalized it.[34]

Margaretta, Elizabeth, and their family had lost so much— their father, their childhood home, their proximity to schools, museums, family, and friends. However, between their mother and their great-aunts, they had no shortage of strong, self-assured women to serve as their role models. Their home was a matriarchy. Their great-aunt Elizabeth Powel encouraged teenaged Margaretta's and Elizabeth's interests in gardening, praising them heartily when they turned the raspberries they planted and tended into raspberry syrup. Not only were the girls supported in their interests, they were constantly in the company of well-read women who never hid their intellectual strengths from the wider world.[35]

Still, Ann had some hesitations about their life in Germantown, so far from everything she once had known. She worried that she might regret removing her children from the center of social life in Philadelphia. She worried about who her children would someday marry. Along those lines, she was also concerned that her daughters seemed to prefer spending long days alone in the garden.[36]

Margaretta and Elizabeth, however, relished that seclusion and the time it afforded them to explore and tinker with experiments outdoors surrounded by the insects and plants that consumed their attention. And, while they were far from the center of Philadelphia where the pulse of science beat loudest, Germantown would ultimately provide a surprising number of opportunities.

Two

In a Tangled Wilderness
Without a Guide

HIKING PAST THE PAPER MILL, ON NARROW, UNEVEN PATHS woven with the roots of chestnut, oak, and hickory trees, Margaretta and Elizabeth watched their step as they reached the Wissahickon Creek. The young women brought picnics that they unwrapped as they found space to sit on rocks shaded by the dense canopy overhead. Even on the sunniest days, they were hidden in dappled light as they sang songs, sketched landscapes, and scribbled lines of poetry, meandering down paths for miles to collect treasures like lady's slipper orchids (*Cypripedium acaule*) and regal moths (*Citheronia regalis*) to put in their tin vasculum cases and pillboxes in order to study them back home.[1]

The Wissahickon, just a mile from their home in Germantown, was one of Margaretta and Elizabeth's favorite places to go in search of specimens and adventures. Before the actress Fanny Kemble alerted tourists to the beautiful forest so close to Philadelphia, before Edgar Allan Poe waxed on about floating down the "lazy brook," Margaretta and Elizabeth were venturing out into the forest, climbing boulders and hiking for miles alongside tutors, neighbors, siblings, friends, and lovers.[2]

These adventures in the woods were social events as much as they were scientific explorations. Margaretta and Elizabeth benefited from a network of relatives and neighbors who connected the girls to mentors and taught them to read their environment. If their garden was a space where they could tinker and experiment, the forest around the Wissahickon Creek set the stage for exploration alongside their ever-growing community of scientists and friends. The young women who joined them to watch meteor showers, the mentors who taught them to closely observe and draw fuchsia, shield ferns, and damselflies, and the neighbors who shared their books and invited the young women on hikes with visiting scientists all helped Margaretta and Elizabeth navigate their youth.

Inspired after a day with friends along the creek, Elizabeth returned home and sat at her desk to write a poem. Science, poetry, and art were closely connected for the Morris sisters and their contemporaries in both the United States and Europe. Early nineteenth-century scientists were often also artists and poetry lovers, if not poets themselves. They saw the connections between close observation and the celebration of nature in ways that modern scientific training sometimes misses. Being able to illustrate what they were studying was as important as being able to describe it in technical terms. To also work their subjects into poetry was a way to embrace the wonder of the natural world while continuing to study it empirically.[3]

Friendship albums, so popular with educated women in this period, showcase this intersection of science and art. Margaretta and Elizabeth both kept such albums, inviting friends to make contributions alongside their own poems and paintings. They were hardly alone. Sarah Mapps Douglass, the Black botanist, educator, and abolitionist, contributed botanical art and poetry to the albums kept by her friends and students in Philadelphia. These albums that highly educated women created, even within racially segregated communities, were nearly identical in form and content. They celebrated affectionate friendships while merging the science of botany with sentimental flower culture.[4]

In her poem "To the Wissahiccon," Elizabeth reveled in the memories she, Margaretta, and their friends had created there:

> *Sweet silent stream! how oft along*
> *Thy rocky banks with tale and song*
> *We've cheated time!*
> *Or danced upon thy yielding sand*
> *A laughing, joyous thoughtless band,*
> *Or o'er thy hills with vent'rous hand*
> *Have dared to climb!*

To the Wissihiccon.

The Wissahickon Creek was one of Margaretta and Elizabeth's favorite spots to search for specimens and take friends along for picnics and hikes. This image was drawn by their younger sister, Susan Sophia Morris. (*Special Collections, University of Delaware Library*)

These were not elegant strolls, free from exertion and sweat, but hikes that required scaling slippery moss-covered schist boulders and outcroppings in long dresses to reach "the most sequestered glen / untrodden by the feet of men." The young women rowed down the creek, scrambling to the shore to make their way home as the night grew dark.[5]

At the Wissahickon, Elizabeth studied the wildflowers that thrived there amid the rocks and shade, while Margaretta sought out uncommon insects like stalk-eyed flies (*Sphyracephala brevicornis*) hiding in the skunk cabbage leaves on the muddy banks of the creek. As they hunched over in their search, blue jays, swallows, and robins squabbled, whistled, chirped, and clicked in the branches overhead. Though the sounds of dozens of mills on the creek and its connected rivulets were never distant, sometimes even drowning out the birdsong, the mills did not make it into the Morris sisters' poetry. Neither did the juxtaposition of industry and forest inspire them to complain. Even when Fanny Kemble famously wrote about the Wissahickon in 1832, she referred to the mills as "picturesque," accepting that the creek welcomed multiple uses. A generation later, Philadelphians would rethink that perspective as they began removing the mills in hopes of keeping their watershed clean.[6]

Regardless of whether Margaretta and Elizabeth might have wanted to wander alone through the woods, they were always accompanied, if not by a large group of friends, at least by each other. Nineteenth-century Americans might have raised an eyebrow at a young woman of their social standing exploring the woods alone. Turning their scientific adventures into social outings insulated them from both danger and the danger of judgment. As Elizabeth put it in her poem: "How often through thy shades I've rov'd / With school companions fondly lov'd." They brought their dearest friends there, the bright young women with whom they traded witty banter and shared their love of scientific discovery. These friendships structured Margaretta's and Elizabeth's lives, as they locked elbows with other young women, moving from astronomy and chemistry lectures in the evenings at the Germantown Academy to the Wissahickon the next morning for another adventure on the mossy banks.[7]

The Morrises had made themselves at home in Germantown. Their mother developed friendships with the women around town, visiting for tea and sharing meals, while raising her children, cultivating her garden, and monitoring the increasing amount of real estate that she inherited from family members. Becoming a landlady and collecting rents around Philadelphia made it possible for Ann to maintain a lifestyle that included an elite education for her children and the luxury of roaming local forests at leisure.[8]

By 1815, Morris Hall had begun to grow quieter, with only the four youngest siblings—Ann, Elizabeth, Margaretta, and Susan—at home with any regularity. Their mother remained devoted to educating them, just as she had her older children, who were now making their way in the world. Tom, who had studied law like his father, enlisted in the army during the War of 1812, becoming the secretary for General Thomas Cadwalader at Camp Brandywine in Delaware. A month after the war ended in 1815 and life seemed to feel more secure, Margaretta and Elizabeth's eldest sister Abby married Justus Johnson, the son of a well-established Germantown family. Margaretta and Elizabeth gladly embraced their role as aunts once Abby began having children of her own.[9]

The Morris sisters who remained at home referred to themselves as the "Independent Taciturn Society of Morris Hall," but the young women were anything but quiet. They loved gossiping about their "own affairs, and the affairs of our friends," weaving in what they had been learning about astronomy, botany, entomology, conchology (the study of mollusk shells), and all other kinds of sciences. Friends who joined in these conversations relished the "good female society" they had in the cozy parlor, chatting with Margaretta and Elizabeth.[10]

While the Morris sisters were clearly well educated, writing in carefully crafted handwriting, reading French poetry as easily as they deciphered scientific jargon, sharing their extensive library of books with neighbors and friends, exactly where or how they were educated in Germantown remains as unclear as it had been when they lived in Philadelphia. It is possible that they attended a boarding school like Madame Grelaud's, where wealthy white American girls learned history, French, and science, as well as painting and dancing. While the school operated in the heart of Philadelphia during the better part of the year, over the summer the instructors and students moved into Loudoun, a large mansion in Germantown, just down the street from Morris Hall. During the War of 1812, the school remained in Germantown semipermanently due to Madame Grelaud's fear that Philadelphia was in danger of an attack.[11]

In the early nineteenth century as more schools for young women opened up across the United States, the sciences became a major part of the curriculum. In general, young women and men received relatively similar educations, but the emphasis on science courses at girls' schools stood out. In part, schools for young men remained centered on the classics for the sake of college entrance exams. While girls were prohibited from attending those colleges, there were also benefits to being untethered from such expectations. Their curriculum, for instance, had space for more focus on astronomy, geography, geology, chemistry, and botany. Thanks also to the cultural embrace of natural theology—or the idea that you could better understand God through a close understanding of nature—religious schools introduced robust science curriculums as well. Combine that with the patriotic idea that educating girls meant training future "republican mothers," and the sciences found fertile ground in schools for girls, even more so than for boys.[12]

Science classes—the more specialized the better—also became a status symbol. They signaled that a child had access to the best education. It was mostly middle- and upper-class young women who had those opportunities. This was true for white girls, like the Morris sisters, but also for elite African American and Cherokee girls who attended schools like Philadelphia's Institute for Colored Youth or the Cherokee Female Seminary. Visitors from Europe, including Alexis de Tocqueville, Frances Trollope, and Fredrika Bremer, found it remarkable that American schoolgirls received such a rigorous scientific education. The English author Frances Trollope, for instance, was taken aback during a visit to a girls' school in Cincinnati when she found "that the higher branches of science were among the studies of the pretty creatures I saw assembled there." Still, in the early nineteenth century, such a thorough education in the sciences was not yet universal and remained restricted to wealthy or at least lucky Americans who had access to these schools.[13]

WHETHER ELIZABETH, MARGARETTA, AND THEIR SISTERS enrolled in Madame Grelaud's school or another academy, they quickly found themselves connected to the growing community of scientists in Philadelphia in the 1810s. In part, this was because their cousin Robert Hare was establishing himself as a notable chemist, gaining membership in the American Philosophical Society in 1803 when he was just twenty-two years old. Hare had invented an oxy-hydrogen blowpipe that produced a heat so intense it allowed for new chemical reactions and discoveries. The American Philosophical Society, which had been founded in 1743 by Benjamin Franklin, was arguably the country's most elite scientific organization at the time, and one of the best known among European scientists. The Philosophical Society counted the country's leading "men of science" as members, in addition to a number of eminent Philadelphians. This membership was an important honor and achievement for Robert Hare, both socially and professionally, as he established his career. He was very close with the Morris sisters, and even closer still after he married Harriet Clark in 1811, as Harriet became particularly fond of teenaged Margaretta and Elizabeth. He would have likely shared news of what happened at the meetings with his scientifically inclined cousins during their frequent meals and visits.[14]

In addition to their cousin, Margaretta and Elizabeth's neighbor, Reuben Haines, connected them to a rising group of passionate naturalists. A decade older than the Morris sisters, Reuben was a regular presence in their parlor since his mother's summer home, Wyck, was just across the street from the gardens where Margaretta and Elizabeth dissected flowers and studied the life cycles of insects. Reuben had a hunger for scientific knowledge that could at times seem frenetic. His exuberance was often coupled with extended periods of lethargic depression. After marrying in 1812, Reuben and Jane Haines gradually spent more and more time at Wyck, eventually making it their permanent home. The neighbors got to know each other well. Reuben's younger cousin Ann—who often lived with them for extended periods of time—became particularly close with the Morris sisters. Reuben's scientist friends from Philadelphia often came to stay at Wyck so they could explore the more rustic environment around Germantown. Reuben, together with Thomas Morris, brought these men on hikes along the Wissahickon in search of new adventures and species, occasionally allowing Thomas's sisters to tag along.[15]

Reuben's friends—Thomas Say, Thomas Nuttall, Charles Alexandre Lesueur, and others—were some of the leading naturalists living in the United States, though they struggled to make a living. Reuben managed his family's brewery and had enough wealth that he was able to fit in his wide-ranging interests in science, agriculture, and technology on the side, even delegating the brewery work to others so he could focus more on his side projects. However, when male naturalists did not come from money as Reuben did, they often had a hard time making a career out of their passions. Many moonlighted as naturalists, working instead as bankers, doctors, lawyers, teachers, or businessmen during the day. At a time before colleges and universities developed extensive science programs, and before major institutions like the Smithsonian were established, it was very rare to find a "man of science" who paid the bills that way, let alone someone who was not primarily self-taught.

Margaretta and Elizabeth had more freedom to pursue scientific studies because there was little expectation that they would hold careers or support themselves financially. While their brother Tom may have felt similarly inclined, he decided to take a more practical path, getting a law degree instead. Since Tom was the only son in a fatherless family, his mother, aunt, and great-aunts encouraged him to find a financially stable career and they celebrated when he became a lawyer. Tom continued to send his sisters specimens, but natural history would remain strictly a leisure activity for him rather than his life's work.[16]

Still, the naturalists who surrounded Reuben Haines were so passionate about their subjects that they decided they would rather live hand-to-mouth than take up more profitable careers. Reuben's childhood friend Thomas Say had helped establish the Academy of Natural Sciences in 1812, bringing together a like-minded if motley group of science enthusiasts, many of whom might not have been elected to the more elite American Philosophical Society, where most of the members were socially prominent. The Academy of Natural Sciences was meant to be distinct from other Philadelphia institutions, like the American Philosophical Society, by focusing exclusively on exchanging ideas about natural science. The Philosophical Society, on the other hand, supported discussion on all branches of science, as well as the arts. Eventually, overlapping membership blurred the boundaries between the organizations and muted potential competition. Robert Hare, Thomas Say, Reuben Haines, Charles Alexandre Lesueur, and Thomas Nuttall all came to hold membership in both.[17]

The Academy of Natural Sciences and American Philosophical Society did not explicitly ban women from membership, but by practice they never invited women to join their meetings. While these men built communities around publications, regular meetings, and shared work spaces, the women who were excluded, like Margaretta, Elizabeth, and their friends, formed their own communities traipsing through the forest and searching for announcements of the latest lecture series to attend together.

While the Academy fostered camaraderie among the many men who became members, it did not generate any income for them. Say, who would become known as the "father of American entomology," slept under the Academy's horse skeleton at night, throwing a sheet over its back to make a tent rather than leave the collections or pay for rent. The botanist Thomas Nuttall followed Say's example and camped beneath the Academy's mastodon skeleton. While Charles Alexandre Lesueur, the French artist and naturalist, slept in an actual apartment, he complained to a friend that "fortune shuns me." Natural history was not a lucrative calling.[18]

Despite their struggles, the men were prolific. They filled the pages of the new *Proceedings of the Academy of Natural Sciences* with articles and illustrations, announcing, naming, and describing species they collected on expeditions, including their trips along the Wissahickon while visiting Reuben Haines. Thomas Nuttall used the Academy's library and collections to help complete his *Genera of North American Plants* in 1818, which would win him international acclaim. Elizabeth Morris bought her copy within a month of its release.[19]

Like the Morris sisters, these men were passionate about making sense of Philadelphia's environment. They hoped to contribute to the efforts to wrest control of naming species from European scientists who often beat them to it. With memberships in the American Philosophical Society and Academy of Natural Sciences, as well as access to publishing in their journals, these men could now lay claim to these species and declare themselves the discoverer, even if that declaration simultaneously overlooked and relied on the environmental knowledge of women, the enslaved, or the indigenous. They were pushing back on the colonial past of their country, and colonizing the environment with names of their choosing.

Many of the men in the Morris sisters' orbit took up tutoring as a way to make money so that they could continue their otherwise unlucrative scientific work, and it was in this way that Margaretta and Elizabeth received their more advanced scientific training. Later in life, Margaretta reminisced about her luck in being tutored by some of the leading American scientists at the time. "When quite a child," she wrote, "I had great advantages in the society of Say, Dr. Godman, Lesueur, and Nuttall, who all made a pet of me, and encouraged me onwards." Elizabeth similarly recalled the praise she received from Thomas Nuttall, who told her that her "drawings from nature were *botanically correct.*" At a time before women's colleges and seminaries opened, and before most universities even offered much training in the sciences, Margaretta and Elizabeth were getting one of the best scientific educations possible.[20]

During these tutoring sessions, the young women surveyed the current scientific literature and were taught how to use such texts to identify the plants and insects they were observing. Margaretta and Elizabeth began to fill the garret on the third floor of Morris Hall with bookshelves stuffed with European and American scientific texts. They also subscribed to the *American Journal of Science and Arts* published by Benjamin Silliman at Yale, which was informally referred to as "Silliman's Journal." They took art classes with Lesueur, who walked the seven miles from his apartment in Philadelphia to Germantown several days each week. In a letter home to a friend, he described stopping "three times to give my lessons to the very agreeable young ladies, who speak Greek, Latin, etc. and study botany."[21]

While their interests in the sciences spanned disciplines, Elizabeth was increasingly coming to focus on botany and Margaretta on entomology, no doubt in part because of the influence of their tutors. British-born Thomas Nuttall, in particular, was one of the leading botanists in the United States at the time, following the publication of his *Genera of North American Plants*. Thomas Say was meanwhile defending entomology "against the aspersions cast upon it by some writers and against the ridicule of the inconsiderate." Entomologists like Say struggled to shed the stigma of being eccentric bug catchers. While working tirelessly at the Academy and publishing articles, he was simultaneously drafting *American Entomology*, the first book of its kind intended to catalog North American insects. Margaretta and Elizabeth overheard many of the conversations these men were having both during their tutoring sessions and while visiting their neighbors at Wyck.[22]

The Morris sisters gleaned something even more valuable from these tutors than the practical skills of preserving, identifying, categorizing, and illustrating specimens. Their sense of wonder at how the world worked and their instinct to patiently observe were solidified as their tutors praised them for their detailed work. Margaretta and Elizabeth counted John Davidson Godman as one of their many science tutors, though he was only a few years older than them. He had a medical degree and used the Peale collections at the American Museum to write his multivolume

American Natural History, frequently giving lectures in Philadelphia on comparative anatomy.[23]

Godman wrote for a general audience, not just scientists, and his love of animals animated his writing. He was amused by animals' surprising behaviors, and he humorously described all the difficulties he had tracking them down. He was known for having walked several hundred miles just to observe a shrew mole. Godman was clearly delighted by these creatures, relishing how one popped its snout up from a mound "for the purpose of enjoying the sunshine." Having brought a shrew mole home in a basket to study, Godman wrote comically about having to chase after it as it darted around his living room, pushing furniture aside and eventually hiding behind a pile of books. He was amazed that the tiny "lively and playful" creature could manage to move heavy furniture in order to escape a perceived threat. Godman clearly embraced the technique of slow and patient study, and exhibited a deep love of his subjects—something that Margaretta and Elizabeth took to heart. There was joy to be had in their perpetual, insatiable fascination with the workings of the natural world.[24]

Though many of Margaretta and Elizabeth's tutors struggled to find economic stability, their cousin Robert Hare was having much more success. In 1818, he was hired as a professor of chemistry by the University of Pennsylvania. In many ways, Hare's trajectory stands in contrast to the Morris sisters, particularly because they shared a similar social status as white, wealthy, and well-connected Philadelphians. As a man, he was able to go to medical school and gain the training he needed to become not only a doctor but also a chemist. He gained admittance to scientific societies like the American Philosophical Society and the Academy of Natural Sciences, both of which at the time had no women members. It was through those communities that he connected with leading scientists from around the country and overseas. He shared the same kind of inquisitive nature and persistence as his younger cousins, but these opportunities and networks made it possible to rise to a level of eminence impossible for them. His new professorship would only open additional doors for him as that credential became shorthand for his expertise and respectability. He was intensely devoted to his family, though, and that devotion meant he would use his position to help Margaretta and Elizabeth as best he could.[25]

AT THE SAME TIME THAT THE SISTERS WERE FULLY DEVELOPing their passions for entomology and botany, they were also settling into their twenties when courtships and marriage would have been on their mother's and their minds. Marriage provided economic stability for young women. With so few occupations open to them, it would have been difficult to achieve financial independence, let alone maintain the lifestyles they had grown accustomed to, without marrying. Marriage also carried cultural heft inasmuch as many novels, short stories, and advice manuals touted a good marriage to a worthy partner as a sensible goal for every woman. Still, there were also a rising number of women choosing not to marry in the early nineteenth century, particularly among Margaretta's and Elizabeth's friends and more generally among the daughters of the elite. Their reasons varied but often included wanting to pursue a vocation or simply not finding a good match.[26]

Ann worried about her daughters. Her sister, Elizabeth Gibbs Willing Alleyne, lived in Barbados on an enormous sugar plantation and kept tabs on her nieces and nephew through their correspondence. In 1818, when Elizabeth was twenty-three, and Margaretta nearly twenty-one, their aunt Elizabeth sent a letter addressing Ann's worries. "I can easily infer my fellow Sister," she wrote, "into your feelings on viewing your four lovely girls in a great measure secluded from the kind of Society you would enjoy with them." Ann and her sister had grown up in the "refined

Society" of Philadelphia, as she put it, with plenty of suitors. Implying that it was perhaps better not to marry if they would otherwise have to settle, their aunt suggested that Margaretta and Elizabeth might be better off "in preferring the seclusion they are in." Elizabeth certainly was interested in finding love, as she scribbled hearts in her scrapbooks among cryptic notes and initials, filled albums with poems about marriage and love, and shared handwritten quizzes about possible suitors with friends. Margaretta, however, left little evidence that she was interested in suitors at all. She seemed to enjoy the seclusion her aunt alluded to. At the very least, she was not bothered by it.[27]

Whatever may have happened with potential courtships, Margaretta and Elizabeth structured their lives around their female friendships, weaving their passion for science seamlessly into everything. Margaretta and Elizabeth spent a great deal of time with their older neighbor Deborah Norris Logan, a sometimes cranky but always gracious host who was a magnet for many of Germantown's intelligent and unmarried young women. She regularly hosted Sarah Miller Walker, a young friend of her family, at Stenton, her home, for anything from months to a year at a time, practically adopting her as a daughter. Sarah otherwise lived with her widowed father, sister, and brothers several hundred miles away in northeast Pennsylvania. During her long stretches in Germantown with Deborah Logan, she became close friends with Elizabeth and Margaretta, as well as Ann Haines, Reuben Haines's younger cousin.[28]

Deborah, who wrote several history books, also had a passion for astronomy, so she found common ground with Margaretta, Elizabeth, and her other young friends, searching for comets at night and attending lectures. "Oh if I had a Telescope and an Instructor," Deborah admitted in her diary, "what delight I should take in cultivating some acquaintance with the Sublime Science of Astronomy." The young women would report to her on their findings, sharing what they had learned about comets, eclipses, or other astronomical experiences, using her atlas to read about historic sightings. While Deborah lacked equipment and training, she found a community of young women eager to share her enthusiasm and learn together. "They are sensible, well-informed women, knowing more on many branches of science than is usual in the present day for Belles and beaus to be acquainted with," Deborah wrote.[29]

Deborah delighted in the company of these women, sharing books and advice with them while meddling in their personal lives, even if she sometimes complained that they arrived unannounced when she was not prepared to entertain. She wrote repeatedly about how she loved the company of Margaretta and Elizabeth (though she never enjoyed spending time with their mother, who was part of her own generation). Deborah considered Margaretta and Elizabeth to be "sensible agreeable Girls." Elizabeth kept conversations interesting with her knowledge of science and literature, and Deborah delighted in how she had "more information than well-educated men frequently possess." Margaretta was a "right cheerful companion" even if she loved to gossip with stories of "treachery and disagreements." Their grandmother, Elizabeth Hudson Morris, the Quaker preacher, had long ago served as a similar sort of mentor and confidante to Deborah, and in many ways, she felt she was performing the same role for them. "It seems as if I had a hereditary Right to the acquaintance of these Girls," Deborah wrote, "their Grandmother Morris having been to me what, I suppose, I am now to them."[30]

Just as Robert Hare, Reuben Haines, and all of their tutors connected Margaretta and Elizabeth to the scientific community, so too did Deborah Logan. Though she only occasionally dabbled in astronomy, Deborah was socially connected to many leading figures in Philadelphia, partly because of her late husband George

Logan's political and agricultural activities, and partly because of her own social network. So when an eighty-three-year-old Charles Willson Peale, the famous painter and owner of the American Museum in Philadelphia, came to visit, Margaretta and Elizabeth got to spend time with him as well. At one visit when Elizabeth and Charles Peale joined Deborah for tea, Deborah was tickled that her "old venerable friend" Peale escorted Elizabeth home. "When my younger Guest took her leave," Deborah recounted, "the Old Gentleman waited on her home with all the gallantry of five and twenty, tho he is upwards of fourscore. I think I never saw anyone wear quite so well." Elizabeth was likely amused by all of this. The Haines family, the Morrises, and Deborah Logan frequently hosted each other for meals, bringing in visiting dignitaries like the governor of New York, DeWitt Clinton, or even the Revolutionary War hero General Marquis de Lafayette, as well as visiting scientists and educators. Germantown was fertile ground for young, well-connected naturalists like Margaretta and Elizabeth.[31]

Though the crowd of young women spent a good deal of time in Deborah Logan's parlor reading books and doing crafts, they also left her at home to go on local adventures, taking picnics while exploring the paths along the Wissahickon—what Deborah called "a most wild and romantic spot." These were exactly the kinds of days Elizabeth memorialized in her poem:

> *How soon those happy hours have flown!*
> *Time's feet winged with softest down*
> *In those young days—*
> *Yet still I seek each well known spot,*
> *Nor is one moss-grown bank forgot*
> *Where rest, or shelter we have sought*
> *From Sol's bright rays.*

When illustrators and painters depicted the Wissahickon throughout the nineteenth century, they portrayed it either as a sublime wilderness or as a space for picnics and courtships, as seen here in the silverplate engraving *A Pic-Nic on the Wissahickon* that accompanied an 1844 short story about love at first sight in *Graham's Lady's and Gentleman's Magazine* (vol. 25).

Margaretta and Elizabeth had found true kinship, even their own scientific community, among the young, unmarried women in Germantown who shared their love of adventures.[32]

The women also regularly attended scientific lectures at the Germantown Academy in the evenings together. In the 1820s and 1830s, public lectures became a popular form of education and entertainment for primarily middle-class and wealthy Americans. The Germantown Academy, a boarding school for boys, used public lectures as fundraisers, hosting scientists such as Thomas Nuttall, who gave a series of lectures on botany for students and community members in the summer of 1818. The scientists were able to earn money from the ticket sales, which cost five dollars each. As Nuttall's proved to be a popular series, the school offered other community programs over the years, including lectures on geology, minerals, chemistry, astronomy, and the like. Margaretta, Elizabeth, and their friends devoured them all. With so many schools, especially college programs closed to women, these sorts of lectures were a chance for the Morrises and their friends to get a taste of postsecondary education.[33]

A number of female seminaries had begun opening in the 1820s and 1830s, including Emma Willard's Troy Female Seminary in New York, Catharine Beecher's Hartford Female Seminary in Connecticut, Mary Lyon's Mount Holyoke Female Seminary in Massachusetts, as well as the coeducational Oberlin Collegiate Institute. Elizabeth, Margaretta, and their friends, however, were the age of the teachers, rather than the students, by the time these institutions opened. Having missed out on the early boom in advanced education for women, the Morris sisters had to make do with the public lectures and private tutors. Their ability to hire these tutors, their close social connections with so many of Philadelphia's scientists, and their proximity to the Germantown Academy and other venues that hosted lectures together resulted in one of the best scientific educations possible at this moment in American history.[34]

The Morris sisters also used their connection to the principal of the Germantown Academy to gain access to the school's resources after hours. Deborah Logan logged the comings and goings of the young women in her diary, noting how Elizabeth would come by to take Sarah Walker to Morris Hall overnight "promising her a glance thro' the Telescope at the Academy and a walk to the Wissahicon tomorrow." While Deborah Logan was sometimes miffed to see her houseguest whisked away, she also admired the close-knit friendships she saw among the young women. Both Margaretta and Elizabeth, enchanted with the hope of spotting comets, often took their friends on late-night trips to use the telescope that the Academy purchased with funds from one of the lecture series.[35]

While it would never become their passion, astronomy captivated Margaretta and Elizabeth, and they created social events out of meteor showers. In one of her albums, Elizabeth copied down an article from Silliman's *American Journal of Science and Arts* about the astronomer siblings Caroline Herschel and William Herschel, and their cutting-edge telescope in England. The author lingered for some time on Caroline Herschel, who became one of the first women to publish in the journal of England's Royal Society, winning awards for her research and honorary memberships in scientific societies. Elizabeth must have been particularly taken with the description of how Herschel hosted "parties of ladies" to take tea "and then, as the evening comes on, to gaze at the stars, through the largest telescope in the world."[36]

The Morris sisters and their friends found kinship with their English equivalents, delighting in the ever-expanding science of astronomy. During one meteor shower, the women woke several times in anticipation, eventually witnessing the bright light of the falling stars at three in the morning. "We all dressed as rapidly as possible, and went into the garden, where we had a good view of the heavens," Elizabeth recalled. They then ran across the street to wake neighbors so that they too could witness it, as "it was a sight not to be enjoyed alone." The wonder that they felt for insects and flowers they also felt for the solar system. Elizabeth, Margaretta, and their friends were at a loss for words as they took in the multicolored magnificence of stars shooting in all directions, lighting up the early morning skies.[37]

THE WOMEN WERE THRIVING WITH THEIR TUTORING SESsions and their friendships, but they were simultaneously facing personal losses at home. Their mother's only sibling, Elizabeth

Alleyne, died suddenly while traveling in England in February 1820. Ann, who had not seen her sister in decades, would never have the chance to see her again. The family's complicated relationship with slavery, where one great-aunt funded abolitionism while another enslaved hundreds, was further complicated by the fact that their now deceased aunt enslaved more than four hundred people, while their sister Abby's in-laws would come to host a stop on the Underground Railroad down the street in Germantown. This family encompassed so many of the contradictions and complications of white Americans in the middle of the nineteenth century.[38]

While their mother was still grieving the death of her sister, Margaretta and Elizabeth's older sister Ann died in July 1820 just after turning thirty. How she died is unclear from the family's records, but what is clear is that they were devastated, their mother especially. Elizabeth recalled decades later how deeply "depressed in spirits" her mother had been after losing her daughter. In the year that followed, Ann Willing Morris struggled to regain her own health and happiness, and her aunt Elizabeth Powel and cousin Martha Hare grew concerned. When Ann constructed her own last will and testament, the first directive she gave was that she would be buried with her daughter Ann in St. Luke's churchyard. Her children also struggled with their grief, and they copied poems into each other's albums in memory of their "dearest Spirit" Ann: "Thy place no longer knows thee, beside the household hearth, / We miss thee in our hour of woe, we miss thee in our mirth." Poetry—an outlet for the Morris sisters as they fumbled their way through love and heartbreak, scientific wonder and historic events—provided solace, too, as they navigated their grief.[39]

While the family struggled with these profound losses, there were also reasons to celebrate. In 1823, Thomas proposed to Caroline Maria Calvert, the daughter of a wealthy Maryland family. Her father determined that while Thomas had little property, he was at least "a man of good connections," given his relation to his great-aunt Elizabeth Powel. His sisters served as bridesmaids in the wedding, where they met the Calverts' friends and relatives, including Eleanor "Nelly" Parke Custis Lewis, the granddaughter of Martha and George Washington. In the summer after the wedding, Elizabeth renewed her own efforts to find love, perhaps feeling pressure as the oldest unmarried Morris sibling at twenty-eight years old. While she vowed to her other single friends that she would never marry, she also swooned over crushes, traveling to see friends and attend parties in hopes of finding a future spouse.[40]

In 1826, Elizabeth began to fall for a younger man named John Stockton Littell. Eleven years her junior, John was as interested in science and poetry as Elizabeth was and filled her album with poems he wrote that blurred the lines between love and science. A poem about anemones waxed on about beauty; another about twilight lingered on the love of growing darkness:

> *I love thee, Twilight! as thy shadows call,*
> *The calm of evening steals upon my soul.*
> *Sublimely tender, solemnly serene,*
> *Late as the hour, enchanting as the scene.*
> *I love thee, Twilight! for thy gleams impart*
> *Their dear, their dying influence to my heart . . .*

While John's labored, romantic poems were supposedly about nature, they were also very much about love. In another, he wrote, "I wander at evening and dream of her eye / I call on her name—but I hear no reply." John was enamored with Elizabeth. However, not everyone approved of this match. Deborah Logan recorded in her diary how "Elizabeth's stripling Beau," though clever and virtuous, was so young that he was still in school. He was twenty when

Elizabeth was thirty-one. "It looks too ridiculous," she wrote. She must have expressed this to Elizabeth.[41]

The eleven-year age gap felt significant to Elizabeth's family and friends, and perhaps it was partly responsible for ending the courtship, much to John's dismay. In a poem that he originally kept to himself, titled "The Wissahiccon," John wrote of his heartbreak:

> *Sweet placid stream! as lone I wander by*
> *The hills and dales that have surrounded thy waves,*
> *And as I gaze upon the dark blue sky,*
> *That high o'er hang thy deep, romantic caves.*
> *I think of her who whilome used to tread*
> *Along thy banks with lightsome step, and gay,*
> *Who with her silvery voice would make the mead*
> *Ring, with a sweet romantic tale or lay.*
> *But now alas! alone I wander o'er*
> *These banks of thine, thou lonely beauteous stream!*

The Wissahickon, site of the Morris sisters' science experiments, hikes, and picnics, had now also become a space that reminded John Littell of his pain. Elizabeth was heartbroken as well, and her friends expressed concern that her interest in dressing up was waning. She seemed to find "even the most rational enjoyments of this life" unsatisfying. They worried whether Elizabeth had given up on finding love.[42]

Margaretta, too, harbored crushes and hopes, but unlike her sister left far fewer traces of her longings. Even Deborah Logan, who cataloged her judgments of the young women's suitors in her diaries, never seemed to mention any "beaus" for Margaretta. The few traces of Margaretta's loves can be found in the scattered poems she wrote and copied into her albums and the albums of friends. One was the English poet Letitia Elizabeth Landon's 1822 poem about burning a love letter: "I could not bear another's look / Should dwell upon one thought of thine." Perhaps Margaretta burned her own love letters, given that so little remains. Still, she longed for closeness and the poems she copied relished female affection. In one poem that she may have written and published anonymously in the *Analectic Magazine* (or otherwise copied out of that magazine), she celebrated the love of women: "The breath of spring, to meet / For like morning air, is sweet / And woman's love is sweeter than roses in May." The poem continues, comparing female love to the best part of each of the other seasons: the brightness of summer, the soft silvery ray of the moon in autumn. Even death in winter brings its fleeting thrill.[43]

Margaretta also wove science into her poems. In one poem titled "Arbor Saturni" that she wrote for her friend Louisa Miller, she interlaced her interests in chemistry, geology, meteorology, and Louisa. Describing the magnificent way that a shaving of zinc dissolved in a lead acetate solution manages to form something that resembles a tree, Margaretta's poem imagines the crystals "brightly encircling a beautiful finger / Around it appears like a glory to linger." In these somewhat cryptic poems, Margaretta found space to express her passion. This poem, left as a draft in a collection of Margaretta's papers with lines crossed out and revised, was something she likely copied into Louisa's album or mailed to her alongside a letter. Margaretta assumed or knew that Louisa would be fluent enough in the chemistry involved in Saturn's Tree to understand the beauty and symbolism she meant to convey.[44]

These sorts of allusions and euphemisms meant something to Margaretta and Louisa but likely left most outsiders mystified. Perhaps they helped to obscure a romantic facet of their friendship. In the early nineteenth century, deep, intimate, even sensual relationships between women were common. Before the increased stigmatization of homosexual love and gender expressions in the late nineteenth and early twentieth centuries, such relationships between women were considered acceptable even among relatively socially conservative families like the Morrises. They were not even necessarily considered incompatible with heterosexual marriage. There were certainly relationships that pushed at the edges of cultural acceptability and made it into newspapers, such as when one partner was transgendered, but there were also lesbian partnerships that were open secrets in rural and urban communities alike.[45]

Reading Margaretta's sexuality into her scattered remaining poems, the absence of discussion of beaus, even her mother and aunt's concern about her preferring solitude to suitors means embracing some uncertainty. So much of queer history is erased both by self-censuring writers like Margaretta and the curation of archives by descendants who may have also helped burn letters and poems. Few of Margaretta's personal letters remain, and neither Elizabeth's nor Margaretta's journals ever made it to the archives. But from the breadcrumbs that Margaretta left behind, it is reasonable to assume that she preferred the companionship of women.

WHILE MARGARETTA AND ELIZABETH PINED FOR LOVERS, wrote poems, and experienced heartbreak, the one thing that remained constant was their abiding passion for science. However, the community of scientists that they cherished was beginning to fall apart. In 1825, there were murmurs that many in their circle—their teachers, their friends, their neighbors—were caught up in the excitement of building a new utopia in Indiana.

Just as Margaretta and Elizabeth's tutors gave lectures in order to make a living, they also orbited around wealthy philanthropists like the Scottish merchant and geologist William Maclure, who was energized by the potential that the United States held for educational reform and scientific innovation. After he immigrated, he helped fund modest expeditions around North America for these naturalists, while also contributing books, specimens, and scientific instruments to the Academy of Natural Sciences. The grateful naturalists elected him their president. Maclure enjoyed taking on an avuncular, patron-like role with the young idealistic scientists.[46]

When Robert Owen, a Welsh factory owner, visited the United States on a lecture tour and began to speak about the utopia he planned to start in Indiana, Margaretta and Elizabeth's tutors were immediately enchanted by the idea. New Harmony, as it would be called, was meant to disrupt the inequalities that structured so much of nineteenth-century life. The early nineteenth century was a moment when many Americans, uncomfortable with the way capitalism was transforming their communities and energized by a call for reform, felt they could create a better civilization from scratch. Owen's was one of many utopias intended to design a perfect society to improve human nature. Property would be held in common, women could vote in community elections, and everyone would work. While scientists had a hard time finding jobs in contemporary society, Owen promised that science would be celebrated in New Harmony. In many ways, Robert Owen was much like William Maclure, in that he hoped to use his considerable wealth from industrial pursuits to make good on the revolutionary promises that had buoyed earlier generations. Thomas Say, Charles Alexandre Lesueur, and other scientists were so taken up by Owen's ideas that they convinced Maclure to invest his money in the venture.[47]

As Owen toured the United States in the fall of 1825, giving lectures on New Harmony and his plan for an egalitarian, innovative society that had no need for currency or private property, the reception was mixed. Owen promised so much. A writer for the *Philadelphia Gazette* found Owen's confidence that he had found the one true solution for all social ills offputting. Owen's criticism of the Bible, too, made many in the audience raise an eyebrow or even leave, while still others applauded. Positions such as these made Owen and New Harmony seem too radical, even scandalous.[48]

Margaretta and Elizabeth heard about this utopia regularly when visiting the Haines family at Wyck. Reuben Haines was obsessed with it, which dismayed Deborah Logan. Practically rolling her eyes, Deborah described how he was "altogether absorbed at present in a contemplation of Robert Owen and his schemes," but she was relieved to hear that his wife and cousin had blocked his attempts to participate.[49]

It is unclear if the Morris sisters considered leaving Philadelphia for New Harmony. Owen's ideas promised that their own trajectories would transform in a society that embraced the equality of sexes and an educational system that emphasized the sciences and arts so directly. Elizabeth might finally have been able to go on the kinds of botanizing voyages she had long dreamed about; Margaretta might have been able to focus more of her time studying insects. Still, it would have been risky for them to leave their community, even if they longed to devote all their waking hours to the sciences. Margaretta and Elizabeth would have also agonized about leaving their mother, who was still reeling from the death of their sister Ann. Regardless, their family would not have approved of the idea, entrenched as they were in Philadelphia's elite society. Deborah Logan, of course, had little patience for the scheme and would have also warned her young friends against it. Logan mused that she saw "symptoms of this folly breaking up pretty soon."[50]

Even though Margaretta and Elizabeth decided to remain in Germantown, the creation of New Harmony had a very real impact on their lives. Margaretta felt abandoned as her instructors embarked down the Ohio River in December 1825 aboard the *Philanthropist*, which would be nicknamed the "Boatload of Knowledge." "Their removal left me in a tangled wilderness without a guide," Margaretta recollected. It was a low point in her training.[51]

During this moment of crisis, Elizabeth took Margaretta's friendship album and copied a poem into it titled "A Sister's Love":

> *When o'er my dark and wayward soul,*
> *The clouds of nameless Sorrow roll,*
> *When Hope no more her wreath will twine,*
> *And Memory sits at Sorrow's shrine*
> *Nor aught to joy my soul can move,*
> *I muse upon a Sister's Love.*

The poem, anonymously penned by Sara Coleridge and published in the British poetry journal *Etonian*, captured how stabilizing a sister's companionship could be during a period of upheaval. It was a kind of love that was "changeless." It continued:

> *When, tir'd with study's graver toil,*
> *I pant for sweet Affection's smile,*
> *And sick with reckless hopes of fame,*
> *Would half forgo the parting aim,*
> *I drop the book,—and thought will rove,*
> *To greet a Sister's priceless love.*

Just like the poetry that filled their albums, Margaretta and Elizabeth's sisterly affection would see them through these moments when they felt unmoored. They could discuss new scientific articles together, and they could continue to experiment in the garden or search for specimens in the woods together.[52]

Margaretta and Elizabeth were not the only ones feeling abandoned in the wake of the New Harmony exodus. Thomas Say, Charles Alexandre Lesueur, William Maclure, and the other scientists who left for New Harmony were some of the most active participants at the Academy of Natural Sciences. Even though Maclure stayed on as the president, running business remotely, members worried that the institution might be set adrift. Many believed that Thomas Say, always amiable and continuously present, had held the Academy and its morale together. In the absence of several key members, chaos ensued for some years. Thomas Nuttall, who had recently moved to Cambridge to run Harvard's Botanic Garden, found little reason to return to Philadelphia and the Academy now that his close friends were gone. New Harmony, in short, shook the foundation of Philadelphia's scientific community.[53]

As the utopia began to falter, those left behind savored stories of missteps and mismanagement. When a defector, Helen Fisher, stopped in Germantown and filled tea table conversations with tales of radical outfits ("Lesueur looked most inexpressibly ugly in this costume"), wild beards, unmilked cows, and neglected gardens, Deborah Logan was smug. She had predicted it all. Many women at New Harmony were dissatisfied with their workload and loss of autonomy, juggling far more labor than they had in their previous lives, as they were considered the wives of the entire community. Margaretta and Elizabeth had dodged a bullet.[54]

Though the utopia dissolved within a few years, the Morris sisters never regained their tutors. Despite considering New Harmony a "singular & unique drama or farce," Say chose to remain at New Harmony, as did Lesueur, in order to keep their scientific work going. They had few other options financially or otherwise. Thomas Say, who married Lucy Sistare, a fellow traveler on the Boatload of Knowledge, would end up dying of some combination of liver failure, typhoid fever, dysentery, and perhaps even malaria in New Harmony a few years later. His untimely death at age forty-seven, however, secured his fame among scientists.[55]

THESE YEARS IN THE LATE 1820S WERE A MOMENT OF RECK-oning for Margaretta and Elizabeth, who now were both in their thirties and trying to determine what their future might hold. With their art and science teachers gone, it was unclear how to continue their studies. Unmoored but determined, Margaretta and Elizabeth continued on with the work that they loved, using books and journals they bought and borrowed to structure their own courses. They continued to collect and write and attend lectures. Perhaps being in a "tangled wilderness without a guide" would provide opportunities they had not yet come to realize.

Still, Margaretta was devastated. She had loved having intelligent conversations about the habits and transformations of insects, and with the loss of Thomas Say and the other naturalists, she was left "ardently longing for information and instruction." This did not mean she had to stop studying the world around her. Losing track of time outside while studying insects and their worlds had brought her "inexpressible happiness." She used what she had learned from her tutors as well as what she read in books and journals to inform how she made sense of the world. Without a structured curriculum, she certainly may have taken some wrong turns, but she also learned to trust her instincts and follow what intrigued her. She found her calling, or vocation, as she put it, in entomology. "I can scarcely withdraw my attention from a glass jar when two chestnut worms are wandering up and down instead of quietly making their nest in the earth as their brothers have done," she mused.[56]

Elizabeth, too, dove headfirst into botany, continuing to collect specimens in the woods along the Wissahickon Creek, transplanting some into the Morris Hall garden and pressing others to add to her herbarium. She was building a collection that she could return to again and again to study. The women cultivated their garden, filled as it was with plants and fruit trees, as a space for observation. These were spaces they inhabited and manipulated. They hired laborers to do the physical work they could not or did not want to do, instructing them in methods, and arguing with them over techniques. Their garden, and the Wissahickon, too, were scientific spaces where they honed their research practice.

Like many of their contemporaries who were well educated and unmarried, Margaretta and Elizabeth tried their hands at becoming teachers, albeit briefly. Thanks in large part to Reuben Haines, Germantown had become a center for early childhood education, featuring one of the first kindergartens or "infant schools," as it was called, intended for working-class boys and girls. Reuben even helped recruit Bronson Alcott among other educators to run a school in the neighborhood. The Morris sisters and their circle became friends with Bronson and Abba Alcott, who gave birth to two children, Anna and Louisa May Alcott, during their brief time in Germantown. While Margaretta, Elizabeth, and many of their friends helped to run the Infant School, fundraising and occasionally assisting in the classroom, Margaretta's heart was not in it. She preferred to be spending time at Carey's bookstore in Philadelphia or in her garden taking notes on insects—not handling the day-to-day logistics of running the school. This would not be her path.[57]

In the midst of all of this, Elizabeth faced fresh heartbreak. Her former love, John Stockton Littell, who had written her poetry years prior, had begun courting her youngest sister, Susan. He even proposed, but Littell had few job prospects, and Tom Morris did not approve of this pairing. Elizabeth helped to fix this for her sister, using her social connections to find John a job at the railroad. In spite of this generosity, Elizabeth still ached. On the day of the wedding, Deborah Logan mentioned how Elizabeth stood out for being "abominably attired." Having borne witness to the drama surrounding Littell and the Morris sisters, Deborah added, "Something I am sure sits heavy upon Elizabeth."[58]

While Elizabeth always longed to find love, in some ways marriage might have hampered her ability to devote herself completely to the practice and study of botany. This was not true for everyone. Some scientific women married male scientists and essentially managed their husband's research, illustrating articles for them, organizing data, handling their correspondence, and expanding what the men could have produced alone. Despite receiving little or no credit for their contributions, "science wives" were often instrumental in defining not only their husbands' careers but also their legacy after death. Susan had some of the same training as Margaretta and Elizabeth and specifically loved ornithology, botany, and astronomy. Marriage, however, along with the expectations that she would bear children, educate them, and run the household, meant that her scientific interests were pushed to the periphery of her life. When John Littell wrote his wife's obituary several decades later, he honored her "self-sacrificing devotion," praising how her "scientific achievements were as extraordinary as they were modestly and meekly borne." Susan's path could have easily been Elizabeth's had things gone differently, and Elizabeth might have even preferred it.[59]

DEATH WAS NEVER VERY FAR FROM MARGARETTA AND ELIZabeth's lives. When their great-aunt Elizabeth Powel passed away in 1829 at the age of eighty-five, it was not unexpected, but it was a loss nonetheless, particularly for their mother, who had relied so dearly on her aunt for not only financial support but also advice at her own crossroads. Elizabeth Powel set up her niece with an annuity, and sensing that Margaretta and Elizabeth might never marry, willed Morris Hall to them.[60]

More tragic was the loss of their thirty-five-year-old tutor, John Godman. One of the few of their tutors who had not left for New Harmony, John Godman had been sick with tuberculosis for years, causing him to scale back on his lecturing as well as his adventures tracking creatures like shrew moles. Though married with three children, he seemed to grow particularly close to Elizabeth while in his sickbed. He copied a number of love poems into her album, with lines like "I think of thee, when the young morn is breaking / In radiance bright; / Thou art the earliest thought of my awakening— / My last at night." Whether they had an affair that went beyond the page is impossible to know, but they certainly shared some affection. He penned some of his own poems in her album, including one that dealt with how scientific knowledge awakens observers to both the beauty and mortality before them. After he died, Elizabeth penciled in on the page of one of his poems: "Long shall my care this sweet memorial save / The hand that traced it rests within the grave!" His death was painful, not only for Elizabeth, but also for the scientific community at large.[61]

Then Reuben Haines, so central to so many different Germantown institutions and endeavors, not to mention the lives of his family and neighbors, died in 1831. He regularly spent the night in Philadelphia after attending the weekly meetings of the Academy of Natural Sciences. One October evening after returning to his room, he apparently died by suicide after taking an excessive amount of the opiate laudanum. The story of what happened to him, so filled with shame for his family, shifted again and again, and what was initially reported to neighbors as suicide was reframed as apoplexy, or a stroke, but then switched back to laudanum. Reuben had written his will four days prior and acquired a large amount of the drug from his uncle on the day of his death. Deborah Logan wrote in her diary, "The very Demon of

Suicide seems to have taken possession of that family, otherwise so fortunate."[62]

A cholera pandemic ravaged Philadelphia and other American cities in 1832, and with its causes then unknown, many feared their lives could be at stake whether the winds changed or they ate a bad melon. Life seemed dizzyingly tenuous and fragile for the Morris sisters, who had lost still other friends to cancer and tuberculosis in quick succession. These kinds of tragedies have a way of shaking complacency out of one's life. Elizabeth came to accept that she might not marry, cracking jokes with friends and sharing humorous poems about spinsters, even if she sometimes wallowed in her loss. She was not alone, though, because she had her sister. With their housing secure, Margaretta and Elizabeth began to shape their futures together.[63]

Both Margaretta and Elizabeth found comfort with each other and in the closeness of their friendships with other women, filling parlors with uproarious laughter and letters with sly nudges and humor, referring to each other as "your whimsical ladyship." They traveled sometimes hundreds of miles with their friends, structuring their years around cherished visits. When Margaretta and her friends traveled together, one of her neighbors called the group "stout dames" in reference to their bravery. They pushed at the edges of what was considered safe for women, traveling unaccompanied by men across Pennsylvania as they visited each other's homes on various adventures.[64]

United, Margaretta and Elizabeth formed their own household. Though they were not responsible for caring for a husband or children, they were primary caregivers for their aging mother. They helped their mother run the household, and increasingly, Margaretta took over the finances and renovations of the home while Elizabeth handled the cooking, gardening, and the management of hired help. They had their roles and routines. And wherever they found time, they filled it with science.[65]

Margaretta's and Elizabeth's lives might not have been unfolding as they had once imagined. They had lost mentors, friends, relatives, and lovers. They were unmoored but they were also untethered. After clearing Morris Hall's garret of old furniture and odds and ends, they set it up as a laboratory with a library full of scientific journals and books as well as the art supplies they needed to illustrate what they found. Their tables were covered in experiments, with bell jars, a microscope for observing larvae and plant stems, and cabinets for Elizabeth's growing herbarium and Margaretta's pinned and bottled insects. While they had feared that their studies were doomed when they lost their tutors, they were only just beginning. The sisters were forging their own paths forward, and embracing their passions for science no matter the adversity they would come to face. And on the best days Margaretta and Elizabeth headed out, as they always did, to the Wissahickon Creek. With coarse dresses that could handle thorns from plants like devil's walking stick (*Aralia spinosa*) that lined the trails, and sensible shoes for traversing paths tangled in roots, they packed lunches into their vasculum cases and set off together.

Three

An Object of
Peculiar Interest

Shifting and hoisting her skirts to make her way through the narrow rows of wheat with her net, knife, magnifying glass, and containers, Margaretta was on a mission. Her neighbors' field showed signs of a fly infestation. "Prompted by no other motive than a love of study," she recalled later, "this, to me, new insect was an object of peculiar interest." She wanted to observe the fly with her own eyes, to study its behavior, to collect specimens, and, most importantly, to devise a strategy to stop its spread. Cutting off sections of infested wheat with her knife, she carried the awkward bundle home to place under bell jars in the library that she and Elizabeth had created on the third floor. She hoped to witness the flies mature and transform. As she was fond of saying, "an evil investigated and understood is half remedied."[1]

This was the summer of 1836 and Margaretta and Elizabeth, now thirty-eight and forty, respectively, had been running their household together for several years, though with their aging mother's continued involvement. Ann was not always willing to cede control of the kitchen or garden, which sometimes frustrated her daughters. There were surely growing pains, amplified by Margaretta's and Elizabeth's struggles to strike out on their own to sustain their studies after the collapse of their intellectual community. But strike out on their own they did. Each sister embarked upon her individual but parallel journey into her respective field, forging a new path.

Margaretta realized that something was amiss once she noticed that her neighbors' field was not the only one suffering. Newspapers and a few agricultural journals were beginning to report on how Hessian fly larvae were gorging themselves on young wheat plants. Margaretta noted that the fly was appearing around Philadelphia in "appalling numbers" that had not been seen for a generation. The Hessian fly, or *Cecidomyia destructor*, was capable of devastating farmers' wheat yields, creating cascading consequences for farmers and consumers alike in a culture so dependent on flour and bread.[2]

At the front lines in the fight to stave off the pests were American farmers, who had been disagreeing about the behavior of these flies and the best methods for protecting their crops for decades. In a primarily rural country, farmers spanned social classes and political parties. They filled pages of local newspapers with descriptions of infestations and tips for how to handle the problem: planting wheat after certain dates or planting certain kinds of wheat, burning stubble, or even seeking out and destroying the countless eggs by hand. Some overwhelmed farmers admitted that nothing they did seemed to help.[3]

Agricultural journals—which supplied farmers with the latest information to support their work—provided a forum for them to debate how the wheat fly behaved and how they might be able to get a handle on the problem. These journals had recently begun to publish articles by a few entomologists who specialized in agricultural pests. Some farmers appreciated this, while others balked at the prospect of men, who often lived in cities and were not farmers themselves, presuming to instruct them in ways that countered their own experience and observations. While some embraced "book farming," as it was often called, as a way to bring scientific methods into their fields, others mocked it as an unnecessary intrusion. Emotions tended to run high in the pages of agricultural journals when wheat flies came up, not only because experiences across regions differed, but also because it invoked a cultural clash between academic expertise and lived experience. So much was at stake, including farmers' livelihoods, particularly since wheat was the primary grain crop in the United States at the time.[4]

Today, entomologists recognize just how little they know about flies in the *Cecidomyia* family. Only a tiny fraction of them have been identified, described, and named—and by some estimates there are close to two million species of *Cecidomyia* worldwide. In the early nineteenth century, however, most publications on North American pests listed just three or four *Cecidomyia* at best. In other words, agricultural science had not yet advanced to a point where entomologists and farmers could readily distinguish the differences among various species of the wheat flies.[5]

As Margaretta peered into her bell jar full of larvae and wheat stalks, she was about to wade into the middle of this tumultuous landscape of active discovery and rampant contradiction. Margaretta was neither a farmer nor a man of science and she found it difficult to prove to either of these groups that she was qualified to speak as an expert. No matter her training or even her close connections to other naturalists, simply being "Miss Morris" would come to hurt her credibility.

THOUGH SHE HAD NEVER SEEN ONE BEFORE SHE WENT LOOKing for them in the wheat field, Margaretta had substantial knowledge about the Hessian fly. American farmers had called the notorious wheat fly "Hessian" because it seemed to arrive with the mercenary soldiers who fought alongside the British in the American Revolution. For decades to come, entomologists would debate whether the Hessian fly was truly Hessian—in other words from the European region that would later become Germany. Regardless of their origin, Hessian flies had devastated American wheat fields to varying degrees from the 1780s onward, flitting around the leaves, laying eggs that matured into ravenous larvae. They would then mature into pupae, ultimately emerging from their chrysalises into flies that would lay their own eggs and continue the cycle until cold weather or the end of the planting season brought it all to a halt. Seemingly fragile and resembling mosquitos, these tiny flies could wreak havoc on a devastating scale. Margaretta's tutor, Thomas Say, had observed and described the fly's life cycle in the very same fields that Margaretta was exploring. He named it *Cecidomyia destructor* (Say), describing its destructive nature in its Latin name, though it would continue to be known colloquially as the Hessian fly. Say published his findings in the first volume of the *Journal of the Academy of Natural Sciences* in 1817 with illustrations by Charles Alexandre Lesueur, back when Margaretta and Elizabeth were still students of theirs.[6]

Margaretta had set out to examine the Haines family's field fully aware that Say and Lesueur had studied the fly there. While still a teenager taking art classes with Lesueur, she had overheard the men discussing their findings. All these years later, Margaretta and her family had remained a regular presence at the Haines family's tea table, strolling through their gardens at Wyck, and going on local botanizing adventures with them. Young John Haines had been managing the small farm after his father's suicide a few years earlier, and, appreciating its impact on agriculture, he was eager to learn more about entomology. Fortunately, he had a seasoned expert for a neighbor who was happy to indulge and

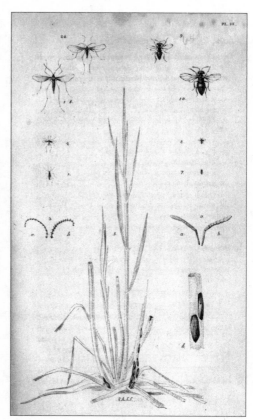

Margaretta's visual reference as she observed the flies was Charles Alexandre Lesueur's 1817 drawing of the Hessian fly and its parasite from the *Journal of the Academy of Natural Sciences.*

nurture his curiosity. John would have given permission for Margaretta to study their struggling wheat without a second thought.[7]

Back in her library, Margaretta reread Say's description of *C. destructor* in her copy of the *Journal of the Academy of Natural Sciences.* Say wrote the kind of clipped descriptions typical of entomological studies from that period: "Head and thorax black; wings black, fulvous at base; feet pale, covered with black hair." She noted that these physical traits were consistent with her specimens, but that parts of his description were imprecise: he had recorded the size of one body part relative to another instead of measuring them, and had described something as vaguely "brownish" or "whitish."[8]

It was Say's vague description of the fly's behavior, though, that gave her pause. He had begun his explanation of their life cycle with a statement that seemed to lack confidence: "The history of the changes of this insect, is probably briefly this—." That phrase—"probably briefly"—suggested that Say had made assumptions based on a limited study. He went on to describe how the female fly deposited the eggs "between the vagina of the inner leaf and the culm nearest to the root of the plant." And there, Margaretta noted a discrepancy. The flies she observed laid their eggs in the head of the wheat, amid the seeds—not in the groove between the leaf and the stem.[9]

Perhaps Say had been mistaken. Or else this was not the same fly. But given that she had collected her specimens from the same fields as Say and Lesueur, Margaretta suspected the former. By avoiding specifics, Say had "left room for these doubts." However, Thomas Say had died a premature death less than two years earlier in New Harmony, and his old friends had celebrated him as the founder of American entomology and zoology ever since. He was practically a scientific saint. Contradicting him would ruffle more than a few feathers.[10]

By the fall of 1836, it was becoming clear that Margaretta was not the only one who had observed the flies laying their eggs among the wheat seeds. As they had in previous years, farmers gave contradictory reports of the flies' behavior. Some farmers found flies laying their eggs just as Say had described, right at the point where the leaf met the sheath. But others found eggs in the grain itself. This discrepancy mattered. Farmers could only prevail against the flies if they knew where and how to target them, whether by letting fields lie fallow, planting different varieties of wheat, burning the infested plants, or some other method.[11]

More troubling, it was becoming clear through the reports that the farmers' plight was not just restricted to Philadelphia or even Pennsylvania. The infestation stretched from Virginia to upstate New York, and wheat fields that showed promise in May had withered by July, producing paltry harvests, if any. As this plight multiplied from field to field and state to state, farmers' misfortune transformed into a national problem. The price of flour in Philadelphia jumped from $6.50 per bushel in January 1836 to $11.00 just a year later. Similar increases occurred across the country. Consumers—already shaken by the other constrictions on their personal finances during what would become known as the Panic of 1837—felt this price hike in a very real way. Anger erupted in newspapers and on city streets, the most extreme example being New York's Flour Riot that winter.[12]

Spurred by the rising national anxiety, Margaretta shared her findings with her cousin, Robert Hare, explaining that contrary to Say's description, she had seen the Hessian fly laying its eggs in the head of the wheat, among its seeds. He strongly urged her to send a report to the members of the American Philosophical Society (APS), assuring her that he would even present the work himself. Women rarely sent reports to the APS, but when they did, a male member of the society typically read the paper on their behalf. Margaretta hesitated. She had not ventured into the fields at Wyck with the intention of formally publishing her observations, and if she submitted a report to the APS, her name would be in their *Proceedings* and *Transactions*.

Margaretta took her work seriously, but she had never anticipated taking such a radical step into the public sphere. She may have wanted to publish but perhaps thought the possibility was slim, given her lack of membership in scientific associations, not to mention her gender. Women authors in the mid-nineteenth century sometimes gained acclaim but were subjected to prejudice and hostility—even if they wrote for an exclusively female audience. If Margaretta agreed, her research would be scrutinized by a very masculine audience of naturalists and farmers. The APS publications did not allow for anonymity, either, which might have protected her from gendered criticism. Publishing a report was a bold step and one she was not ready for. Margaretta decided she was "unwilling to appear before the public" and refused her cousin's repeated requests.[13]

Her confidence grew, though, as she returned to the Haines family's perpetually infested wheat field in 1837 and 1838 to observe the wheat fly's behavior. What had started as a project to satisfy Margaretta's curiosity had transformed into a mission. Her passion for and obsession with the insect also grew. As her friend would later put it, "Margaretta's heart is as full of Hessian flies, as ever was a wheat field." Witnessing firsthand how her neighbor and other farmers struggled mightily with this fly, and recognizing how the crisis persisted nationally, Margaretta was driven by a sense of duty to help. Finally, in 1840, she succumbed to the persistent pleas of her cousin and allowed him to present her findings to the APS. While she may have regretted not stepping forward during the Panic of 1837 when the crisis was at its worst, her entomological knowledge still had the potential to help farmers avoid a similar economic and agricultural crisis in the future.[14]

Standing before the twenty-nine men present at the meeting on October 2, 1840, Robert Hare read his cousin's report on the Hessian fly. Margaretta was careful not to contradict Thomas Say's authority, but she pulled no punches, suggesting that Say's lack of precision was a result of incomplete information. "Had the information of Mr. Say, respecting the history of the insect, been as accurate as his knowledge of its appearance, he would not have left room for these doubts." At issue was not just *where* the wheat fly laid its eggs in June but also *how* the larvae would come to be found on a new crop of plants sown in the fall or spring, given that the fields had been cleared in between. Finding a gap in Say's description of the fly's life cycle, Margaretta wrote, "We are here left in the dark respecting the home and food of this second brood, since there is no wheat growing from June to September, when the grain is planted again." If what Margaretta proposed was true and the fly laid its eggs at the head—or grain—of the wheat, then it followed that they would be collected with the seed that farmers carefully preserved for future crops.[15]

Margaretta's argument was grounded in evidence. "The fact of the egg being laid in the grain does not, however, rest on inference," she contested. "I have actually detected the larva in the grain, when peculiar circumstances had prevented it from leaving its birth-place in order to ascend the stalk, as it is prone to do." She went on to describe several crops and seasonal plantings she observed closely, noting that the only field that was unaffected was one where the farmer obtained Mediterranean seed—and even then, it was only immune for a year before the flies claimed it. She had not found ways to destroy the eggs in the seeds without destroying the seeds too. "The only remedy," she suggested, "is to procure seed from an uninfested district."[16]

After Robert Hare finished reading his cousin's report to the fraternity of male scholars, a committee of the American Philosophical Society—consisting of the naturalists Benjamin Hornor Coates, Isaac Lea, and Thomas Nuttall—reviewed Margaretta's findings and deemed them plausible and significant enough to publish her report in the society's journal. Knowing, however, that it would be another few years before the volume would be printed, they decided to immediately alert farmers about its potential significance for their harvests. While there was little consensus among farmers, confirmation from a careful, multiyear study rather than contradictory and anecdotal reports might help to move the conversation about wheat flies forward.[17]

Benjamin Coates, a medical doctor who had interest in but no specific expertise in entomology, took the lead with the American Philosophical Society's endorsement and wrote an article on Morris's behalf for the *Farmers' Cabinet*, a relatively new but popular agricultural journal published out of Philadelphia. He described her careful method, what she had observed each year, and then offered her solution to farmers: avoid planting seeds from districts where there had been Hessian fly infestations, since the seed may be infested. Coates's article was three times the length of Margaretta's report to the APS, belaboring the basics in ways that were at times condescending to his readers. The article read like a rudimentary course in entomology rather than a warning to a community with extensive experience in pest prevention. This was not the best way to win the hearts and minds of farmers.[18]

Coates also took many liberties with Margaretta's research, describing things in ways she never had. He went on a long-winded tangent comparing plant seeds and the eggs of "lower animals" like insects and birds, before later writing, "The farmer who uses seed-wheat from a district ravaged by this animal, actually commits the absurdity of *planting Hessian flies* for the benefit of his next year's crop!!" While Margaretta did not make it fully clear in her report how the eggs transferred from the stored seeds to the new plants, she never implied that farmers were planting a crop of flies. She certainly never accused them of committing an absurdity. Given how Margaretta was not a member of the APS, it is unlikely that she had a chance to read or correct a draft of the article before it was published. She had lost control of her message at just the moment when she was stepping into the middle of a contentious debate.[19]

Reactions from readers were swift, unkind, and mostly anonymous. They were also much harsher than the usual disagreements about flies that filled the journal's pages. Perhaps the most vicious came from a writer who identified himself only by his town of "New-Garden." Accusing Margaretta of reviving a long-settled debate, he called into question her scientific skills, claiming that her findings were "opposed by the every-day experience of thousands of observant farmers." What he found most upsetting, however, was that her study had the endorsement of "imposing names" from the American Philosophical Society. In an article for the *Southern Planter*, based in Richmond, Virginia, an author reported on Margaretta's findings but mostly focused on her gender: "notwithstanding [the Hessian fly's] cunning, he has been unable to elude the feminine curiosity of the lady." Mocking Morris and discrediting her work throughout, he made a plea to readers: "As we know that some of our male friends . . . entertain different views on the subject, we invite them, if their gallantry will permit, to entertain the lists with the lady." Though encouraging debate, the author implied that engaging with Margaretta at all might come off as unchivalrous and damage the reader's honor. He used the rules of honor culture to dismiss her from the conversation. Through humor, he was calling out the fact that Margaretta had deviated from the prescribed ideals of antebellum white womanhood, laden as they were with a type of submissive domesticity, by stepping boldly into this public scientific debate.[20]

Not everyone's responses were negative, but even those who took Morris's work seriously could not get past her gender. In the *Cultivator*, an editor wrote, "Few things can more forcibly demonstrate the importance of entomological research than such discoveries, and it is honorable to Miss Morris, that she has unraveled a mystery which has long perplexed men of science." An anonymous writer to the *Farmers' Cabinet*, "W.L.H.," made the unusual suggestion that women were particularly well suited to entomology—a comment typically reserved for botany—as they were "endowed with more delicate nerves, and more acute perception, than commonly belong to our sex." These authors' reflexive need to comment on Margaretta's gender perpetuated the idea that she did not belong. She read every one of these critiques, cut them out of the journals, and put them in a file.[21]

Still, there were others who trumpeted Margaretta's discoveries. William Darlington, a botanist who lived in West Chester,

Pennsylvania, and ran the Chester County Cabinet of Natural Science, delivered a speech to his community about wheat and other grasses in February 1841. As part of his lecture, he went on a tangent about the importance of Margaretta's conclusions. He proudly introduced her as a "Lady of our own State" before citing details from her report to the American Philosophical Society. "Should the history of this little animal, as presented by MISS MORRIS, stand the test of future observation and scrutiny," he declared, "we may yet learn to take advantage of its habits, and thereby arrest its destructive career." He continued to sing her praises, proclaiming that "the researches of our ingenious countrywoman . . . will place her name on the roll of national Benefactors." More so than those who wrote anonymous letters to the *Farmers' Cabinet*, William Darlington seemed to truly appreciate the impact of her research on the nation's agriculture, citing her work alongside that of esteemed botanists and college professors.[22]

After all of the criticism Coates had seen in the *Farmers' Cabinet*, he was relieved to read a printed copy of Darlington's speech and thanked him for his "liberal and kind expression." The poor reception of his article on Margaretta's findings had been a turning point for Coates. He complained to Darlington that he had been "dragged" into the wheat fly controversy and disliked how he now had to defend Margaretta's research. He was cripplingly shy among strangers and uncomfortable in the spotlight. Still, Coates rose to Morris's defense, writing an aggressive rebuke. He may have regretted how he had presented her research, given how many of the critics had fixated on moments where he took liberties describing it. One article, for instance, had called Coates's (and by association, Morris's) idea about planting flies as absurd as a farmer planting beef "for the purpose of raising a crop of oxen." Coates was not a particularly skilled science writer, and Margaretta was bearing the consequences.[23]

Coates's response, published in the March 1841 issue, berated the pseudonymous critics for taking offense at "efforts to do fellow-citizens a service," signaling that the attacks were not only ill-placed but an ungentlemanly treatment of a woman. "Any one disposed to be severe with a lady," Coates wrote, "should be told that the observations were not designed by Miss M. for publication." Coates invoked Morris's gender, but this time as a rebuttal. The etiquette rules that policed gender could be used as both a sword and a shield.[24]

FARMERS MAY NOT HAVE TRUSTED MARGARETTA, BUT HER fellow entomologists were little better. In the mid-nineteenth century, entomologists depended on their ability to trust their peers to identify and report on species in places they themselves could not visit. Careful drawing and painting skills, as well as specialized jargon and descriptions of anatomy, were integral to communicating with other naturalists if you could not send specimens. Peer review and the support of major institutions like the American Philosophical Society were even more vital to establishing one's credibility. But even if a naturalist's drawings were perfect and the description precise, even if he or she used a microscope to amplify observational skills, social trust was the currency of America's burgeoning scientific community. And most entomologists did not trust Margaretta.

The entomologists most intrigued by and critical of Margaretta's findings were those who focused on agricultural entomology, or the study of pests that threatened farms, orchards, and gardens. Relatively few focused on the field, which was also known as economic or applied entomology, especially compared to those who studied systematics, or the classification of insects more generally. Given how predominant farming was in the antebellum United States, American entomologists were taking the lead internationally, even if their numbers were initially small. Two fastidious agricultural entomologists trying to advance the field were Edward Claudius Herrick and Thaddeus William Harris, the librarians for Yale University and Harvard University, respectively. In this messy moment early in the professionalization of entomology, there was a blurry line between enthusiasts and experts, and both were essentially one and the same. There were no graduate programs in entomology. Rare medical schools provided natural history or zoology classes. Rarer still was the employed entomologist. Herrick and Harris were amateurs in the original sense of the word—they pursued this work because they loved it. However, as two men looking to establish themselves and their favorite field, they had to repeatedly prove that they were more than just hobbyists and collectors. The stakes were high and in their desperation for their field and themselves to be taken seriously, they were part of the shift that redefined amateurs as nonexperts, even though they themselves were amateurs. They needed to distinguish themselves in order to seem professional.[25]

In many ways, Edward Claudius Herrick and Margaretta Hare Morris lived parallel lives. While Herrick was a dozen years younger, both entomologists grew up in families that prized education, both never married, and both lived with their widowed mothers late into their lives. And, most significantly, they both fell in love with insects while young. Herrick's self-initiated training also paralleled Morris's as he did not go to college because of health issues. Instead, he gained most of his knowledge of insects from books and discussions with Yale professors. By all accounts, Herrick and Morris should have been kindred spirits, but the landscape of rapidly professionalizing American science would put them very much at odds. They were rivals. Herrick was drawn to studying the Hessian fly specifically after reading an article he believed to be full of errors in 1832. He remained so devoted that he decided to cultivate a laboratory of wheat in his mother's small yard in New Haven, Connecticut. While he kept notebooks

Thaddeus William Harris, librarian at Harvard, was an agricultural entomologist and peer of Margaretta Morris. (Samuel H. Scudder, ed., *The Entomological Correspondence of Thaddeus William Harris*, 1869)

stuffed with his observations on wheat, barley, and rye pests, his interest in the natural world extended beyond disciplinary boundaries—he also swept the skies at night, fascinated by shooting stars, meteors, and comets.[26]

Thaddeus William Harris, who earned both his bachelor's degree and medical degree from Harvard, would have been delighted to be a full-time entomologist. In 1823, he confessed to Thomas Say—the chronicler of the Hessian fly—that he had "an ardent love of Natural Science." He occasionally lectured on the subject to Harvard undergraduates, though he did not hold the coveted natural history professorship. Instead, library work filled his days, and he spent time with insects whenever he could. His note-taking and collecting protocols were as meticulous as his writing. He wrote drafts for nearly every letter he sent, and his handwriting was exquisite. Although Harvard repeatedly turned him down for the professorship he sought, Harris was embedded in the intellectual community in Cambridge, which granted him privileged access to connections and resources that other naturalists could only dream of.[27]

Harris and Herrick were in close contact as Harris prepared a book on the insects of Massachusetts. The two friends exchanged dozens of letters discussing the finer points of Herrick's research on the Hessian fly: the color of its larvae, the number of segments on its antennae, and how other entomologists had described the tiny winged creature, rightly or wrongly. Nineteenth-century entomologists—without the benefit of twenty-first-century genome sequencing, let alone photography—grew frustrated identifying this notoriously indistinguishable group of flies. Given the paucity of books and articles on North American insects, Harris relied heavily on his correspondence with Herrick to publish a definitive account of Hessian flies. Herrick, an equally meticulous note taker, was essentially the regional expert on grain flies and a great resource for his fellow librarian.[28]

We can only imagine what Thaddeus William Harris might have felt when he sat down at his desk to thumb through the latest issue of the *Proceedings of the American Philosophical Society* in December 1840 and encountered Margaretta Morris's theory. The *Proceedings* did not include her full letter—it would not be printed until the Society's *Transactions* came out in 1843. He would have to settle with reading the abstract, along with the endorsement of the APS committee. Immediately, Harris recognized the threat that Morris's observations posed to his work: a significant section of his book, due to publish shortly, would be incorrect if she proved right. His years of work, study, writing, and rewriting would be for nothing. If he had an embarrassing error in his book, his authority among naturalists might prove shaky. If he ever hoped to win that professorship he sought, his work needed to be unimpeachable.[29]

Harris reached out to Massachusetts's agricultural commissioner, who confirmed that "among the farmers the notion *generally* prevails that the fly glues its eggs to the grain in the fields." That was not the answer Harris had hoped he would receive. Breathlessly—and with uncharacteristically minimal punctuation—Harris penned a hasty letter to Herrick to alert him to the brewing controversy. He needed an ally, and knowing that Herrick's interests had shifted toward astronomy, Harris also hoped to pressure him to publish an article on wheat flies and reengage with his own favorite field of science, entomology. Assuring Herrick that he trusted him over Morris, Harris wrote, "I do not wish to give currency to erroneous statements, & have hesitated about mentioning those of Miss Morris; but I am inclined to think that her observations coinciding with the peculiar notions, are worthy of notice." He urged Herrick, however, to publish something and quickly—perhaps in Benjamin Silliman's *American Journal of Science and Arts* based out of Yale—that would put the egg-laying controversy to rest, which Harris could then cite in his book when it came out. Both Harris and Herrick were familiar enough with the systems of academic legitimacy that were forming that they could deploy them against Margaretta. Harris's goal was a "*correct* history of the destructive insects." He was so convinced that Margaretta's alternative theory was wrong, he worried she and her endorsers would mislead farmers.[30]

Following Harris's advice, Herrick sent a letter to the Yale Natural History Society on April 28, 1841, that was then published by Silliman in his journal. Herrick's privileged position within male scientific circles at Yale enabled him to ask a colleague to publish something that would repress Morris's findings. Ever the perfectionist, Herrick gave readers the disclaimer that the letter was merely a "Brief, Preliminary Account" and promised to submit something more complete soon. The urgency, of course, in getting this incomplete account published was to counter Morris's theory in time to help Harris's book.[31]

While Herrick avoided naming Morris or the publications of the American Philosophical Society, to knowledgeable readers it would have been clear that he wrote to dismiss her theory and her resulting recommendation not to plant with seed from previously infested fields. Herrick took an immediate stance, addressing the placement of the eggs by noting that the female "deposits her eggs on the upper surface of the leaf (i.e., the *ligula*, or strap-shaped portion of the leaf) of the plant." Without going so far as to challenge the credibility of Morris or her endorsers at the American Philosophical Society, Herrick refuted her theory that flies lay their eggs in the grain of the wheat. He excluded her from the fraternity of entomologists by never naming her and refusing to engage with her directly.[32]

Morris and Coates were understandably annoyed with Herrick's quick dismissal of her observations. Coates had previously written a letter to Benjamin Silliman that described Morris's findings with the intention of having it published in the same journal. Yet the letter never appeared. Since her cousin Robert Hare was a longtime friend of Silliman's, Margaretta felt comfortable writing directly to him to ask why he had decided to dismiss her research. While the letter that she wrote to Benjamin Silliman has not survived, his response implies that she was concerned that the reason why Coates's letter had not been published was because she was a woman. "I do not think it was stepping beyond the line of female propriety to have your observations communicated to the philosophical society," Silliman responded. He then confirmed that he would be happy to publish any communication under her name or anonymously as from "a Lady." Increasingly aware of the gendered boundaries restricting her scientific authority, Margaretta identified and challenged Silliman's decision to privilege Herrick's perspective over her research. As it was, few women contributed to Silliman's journal prior to the 1840s, despite their rising numbers among naturalists. Though he personally welcomed women into his lectures, Silliman had, decades earlier, also pseudonymously published a series of essays lambasting Mary Wollstonecraft's argument that women and men should be treated and educated equally. Perhaps he had changed his mind by 1841, but he never did provide Margaretta with a platform to respond to Herrick or otherwise tell readers about her findings.[33]

Meanwhile, Thaddeus William Harris finished writing his *Report on the Insects of Massachusetts* in the summer of 1841, feeling all the more confident about his section on Hessian flies now that Herrick's supporting research had been published in a prestigious journal. In his extensive section on wheat insects, he praised Herrick, who "kindly permitted me to make free use of his valuable account of this insect, contained in the forty-first volume of 'The American Journal of Science,' and of other information communicated by him to me in various letters." In addition to quoting Herrick extensively, he celebrated him as a leading expert and "careful observer." Harris, throughout his entire volume and even casually in his letters to fellow entomologists, was fastidious in giving credit to others and including all possible citations. This practice highlights his subtle disdain for Margaretta Morris when he mentioned her in his book. Gone are his typical offerings of praise, gratitude, and respect. Like earlier critics in the *Farmers' Cabinet*, Harris described Morris as reviving an "old discussion," implying that it had long ago been settled and discredited. After describing her intervention briefly, he dismissed it, writing, "The fact that the Hessian fly does ordinarily lay her eggs on the young leaves of wheat, barley, and rye, both in the spring and in the autumn, is too well authenticated to admit of any doubt." Protecting his reputation should Morris's theory also prove to be correct, he allowed that perhaps the impressively adaptable fly could behave differently in different environments.[34]

Harris decided to mention Margaretta in his book, but not out of any respect for her contributions or intelligence. Anticipating that Herrick would be confused upon seeing that Morris was not completely omitted, after their lengthy exchange over discrediting her, Harris wrote him a note of explanation when he gifted his friend a copy. He had only mentioned Morris's "pretended discoveries," he explained, because they had been "warmly defended" by Coates. Harris was not willing to engage in a battle with the men of the American Philosophical Society, whom he clearly respected, without seeing the specimens himself. Challenging the legitimacy of her claims or even fully ignoring her would have meant potentially making enemies with professional colleagues. It was too much of a political risk when his own standing in the community was insecure. Still, his tone and emphasis throughout the section of his book worked to support Herrick's claims and dismiss Morris's, disregarding the possibility that Morris had discovered something that needed addressing in the wheat fields of Pennsylvania.[35]

While Harris and Herrick were busy trying to swat down the "Hessian-fly controversy" stirred up by Morris, Morris and Coates tried to further bolster their claims by presenting specimens and new findings to the Academy of Natural Sciences of Philadelphia. Coates was a member of both the elite American Philosophical Society, where Margaretta initially presented her findings the previous year thanks to her cousin Robert Hare, and the more up-and-coming Academy of Natural Sciences. Like Coates, many local naturalists held memberships in both groups, but the Academy members would have perhaps been more interested in an extended conversation and multiple publications about the flies and their life cycles, given their exclusive focus on natural history, thus encouraging Morris and Coates to direct their efforts there.

When the summer of 1841 came, Margaretta was back in the Haines family's fields to replicate her earlier findings and collect further evidence to combat her critics. Herrick's "formidable opposition roused me to renewed exertion," Margaretta would later recall, "and not for one moment doubting the truth of his observations, [I] determined to prove the correctness of my own." She could have shrunk back, but she was determined. While she said the opposite, she did in fact question Herrick's accuracy and was eager to prove that his bug was actually something more like a cutworm or beetle, rather than a Hessian fly. Luckily for entomologists (though not for farmers), wheat flies had returned that summer to wreak havoc on the crops once more. Her critics may not have trusted her, but they had to accept specimens that they could see for themselves. She just needed to collect specimens of the fly in all its forms, which was no small task, given how hard it was to get the flies to mature in captivity.[36]

Taking advantage of a summer of infestation, Margaretta continued collecting and observing, and Coates, too, went into the fields to help. Coates also continued presenting information on Morris's behalf to the Academy of Natural Sciences, wanting to give her support. At one such meeting in July, his discussion of flies spurred a larger conversation about the conflicting accounts of the entomologists at Yale and Harvard. After Coates introduced a partial set of new specimens Margaretta had collected, an extended discussion broke out before Coates had finished his presentation. One member, clearly concerned that Morris's work conflicted with Herrick's, began describing "the beauty, learning, and careful preparation of Mr. E. C. Herrick's recent memoir" among other observations made by "that gentleman." The members of the Academy were not willing to disregard Herrick's report in Silliman's journal—a thorn that would remain in Margaretta's side. The enthusiasm that the Academy members once had for Margaretta's cutting-edge research seemed to wane in the shadow of these well-known and well-connected New England men.[37]

But the tide was turning in Margaretta's favor. That very month, while busy in the wheat fields collecting specimens and observing flies, she caught sight of a female wheat fly laying her eggs among the grains of wheat. The fly, interrupted mid-ovipositioning, continued laying her eggs on Margaretta's finger. Excitedly, she wrote to Coates: "I have seen a Tipulous fly in the act of placing her eggs on or in a grain of wheat.—This fly and these eggs I have in good preservation." When Charles Alexandre Lesueur had drawn the Hessian fly in 1817, he specifically depicted both the male and the female of the species to show their differences. Comparing her specimens to Lesueur's drawings, Margaretta found that the male version of the insect that "for years destroyed the wheat in the neighbourhood of Philadelphia" looked practically the same, but the female stood out as different. Her body was entirely black or blackish-brown and her wings were "destitute of the hair fringe so conspicuous in the male." It seemed that the solution to this mystery could be found with the female flies.[38]

This was a significant discovery. While initially she had believed she was observing the same Hessian fly that Say had once described, she now realized that she had actually discovered an entirely new species of wheat fly—barely distinguishable but with different egg-laying behaviors. She had stumbled onto the fact that there were still more *Cecidomyia* flies than were known and named. Regardless of whether or not her fly was the same as Say's, hers was at least partially responsible for the recent infestations. If she had realized all of this from the start, she might have been more warmly received by farmers and entomologists. Still, the sexism that underlay so many of their critiques also shaped their reception of her research. Margaretta would need to find other ways to establish herself in their circles.

Recognizing that these men of science did not trust her and would therefore remain skeptical of her work despite the evidence, Margaretta made the strategic decision to enlist other authorities to endorse her findings. In a letter to Coates, she suggested that he present her specimens and observations to the Academy of Natural Sciences. Perhaps he could also "induce" Thomas Nuttall, the famed naturalist, to comment. Nuttall was a prominent member of the Academy of Natural Sciences, as well as a member of the committee at the American Philosophical Society that approved publication of Morris's findings the previous year. It seemed possible to curry his public support. "His name as well as knowledge I consider of importance," Margaretta explained, "particularly at this time when Mr. Herrick is so decidedly asserting that the history of the Hessian fly is so well known, and so positively assures the world that we know nothing about it."[39]

An endorsement from someone like Nuttall—with ties to both major Philadelphia scientific institutions and New England institutions like Harvard, and plenty of scientific fame and respect besides—might have silenced her critics up north. She had been tutored by Thomas Nuttall in her youth, though for whatever reason did not directly leverage her personal connection. Maybe she thought it uncouth. Nuttall never came through—likely for a host of personal reasons that kept him distracted, such as the deaths of his mother and aunt that summer, and his impending move to England. But regardless of the outcome, Margaretta was clearly considering new strategies to gain the respect of her male peers.[40]

Within days of receiving Morris's letter detailing her latest discovery, Coates read an excerpt of it at an official meeting of the Academy. And then Margaretta, after assembling all of the specimens and spending the rest of the summer observing in the fields, prepared an even more detailed account. On August 10, 1841, she submitted it to the Academy. Along with the letter, she donated what she had for years sought to capture: "specimens of the insect in all its forms, from the egg to the perfect fly." After several summers of collecting infested sheaths of wheat and putting them under bell jars in her library, only to find that the flies had failed to mature or died, she finally had a complete set.[41]

"I was eminently successful," Margaretta boasted. She had proof now that she had identified a new fly, giving credence to what so many farmers had also been seeing. This fly, which so closely resembled the Hessian fly, behaved differently enough that a more fine-grained approach might be necessary for farmers to suppress these different pests. It was a case of mistaken identity—and by giving her specimens to the Academy, she made them accessible for anyone to study, hopefully helping entomologists and farmers alike. Perhaps someday, with affirmation and support from her peers at the Academy, Margaretta would name this unnamed fly.[42]

Amid the flurry of critiques by farmers and entomologists, Margaretta never lost sight of her mission, or her passion for learning about these flies. That fall in 1841, Margaretta stopped by the Wyck house across the street for tea. At the time, only nineteen-year-old John Haines and the younger children's governess, Mary Ann Donaldson, were home. Mary Ann confided to a member of the Haines family who was away at the time that while Margaretta was "always an agreeable visitor," she could have done without "the interminable discussions about the Hessian Fly." John was so interested in entomology that he and Margaretta could talk for hours about her discoveries. While Mary Ann wished for a change of topic, she still admired Margaretta and her passion: "Her name and fame have traveled far and wide—exciting the ire of some and approbation of many—but whether her theory be true or false, her observations have at least awakened an interest in the subject and a spirit of investigation which may lead to important results."[43]

WHILE MARGARETTA HAD REMAINED QUIET AS SHE RECEIVED critique after critique, that would not last. When she had finally gotten hold of Harris's report on Massachusetts's insects in 1843, she must have flipped immediately to page 422 where the discussion of the Hessian fly began, to see if he had incorporated her findings. Disappointed to find her theory mostly dismissed, she wrestled with whether she should write to Harris to defend herself. Perhaps Elizabeth encouraged her to write and tell him about her more recent discoveries and specimens, so he might update a future edition of his book. Perhaps the file she kept filled with critiques had gotten so thick that her patience broke. Whatever the case, on August 31, 1843, Margaretta sent off a polite letter to Harris, to share her findings since the publication of his book, as she was still awaiting their publication by the Academy of Natural Sciences. Realizing that she was initiating a correspondence without any formal introduction, Margaretta stepped lightly, apologizing for her intrusion. Still, she was not meek. Within a few lines, she declared herself an entomologist and his peer, requesting the "equal privilege" of having her observations taken seriously. After copying the text of her yet-to-be-published report, she offered to send specimens to Harris so he could judge for himself.[44]

Harris never replied to Margaretta's letter. At the time that it arrived, Harris had once again been passed over for the professorship at Harvard, a professional slight that hit him hard. The publication of his book had seemed to make no difference. He also suffered from eye inflammation that made it difficult for him to use a microscope. Facing these financial, physical, and emotional hardships, Harris paused most of his entomological correspondence for several years and put Morris's letter aside. Still, his silence stung, and Morris felt "doomed to disappointment" by the implicit rejection.[45]

More importantly, however, Margaretta's decision to defend herself in this measured way prepared her for an even more direct challenge when Asa Fitch, an entomologist based in New York State, critiqued her in ways that she could not leave unanswered. Fitch, who like Harris was trained as a medical doctor but preferred natural history, was devoted to bugs. Self-consciously shy and always feeling like an outsider, Fitch was nicknamed "The Bug-Catcher" by his neighbors. His big break into the world of entomology began when he started writing articles about New York's agricultural pests for the *American Journal of Agriculture and Science*. His first assignment: the Hessian fly.[46]

In a series of three articles in 1845 that were republished together as a pamphlet in 1847, Fitch cataloged not only the habits of the fly but also the various ways Americans had described the fly over the previous half century. He included some of Margaretta's publications in the list, but toward the end of the pamphlet, he made space to dismiss her account completely, claiming, erroneously, that she must have known she was wrong because she stopped publishing on the topic. She had not stopped publishing. In fact, she had not only continued to publish, but also deposited a full set of specimens at the Academy of Natural Sciences of Philadelphia.[47]

Margaretta was furious. After all the frustration—the constant distrust, Herrick's dismissal of her work, Harris's silence—she had wanted to put the wheat fly research behind her and move on. "Mortified and disgusted, I determined to let the subject rest forever, and would have done so," she recalled. Fitch's indictment of her research, however, pushed her over the edge. She fumed that he had noticed every other publication on Hessian flies ever written but conveniently, pointedly overlooked many of hers. He was writing her out of the history of agricultural entomology. "At first I was disposed to pass him by in silence as I had my other opposers, but was overruled by my friends." Elizabeth, and the many women who had hiked with her on the Wissahickon, listened to her accounts of it all and came through. They urged her to defend herself. Emboldened, Margaretta decided to fight back, and publicly.[48]

In a polite but fierce open letter published in the April 1847 issue of the *American Journal of Agriculture and Science*, Margaretta wrote that she craved Fitch's indulgence to "point out a slight error in his statement, which has arisen from misinformation." Her tone barely concealed the rage hidden beneath genteel language. After defending her research and publication record, she put the onus on Fitch to determine what species the fly actually was, if he was so sure she was mistaken. Well aware that Fitch and other entomologists like Harris and Herrick questioned her observational skills, Morris pushed back. She emphasized her close inspection of her subjects, writing: "If Dr. Fitch will prove that the flies I so carefully watched for so many years, whose larva feeds in the centre of the straw, as seen by hundreds in this neighborhood, is 'the fly he suspects it to be,' I will acknowledge my error as frankly as I now maintain my difference of opinion." Margaretta not only emphasized her professionalism—how carefully she observed the specimens, how she used a microscope to augment her sight—but also emphasized that her observations were corroborated by "hundreds in the neighborhood" who saw the phenomenon for themselves. Fitch had challenged not only Morris and her scientific authority, in other words, but also local environmental knowledge more generally.[49]

Margaretta recognized the inequities inherent in why men of science would not engage with her in good faith. Even with the support of a committee of peer reviewers from arguably the most elite scientific societies in the country, men still sought reasons to dismiss her research. In a statement similar to the one she made in her 1843 letter to Harris, she made a plea for the trust given so freely to other entomologists:

> I do not, nor have I ever doubted the statements of gentlemen so learned in the science of Entomology . . . their assurance that they had seen the insect in its different states of egg, larva, pupa, and perfect fly, was sufficient to satisfy me that it was so; I therefore, in all fairness, claim the same indulgence from them and others, when I state that I saw, captured, and glued to a piece of paper, a fly, while in the act of depositing her eggs on a grain of wheat, so like the drawing made by Lesueur, of Say's Cecidomyia destructor, that it not only deceived me, but all to whom I showed it.

Margaretta, asking her critics to push aside any issue with her gender, called on them to treat her as an equal. And should they still not believe her, they could always see her specimens for themselves.[50]

She wrapped up her powerful rebuke of Fitch by describing the various ways over the last seven years—since her very first presentation of her observations of wheat flies—that farmers and entomologists had justified their disbelief, explaining that she must have observed insects ranging from a weevil or a curculio to a moth, anything but what she knew it truly was: a *Cecido-myia*. "I feel some surprise at being expected to confess myself in error, before I have equally strong evidence that I am so," she fired back. What was at stake was not just whether farmers and naturalists would take Margaretta and her research seriously, but also whether her critics' inability to do so might mean that a tiny, powerful pest would continue to gorge itself in American wheat fields without farmers being able to muster an educated defense.

After all that she had observed in the fields, she had felt it her duty to speak up when the problem persisted.[51]

Meanwhile, in the Academy of Natural Sciences building, Margaretta's specimens withered. These were the specimens meant to silence her critics. When these critics had questioned her observations, she had been able to point them to her publicly available specimens to examine for themselves. No longer. Parasites were common among collections at natural history museums, and they feasted on the specimens of flora and fauna filling cabinets and spilling out onto tables. Neglected and forgotten, the eggs, larvae, and insects Margaretta painstakingly collected, the ones that gorged themselves on Pennsylvania wheat, were in turn gorged upon by these moths, beetles, or silverfish. "I then believed there were two species under observations, which I hoped my specimen would prove to the satisfaction of all," Margaretta wrote, "but much to my disappointment and distress, they too were disregarded, and so neglected that they were never even subjected to a scientific observation, but suffered to lie until both the fly and eggs were destroyed." They were erased. And while Margaretta continued to seek out a new set, praying for yet another summer of infestation, this loss made it even harder to prove her credibility.[52]

MARGARETTA HAD BEEN RIGHT TO WORRY ABOUT HER TRANS-formation from a private enthusiast to an entomologist on the national stage. Farmers and entomologists did not trust her. She had stepped into an active and heavily politicized debate, and farmers saw her as an interloper who had no right to tell them they were wrong. While she had been supported and endorsed by experts, other agricultural entomologists showed her little respect. Harris, Herrick, and Fitch—insecure in their own professional standing as they struggled for agricultural entomology to be taken seriously—saw her challenge as a threat. In their eyes, Margaretta was the kind of amateur who, by association, would make them look less professional. It was easier to dismiss her, better to silence her, than to allow for the complications that her discovery of another wheat fly might mean.

Margaretta could have retreated. Instead, she refused to let the chorus of critics silence her. Her measured and defensive public letter to Fitch helped defend and build her reputation among entomologists and farmers. The influence of the strong female role models in her family as well as her tutors, who had taught her to pursue scientific truth no matter the personal cost, was clear. Less than a decade earlier, Margaretta had been reluctant for her research or name to appear in public, but now, so enraged by the disrespect of her critics and emboldened by the support of her community, she had waged a fierce battle with New York's leading entomologist in the most public way possible. In her mind, there was no turning back.

Four

WEBS OF CORRESPONDENCE

IN THE EARLY MORNING, AS ELIZABETH WALKED THROUGH the garden behind her house, she paused to admire the forget-me-nots next to the barn. The seeds came from the Pontine Marshes outside of Rome and with some tending, the tiny blue flowers with yellow centers were thriving, even spreading, in her Germantown soil more than four thousand miles away from their original home. The moist, shady spot next to the barn was perfect for the delicate, leggy flowers that stretched upward next to partridgeberries and fragrant violets, as well as a variety of tall irises. She had worked hard on this little "Eden for the birds and bees," as she called it, sometimes planting seeds gifted by traveling friends, other times finding spots for a variety of plants she collected on walks in the forest along the Wissahickon, or those she purchased from one of her favorite local greenhouses. Nursing plants to life—especially rare plants such as this *Myosotis palustris*—brought her an enormous amount of joy. For Elizabeth, these forget-me-nots were more than just a cherished, cheerful aesthetic ornament. Her success in cultivating them meant she could share them with other botanists as pressed specimens, seeds, and live plants that they could incorporate into their own gardens.[1]

While Elizabeth loved to begin her day tinkering in the garden for hours, she ended most of her nights at her desk, writing letters as late as midnight to botanists across the country. Under the flickering light of a lard lamp, she would pull out a piece of paper with a muslin-like texture, and use her favorite pen to ink the date on the top right corner. "It is a pleasant reflexion," she mused, "that friends who are too far separated for personal intercourse, have the means of communicating thoughts & feelings by pen and paper." Elizabeth and the botanists she wrote to had so much to gain from the exchange: education, specimens, theories, even companionship with like-minded friends.[2]

Gift exchanges were instrumental for building specimen collections and botanical gardens. It was, in a sense, the main economic system for botanists, particularly before they had the professional standing and institutional support to allow them to pay collectors. At a time when American botanists were striving to rival European naturalists and establish expertise in North American species, having a broad network of local experts and collectors gave ambitious botanists a chance to succeed. Women like Elizabeth Morris who were experts in their local environment were invaluable before the market economy and consumer demand gave rise to a new class of professional collectors.

These exchanges were far from exclusively economic, though. They were gestures of friendship and tokens of acknowledgment. They were a way to share knowledge and build collections when public access to institutional collections was limited. These plant specimens—plucked from fields and forests, jungles and mountains, pressed carefully and labeled, mailed across the country and sometimes even the world—were ways to teach and learn about a new environment. They were also a way to share in the particular delight of botany with other botanists and establish a community. After all, delights expand when they are shared.[3]

Elizabeth understood that she had a lot to gain by participating in scientific correspondence networks, and sought ways to be involved. Naturalists had been connecting with each other through letters for over a century. By the nineteenth century, the scientific journals that the Morris sisters subscribed to had taken on the role of distributing news of cutting-edge discoveries and theories, but letter writing continued to define the community of scholars. The letters were the line separating insiders from outsiders. Being connected meant legitimacy. These letters, with a mix of questions about species names, clarifications of confusing passages in books, requests for plants, as well as personal information about family and health, created friendships between scientists while also establishing a measure of trust and authority. They created an intimacy between members of the community that journal articles and books could never achieve alone. In the United States, none of this would have been possible without the transformative American postal system established in 1775, which spurred a communications revolution and made it inexpensive for American scientists to connect.[4]

ELIZABETH FINALLY GOT A CHANCE TO JOIN THE LARGER world of botanists in 1842, when she met William Darlington and his family while traveling through West Chester, Pennsylvania. Dr. Darlington, as Elizabeth called him, had given a speech trumpeting Margaretta's wheat fly discoveries a year earlier in 1841, and that tribute seems to have inspired the Morris sisters to stop and visit. The sisters quickly learned that he was as devoted to botany as Elizabeth was and was well connected to other scientists throughout the country and in Europe. The son of farmers, William had trained as a doctor but became passionate about politics, enough to serve two terms in the House of Representatives. By the time Elizabeth and Margaretta met him, he was the president of the Bank of Chester County, but his passion remained with botany and botanists.[5]

Elizabeth's eyes must have widened as she toured his extensive herbarium and library of scientific books at the Chester County Cabinet of Natural Science. He collected specimens for the Cabinet as if it were his primary occupation, publishing well-received books about local botany, including *Flora Cestrica*, which surveyed the plants that could be found in his county. He suggested that Elizabeth contact him when she "found botanical difficulties," and she leapt at the opportunity. While her sister had been re-creating some of her intellectual community through her struggles to gain credibility, Elizabeth had not yet found her way back to the kinds of relationships she had once had with her tutors. This was a chance to build a valuable social connection within the world she loved, and she cultivated her friendships with as much care as she did her plants. William Darlington, who referred to Elizabeth as

William Darlington, a botanist from West Chester, Pennsylvania, was not only Elizabeth Morris's mentor but also one of her closest friends.

(Archives of the Gray Herbarium, Harvard University)

"Miss Morris," declared that he regarded "all the zealous cultivators of the science as though they were his particular friends and relatives." He, too, was eager to connect with those interested in botany, no matter their gender.[6]

Elizabeth no doubt realized that an enthusiastic and well-connected mentor like William could open up worlds for her. On the eve of her forty-seventh birthday, when she sat down to write her first letter to Darlington, she had already had a lifetime of conversations with naturalists, attended lectures on botany at every opportunity, and read countless books and journal articles to keep abreast of the latest discoveries. She was, in other words, as well trained as most American botanists of her generation, apart from the medical degrees awarded to many of her male colleagues. She had an extensive herbarium and a well-tended garden filled with local and exotic flora. Discounting her own expertise with the written equivalent of a deep curtsy, she admitted: "You must be too well aware of the advantages which a free communication with one, who like you, has reached the summit of the hill of science, affords to a zealous, though humble aspirant after knowledge, to be surprised at the avidity with which I avail myself of your kindness." Flattering him with sincere admiration and underscoring her eagerness to learn from him, Elizabeth laid the foundation for the friendship she hoped to foster.[7]

Assuming that the power dynamic in the relationship would be unequal, and hoping to reciprocate his generosity, Elizabeth included botanical gifts along with her letter. After carefully reading an essay William had written on the metamorphosis of plants, she found examples of what he discussed in her garden's roses and jasmine, which she pressed and mailed to him inside a copy of a book about Venezuela. Before she had a chance to write her next letter, both William and his wife had made a visit to stay with the Morrises in Germantown. This visit, and others that would soon follow, included discussions of flowers over meals and botanizing trips in the forests around the Wissahickon. Not only was the Wissahickon botanically interesting, but it remained a space where Elizabeth cultivated her deepest friendships. This time together instantly gave William and Elizabeth the intimate familiarity of old friends who could complain about health, housework, and handwriting, while also exchanging books and specimens.[8]

Elizabeth read extensively on botany, and her budding relationship with William meant she now had access to an even larger library. He often lent her expensive European volumes, pulling them from his shelves during her visits to his house, or toting them along when he traveled to hers. She lent him her issues

Elizabeth Morris began her relationship with William Darlington by sending him this pressed specimen and drawing of the "Proliferous Rose."

(West Chester University Special Collections)

of Silliman's Journal, even instructing the booksellers at Carey's bookstore in Philadelphia to hand William her subscription when the journals arrived, so he could read the latest issues first before sending them along to her. That particular setup ended up being more trouble than it was worth, as the bookstore clerks never seemed to remember the arrangement. Still, Elizabeth and William continued to find ways to share books and journals, sometimes tucking pressed specimens or lecture notes inside. Reading the same books gave them more to talk about. Not only were they sharing knowledge, they were deepening their friendship around common ground. To save money on shipping, they often left books to be borrowed or returned at Elizabeth's brother-in-law John Stockton Littell's office in Philadelphia. The railroad facilitated their relationship, not only when they lugged books back and forth, but also when they met regularly at Carey's bookstore or elsewhere in Philadelphia on designated days. By the 1840s, train lines ran from both West Chester and Germantown into the center of Philadelphia, allowing for these frequent meetings and exchanges that might have otherwise been too burdensome to maintain.[9]

Elizabeth was keen to get advice on naming plants in their exchanges. When her brother, Thomas, asked her for help identifying a type of grass with white and green stripes he had found growing in his family's walkway in Maryland, Elizabeth turned to William as a sounding board. In addition to planting the seeds, she drew the specimen and sent it to William, beginning a dialogue about grasses. The botanical reference books for American plants were scarce and incomplete, and it was possible that many of the plants Elizabeth puzzled over did not have scientific names yet.[10]

Identifying, naming, and determining how to categorize species had been central to the study of natural history since antiquity, but even more so since the eighteenth century when Carl

Linnaeus devised a simple organizational system. Faulty as it was, the Linnaean system offered a shared professional language that collectors could use to articulate their observations about different species of plants and organize them based on numbers of pistils and stamens. Many nineteenth-century botanists—Elizabeth and William included—had adopted a new "natural" system of taxonomy that grouped plants by additional characteristics. No single system could perfectly capture the infinite variety of plants they found and studied, but it was integral that botanists establish and adopt a standard language if they were going to make sense of what they were doing.

The more complicated the systems, though, the less accessible botany became. Some American educators like Almira Hart Lincoln Phelps and Amos Eaton argued that the simpler "artificial" or Linnaean system should be taught to beginners, and they continued to use it in their popular textbooks primarily aimed at women and children. The British botanist John Lindley went so far as to say he hoped the new natural classification system would separate serious botanical science from the kind of botany that served as "amusement for ladies." For those like Elizabeth, who gladly referenced the European texts embracing the natural system like Augustin de Candolle's *Théorie Élémentaire de la Botanique*, this would not be the case. Still, taxonomic literacy was becoming one of the many ways that distinguished hobbyists from professionals, women from men.[11]

All of the collections that nineteenth-century naturalists like Elizabeth and Margaretta accumulated and distributed—the shells, the rocks and minerals, the pressed plant specimens, the pinned insects—were part of a methodology to better understand their local environment and ultimately the world. They carefully organized their collections in the garret rooms of Morris Hall and studied them repeatedly. Margaretta had an impressive cabinet of pinned insect specimens and Elizabeth an extensive herbarium full of pressed plants. It was a form of data collection, particularly in the public cabinets, museums, and botanical gardens that became repositories for scientific knowledge. The Morris sisters' close observation of plants and insects helped them become increasingly literate in the language of American environmental science. No matter where they went, nature was no mere backdrop, but an intricate network of relationships alive with meaning.[12]

BUILDING THESE PROFESSIONAL NETWORKS REQUIRED CAREfully tending personal relationships. Elizabeth, who told William that she considered his letters "bouquets to which I often return and feast," continued to effusively express her gratitude. "How shall I thank you for your long and interesting letter, and for all the trouble you took to enlighten my ignorance? Except for your ready help, I am working alone & so completely in the dark, at least when I find any difficulty in *applying* what I read, that it is the greatest relief to know that you are able [and] willing to listen to my doubts, and answer my questions." William was just as interested in the minutiae of plant identification, dissection, and collection as Elizabeth, and she knew the value of having an expert consultant.[13]

Just as Margaretta struggled to gain the trust of entomologists, Elizabeth wondered if she could trust that her relationship was genuine and not obligatory. In the politicized world of developing American science, both sisters grappled with community trust in their own way. Elizabeth was perpetually concerned about becoming a burden. "Thus far," she wrote, "the weight of obligation has been on my side, and here I much fear it must continue." In spite of William's kindness, Elizabeth was acutely aware of the power dynamic that ran as an undercurrent through their friendship. "Words are poor returns, but they are all I have to offer." That was hardly true, though, as she regularly donated books and specimens to his Cabinet. William had to reassure Elizabeth that her questions were welcome and that he truly enjoyed answering whatever she asked. Still, she worried that she was intruding on his work. "Once assured that my letters are agreeable to you," she wrote, "I think I can promise to be a regular and grateful correspondent, leaving it to you to check me, if there should be any danger of trespassing on you too often." Insecure, Elizabeth worried she came across as an amateur when posing questions that might be simply answered—the casual enthusiast that many men of science sought to exclude from their rapidly evolving institutions. She likely understood that she contributed to their friendship, intellectually and materially, but there was a constant dance, especially in early letters, to get confirmation that he recognized that too.[14]

She need not have worried—William knew what a valuable correspondent Elizabeth was. He would actually come to exchange letters with roughly 150 women scientists over his lifetime. He was as fastidious about collecting his correspondence as he was about collecting specimens. He organized all the letters from women in separate volumes from those he received from men. Still, despite Elizabeth being one of his very many correspondents, their relationship was far from transactional. While he exchanged a letter here and there, occasionally a handful, with most male and female correspondents, with Elizabeth he exchanged around 250 in each direction. William grew so comfortable with their exchanges that he did not feel the need to write drafts of his letters to her, as he did to his other correspondents.[15]

While their relationship began as mentor and mentee, it gradually developed a much more personal dimension. Elizabeth felt comfortable enough to complain about the "women's work" that kept her from her "vegetable treasures" and reading the latest botanical scholarship. While unmarried women had more freedom than married women to pursue their passions—and wealthy white women like the Morrises even more so—the domestic labor of caring for relatives and maintaining the household still fell to them. "It will be more good luck than good management," Elizabeth complained to William, "if I lose not the little brains I have, and degenerate into a complete household drudge." The familiar manner in which she spoke with him shows just how close the pair had become.[16]

During one of Elizabeth's visits to West Chester to stay with the Darlingtons in 1842, William shared a letter he had received from Harvard's new botanist, Asa Gray. Harvard had recently hired Asa for the Fisher Professorship of Natural History, disappointing the entomologist Thaddeus William Harris who had angled for that position for decades. A small, energetic workaholic, and far from wealthy, Asa was determined to succeed at his career—despite the fact that his salary was so modest he had to borrow money from his father, a farmer, just to make ends meet. Fifteen years Elizabeth's junior, Asa was becoming one of the most well-known botanists in the United States, especially after the publication of his successful *Botanical Text-book* earlier that year in 1842. After William had mentioned that Asa would be visiting him in the next few months, Elizabeth invited them both to make a short trip to Germantown to visit with her. "Think how proud I should feel to receive Dr. Gray," she wrote. "I should look down upon common folks ever after, I fear."[17]

Ultimately the trip fell through due to Asa's busy schedule, but William was still keen to connect him with Elizabeth. He knew firsthand what a serious botanist Elizabeth was and felt certain she could help Asa expand his reach and collection. When Asa accepted his position at Harvard, he told the university trustees that he wanted to cultivate a rich storehouse of North American native plants. Asa hoped to turn the university's botanical garden into a local attraction and possible source of revenue. He could also gift North American plants that he cultivated or pressed as specimens to botanists overseas and elsewhere, who would in turn

Asa Gray, Harvard University's botany professor seen here around 1841, relied on Elizabeth's specimens and plants from Philadelphia as he built his career. (*Jane Loring Gray, ed.,* Letters of Asa Gray, *vol. 1, 1894*)

send him specimens to expand Harvard's collection. Asa knew the strategic value of scientific gift-giving networks, and an extensive and varied collection would also serve to increase his standing with influential botanists overseas and throughout the United States. Building such a collection, however, would be no minor undertaking, especially faced with the simultaneous demands on his time to both teach courses and write books. Taking on this professional position meant that Asa could no longer find time to venture into the wilds of North America to collect specimens himself. He would have to rely on regional experts scattered around the country like William Darlington to obtain the native North American plants he needed. Asa had sent William a list of roughly one hundred plants and seeds he hoped he could collect for him. It was a tall order and William was clearly taken aback by the magnitude of the request. "You have, indeed, given me a list," William responded. "I cannot honestly promise to procure either roots, or seeds, of several on the list." Always somewhat morbid, William added that if he lived another year, he would give it his best shot.[18]

When Elizabeth returned to West Chester to stay with the Darlingtons, William had an idea. He shared the list Asa sent him with Elizabeth, knowing he could induce her to help Asa and relieve him of some of the responsibility. Flattered to be so trusted, Elizabeth copied down the list to take home with a promise to help in whatever way she could. William reported on all of this to Asa, singing Elizabeth's praises: "I have recently made the acquaintance of a Lady Botanist," he wrote. "Her name is Miss Elizabeth C. Morris . . . and to a zeal & energy of intellect quite unusual in her Sex, adds more extensive knowledge of Plants than any Female I have ever met with." Perhaps wary of potential prejudice—the very same that Margaretta was facing as she published her wheat fly research—William preemptively skirted Asa's possible objection to Elizabeth as a professional correspondent on the basis of her gender by emphasizing how intelligent and exceptional she was for a woman. Further underscoring Elizabeth's material value, William added: "By calling on her, I am satisfied you could engage her to make collections that would be useful to you." He floated the idea to Asa of connecting them directly, and Asa was enthusiastic. "I shall be glad to have so excellent a correspondent as Miss Morris," he confirmed, "and am grateful for your kind offices in my behalf." These connections to intelligent botanists eager to collect materials for him were integral for amassing the volume and diversity of resources necessary to create the kind of garden and herbarium that would advance Asa's career, and, in turn, the international profile of American science.[19]

William was disappointed Asa would not be visiting Pennsylvania for some time, but he kept encouraging him to come and meet Elizabeth. "If we all live until another summer, we must contrive some way to have an interview," he pleaded, "for the true lovers of Botany are too few in our Country, to be kept segregated. We cannot afford to be personal strangers to each other, when an acquaintance would contribute so much to our mutual advantage." William would continue to nudge his young friend at Harvard to build these connections for their sake and for the sake of science.[20]

William also encouraged Elizabeth to reach out to Asa herself, but she bristled at the idea. "Much as I am inclined to believe I deserve all the kind & flattering things you say of me, and highly as I should prize a botanical correspondence with Dr. Gray," Elizabeth responded, "I really want courage to obtrude myself upon his notice by opening a communication with a perfect stranger." She feared coming across as presumptuous. If there was an inherent power dynamic rooted in her friendship with William, the imbalance was exponentially amplified in a correspondence with Asa Gray. She was hinting that she wanted William to broker an introduction. In fact, she went so far as to send the seeds for Asa to William, so that he would forward them along. In the choreography of nineteenth-century epistolary etiquette, it really could come off as bold, even rude, to send a letter without an introduction from a mutual friend or acquaintance, especially if it was a letter to someone of the opposite sex or a different social or professional standing.[21]

William understood. When he forwarded the packet of seeds along to Asa, he added, "if you will acknowledge the receipt of them, *in a note to her*, I think it will prove the commencement of an interesting correspondence. She takes great interest in your operating in behalf of vegetable Science." Less than a week later, Asa followed William's advice and wrote to Elizabeth, mentioning multiple times how "obliged" he felt to her for the seeds. He offered to reciprocate by sending her plants from Harvard's Botanic Garden, and seeds he was receiving from botanists abroad, gifts that would be valuable for plant-loving Elizabeth. Asa knew how his gifts were an investment that would be repaid many times over. "My own heart is very much set upon accumulating here as many of our indigenous plants as I can possibly obtain," Asa wrote, "and Dr. Darlington gratifies me much by the information that I may count upon your co-operation in this matter." Not only would a vast collection of native plants be valuable in his gift exchanges with more powerful botanists abroad, but they might also give Asa the chance to name something new before European scientists could stake a claim—something that had frustrated American scientists for decades.[22]

Elizabeth may have been thrilled to connect with the increasingly famous Asa Gray, but she also made sure William Darlington knew where her loyalties lay. "I am much gratified by the kind interest you feel in my correspondence with your Cambridge friend," she wrote to William, "but be assured that however agreeable I may find it in the future, I can never forget that I am indebted solely to you for that, as well as a thousand other acts of kindness." She must have recognized the value in what she had to offer her correspondents—plants, seeds, favors—and felt the need to remind the not-as-famous William that her first obligation would always be to him. "There is not much danger in my being tempted to neglect my valued old correspondent, for my newer one."[23]

WHILE WILLIAM DARLINGTON WELCOMED ANY SPECIMENS Elizabeth had to offer, Asa Gray's professional standing in global botanical networks teetered on his access to them. He shared his "desiderata" list—the plants he most desired for his garden and herbarium—with her, and she with him, and the two packed boxes full of live plants and pressed specimens and mailed them to each other via Harnden's Express—the country's first express delivery service that operated along the train lines. Not all plants survived the trip, especially if they were not unpacked right away, but most did.[24]

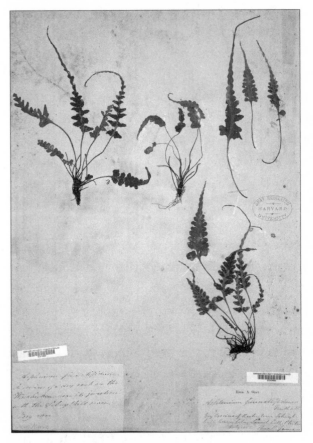

Elizabeth Morris collected this *Asplenium pinnatifidum* (on the left), a rare spleenwort, that helped Asa Gray connect with botanists worldwide. *(Harvard University Herbaria)*

One of the plants Elizabeth thought Asa might appreciate was a rare fern, the *Asplenium pinnatifidum*. Before she had set out to hunt for it, Elizabeth first checked Thomas Nuttall's notes on this lobed spleenwort in his 1818 *Genera of North American Plants*. Nuttall, who was the first to name it, had been Elizabeth and Margaretta's tutor at the time he wrote the book. He often went on botanizing adventures along the Wissahickon just as the Morris sisters had done and continued to do. When Elizabeth flipped to the Cryptogamia (plants with spores) section of the book and noticed that he mentioned finding the ferns in the crevices of rocks on the banks of the Schuylkill, she set off to locate it. Just as they had been for her tutor's generation, the forests surrounding Wissahickon Creek as it cut its way toward the Schuylkill River were a treasure trove of ferns, orchids, and mosses. The ground was covered in fallen trees, duff, leaves, and twigs, making the paths slippery after rainstorms, but also making the Wissahickon a fertile ground for the diverse plants botanists like Elizabeth sought.[25]

Elizabeth typically went botanizing with a tin vasculum strapped across her back. She would have also brought a small trowel to dig up roots, bulbs, or rhizomes and a knife for cuttings, as well as papers to separate the specimens she collected. She might have tucked her leather gloves in with her supplies if they were not too bulky.[26]

After hiking for several miles along the ravine, Elizabeth found the rare fern on the crevices of a dry rock near where the Wissahickon met the Schuylkill River. Removing some specimens, while being careful to leave others so that they could continue to thrive—insurance for further collecting—she began her hike back along the creek. At home, Elizabeth carefully pressed the plants she collected between multiple sheets of paper, arranging the specimens so they looked as natural as possible, turning some of the fronds over to show the spores underneath. Tucked between absorbent drying papers that she changed out regularly as the plants dried, she put the ferns into a press that she tightened twice a day until they were dry enough to adhere with strips of gummed paper—essentially, nineteenth-century tape—to a large, thick sheet of paper. She then labeled the specimen with its scientific name and where she collected it. Like most botanists, she organized the sheets by genera into labeled folios and tucked them in a sealed cabinet with either camphor placed on the shelves or vials of turpentine suspended from the ceiling to keep insects at bay.[27]

It was always wise to press more than one specimen, not only in case pests or mildew destroyed one, but also to share with other botanists. When Asa Gray wrote *Elements of Botany* in 1836, he told prospective collectors about the joy and importance of gifting specimens to friends. "Not only will he have the high gratification of imparting what he knows will be joyfully received, and of contributing to the enlargement and diffusion of correct knowledge, in which all true naturalists have an interest," Asa wrote, "but he will also by such means be certain of receiving, in exchange for his duplicates, the plants of those districts and countries which he might be unable to obtain by any other means; thus advancing his own attainments whilst promoting those of others." The gift exchange may have had political dimensions, but it was also integral for the expansion of botanical understanding. It spread local knowledge broadly, so that botanists like Asa Gray, who had neither the time nor ability to become an expert on the botany of Philadelphia, could benefit from the regional expertise of someone like Elizabeth. These kinds of hierarchies would come to structure professionalized science.[28]

Elizabeth collected plants primarily beneath the large, floppy leaves of the forest canopy surrounding the Wissahickon Creek. However, she dreamed of going farther afield in search of rare treasures. Such expeditions, whether funded independently or through the government, were typically off-limits to women. The eccentric botanist Constantine Samuel Rafinesque described the perils of expeditions in his 1836 book *New Flora of North America*. "Let the practical Botanist, who wishes like myself to be a pioneer of science, and to increase the knowledge of plants, be fully prepared to meet dangers of all sorts in the wild groves and mountains of America," he warned. Rafinesque wanted his readers to know just how taxing and treacherous the work had been. It would take a very strongheaded woman to defy the conventions of elite Philadelphia society and set off on that kind of adventure, even if she had been invited by the organizers.[29]

Elizabeth made do with what was available to her in her neighborhood, and that close attention to her local environment served her well. There was plenty to discover, particularly as many rising male botanists ignored plants that seemed familiar and accessible nearby in favor of exotic new finds from far-off places, marking the development of a certain kind of elitism and a privileging of wilderness. It took a new mindset coupled with these societal restrictions to realize the sublime was not just hiding in some distant landscape. It was in the familiar forest within walking distance, in the garden, and, for Margaretta, it was even in a rotten apple lying beneath a tree infested as it was with a world of creatures otherwise overlooked.[30]

Elizabeth sent a few pressed specimens of the fern to Asa, who was delighted. "Many thanks for the *Asplenium pinnatifidum*, which is the real thing," Asa cheered. "Glad am I to get the specimens." Not only would he keep one for his own herbarium, but he also planned to pass them along to others, including William Hooker, the botanist for Kew Gardens in England. Knowing that Hooker was in the midst of writing a multivolume set on ferns, *Species Filicum*, Asa raced to get the specimen to him so it could be included. Hooker was arguably the most powerful botanist in the world at the time, and the gift would help build Asa's relationship with him. "Pray find me more *Asplenium pinnatifidum* if you can," Asa encouraged Elizabeth, knowing she had struck botanical gold.[31]

While Asa could have taken full credit for the find, he was careful to note in his letter to Hooker that he had received the plant from his "Pennsylvania correspondent, Miss Morris." Hooker was delighted with the rare specimen, telling Gray that he was "exceedingly obliged." In his own muted way, he approvingly admitted, "A good fern is always acceptable to me." When he published *A Century of Ferns* a decade later in 1854, he used

Elizabeth Morris's specimen to create the image illustrating the species. In the text, Hooker did not credit Morris. However, he did emphasize that "this appears to be a rare species, and gathered by very few botanists," a nod to her exceptional talent.[32]

Asa had also written about the *Asplenium pinnatifidum* in his *Manual of the Botany of the Northern United States* (1848). In his short description of the lobed spleenwort, he mentioned that this very rare fern was found on the "cliffs on the Schuylkill and Wissahickon, near Philadelphia, *Nuttall, Miss Morris.*" He gave Elizabeth Morris credit. Asa was not like the entomologists who were so insecure in their professional standing that they were quick to discredit, dismiss, and erase Margaretta's wheat fly research. The different cultures that developed among entomologists and

Having received Elizabeth Morris's specimen of the *Asplenium pinnatifidum* from Asa Gray, Kew Garden's William Hooker was able to incorporate it into his book *A Century of Ferns* in 1854.

botanists might partly explain this, if only because women were a more traditional and accepted presence among botanists. They were not outsiders in the way that many of the entomologists considered Margaretta. But more importantly, Asa Gray had no reason to erase Elizabeth, particularly because he depended so much on her voluntary contributions. It was in his best interest to credit and elevate someone who was so clearly beneficial to his career. Unlike Nuttall's citation, he included the gendered honorific "Miss." Even the Morris sisters' supporters like Asa Gray and William Darlington segregated Elizabeth and Margaretta as women scientists, rather than keeping them on an equal standing with their male peers.[33]

Elizabeth's deep knowledge of the woods surrounding the Wissahickon made her invaluable to the botanists who called her a friend. "I have enjoyed some charming walks, lately," Elizabeth reported to William Darlington, "in search of Orchidaceous plants, Pyrolas, &c.,—which Dr. Gray wants to send to Europe." Elizabeth gladly set out to find the orchids and wintergreens that were hidden in the woods. Every time Asa requested one plant or another, Elizabeth tried to comply quickly. In this case, he wanted to send these plants to the Loddiges, a family that managed a famous and profitable nursery in England. Asa had been getting a number of gifts from them and felt indebted, so he pushed Elizabeth to supply him with sought-after North American exotics that he could send along to them so they might reproduce them for their customers. When she could not find a certain rare orchid in a spot where it typically grew, she had to resort to sending an incomplete box. She fretted, "He is so kindly attentive to me that it pains me to feel there is any request of his which I cannot comply with." However, Asa was immensely pleased with the number of plants she sent and had enough to pass them along not only to the Loddiges, but also to the St. Petersburg Imperial Garden. Elizabeth's plants took root all throughout the world.[34]

While she searched her herbarium, garden, and the forest around the Wissahickon for the plants on Asa's "desiderata" lists, Elizabeth also harvested plants that she wanted to share with him that he had not specifically asked for. In a box full of Asa's requested items, Elizabeth tucked some of her cherished forget-me-nots. Asa understood the message. The forget-me-not, after all, had perhaps the most obvious message in the Victorian floriography. "The pretty 'forget-me-not' you sent us from seeds collected on the Campagna at Rome is much admired," he wrote. The forget-me-nots thrived in Harvard's Botanic Garden, and Asa never lost an opportunity to tell Elizabeth how often they reminded him of her. While some botanists like England's John Lindley were trying to separate the language of flowers from botany as a way to further masculinize the science, Asa found it useful to be literate in both. "Your Roman Forget-me-not—the first of the genus—will long keep you in remembrance," he assured her in another letter. "We are multiplying it very much, and distributing it far and wide." He also incorporated the tiny blue flowers into his *Manual of the Botany of the Northern United States*. Having Elizabeth's plant thriving in Harvard's Botanic Garden made it possible for him to compare it with North American varieties that were often confused for the European forget-me-not.[35]

Asa came to depend on Elizabeth not just for plants but also to review his textbook as he prepared a new edition. Asa's *Botanical Textbook*, which endorsed the natural classification system, needed revisions to make it more accessible. Elizabeth was fully on board with the mission to teach Americans the updated system, but she really did not feel qualified for this role and agonized over how her criticism might threaten their budding friendship. After Asa had coaxed her multiple times, she finally relented in the face of his genuine interest in her feedback. "I am under too many obligations to him, to dare disregard any expressed wish of his if he fancies I

can be useful." She decided to call out the passages in his first edition that had puzzled her, "thus giving him all the advantage to be
derived from my stupidity," as she put it. And when she suggested
some additions he might make, she could only do so "in fear &
trembling." She waited impatiently for Asa's response, knowing
that she could very well have offended him. "Not that I had the
vanity to believe that any remarks of *mine* could be of value to
him," she confided to William. She continued to undervalue herself and her contributions. With something akin to what we might
call Imposter Syndrome today, she admitted, "I have felt a truly
humbling sense of my inability, and have been nervous upon the
subject ever since."[36]

Asa found her input invaluable, however, and the second
edition of his book reflected the thoughtful suggestions she had
made, even if he remained unhappy with the book's imperfections.
"You know I still rely upon your critical remarks," he reminded
her as he shipped a copy of the second edition to her, hoping
she would continue to give him feedback. He worried, too, that
she would be disappointed in the outcome and take it personally
when she saw instances where he could not fully adopt her suggestions. He promised to continue making the improvements she
suggested in the third edition.[37]

Elizabeth used Asa's requests for feedback as an opportunity
to learn more from him. She was interested in perfecting her seed
dissection techniques, and she found Gray's "carpological" (seed
structure) discussion in the textbook confusing enough to ask for
clarification. Glad to take on the role of mentor to a serious student, Gray responded with instructions on how to improve her
precision by cutting small seeds with a penknife on the stage of a
microscope. Almost wistfully, he wrote, "I wish I could give you a
few lessons on the subject. I am apt to forget that many of these
to me simple manipulations have in fact cost me much time and
trouble to learn at first."[38]

Asa often wrote to Elizabeth longingly about how he wished she lived nearby. He wanted her to attend his lectures so that he could know someone in the audience actually understood what he was talking about. "I wish you were one of my auditors," he sighed, "for it is a great satisfaction to feel that there are one or two listeners who will understand and enjoy all you say." He teased her that the images of fruit that he showed in his lectures would tempt her appetite. Elizabeth reciprocated the feeling. She joked to William, "I think I would give a piece of my ears [to attend], and wear caps to hide the deficiency!" Asa also wanted her to be near so she might illustrate his books for him. He even hoped she could design the wax seal he used on letters. There was a wistfulness about the help she could provide but there was something more to it, almost a flirtation. Both Elizabeth and Asa were unmarried, and their shared passion for botany and constant gifts had brought them closer. Men of science often married women who were trained enough to serve as assistants, managing their correspondence, illustrating their books, and more. Elizabeth was already doing much of that kind of work and had proven that she was trustworthy enough. While there is little surviving proof that Asa considered their relationship romantic, it is certainly possible that he saw the potential of a pairing. Personal and professional boundaries blurred and Elizabeth's gender complicated what their intimate friendship could be.[39]

In some ways, it seemed as though Asa lost track of how to categorize Elizabeth. He ended one sloppily written letter with the line "Begging your pardon for this scrawl—for I treat you like my masculine correspondents who are used to my ways." Elizabeth's reaction to this has vanished, if she ever even responded on paper. Perhaps she wondered what it meant to be treated like his "masculine correspondents." For Asa's part, it seemed to indicate that, like William, he felt a different level of intimacy with Elizabeth than with his other female correspondents. It is hard to know if this was a badge of honor for Elizabeth or if it stung, or some combination. Perhaps she just felt the mention of gender was a distraction from the science they discussed.[40]

From the start of their relationship, Elizabeth had been working hard to gain Asa's respect. She had long loved and studied the most humble plants in her garden, including dandelions, that were often overlooked by the famous botanists who considered them garden-variety weeds. She mused with William: "I watch the opening and closing of this too common but curious plant with great interest and yet hire little boys to dig it up & destroy it wholesale!" What she had discovered from observing the dandelions, or *Taraxacum officinale*, was that after they flowered, the scape or stem became prostrate in the grass. It would rise in the morning with the sun only to lay down again at night. It was only after it had fully gone to seed that it would rise higher than before and let "its beautiful feathery ball dig into the sunlight before the seeds are loosened & fly off." She had not read about this in any of her botany books and so was delighted to share her discovery with Asa.[41]

Asa did not immediately accept this dandelion news, though. He rebuffed her, assuming that he, as her mentor and a Harvard professor, knew better. "You have made an inadvertent mistake about the *Taraxacum*," he chided. "Have you not confounded this, with the curious elongation of the *stipe* of the pappus?" He had assumed she had written "scape" but meant "stipe," two different parts of the plant. Elizabeth's lost response must have been firm but polite, defending her observations of the dandelion as fiercely as her younger sister had the wheat fly. Unlike many of Margaretta's critics, however, Asa was quick to admit that he was wrong. "I cry you mercy about the *Dandelion*, where you have me at the hip," he wrote, as if they had been botanically jousting. "I was ignorant of the fact you mentioned; and heartily ask your pardon for supposing you might have made a mistake." Most delightful to Elizabeth, he added: "I shall be careful how I insinuate the possibilities of your making a blunder, after this." Elizabeth had established her credibility and earned his respect. In fact, he even incorporated what she taught him into his description of dandelions in his *Manual*.[42]

Asa depended on women to help with his work and had done so for a long time. "You see that I am likely to be a 'man of many wants,'" he warned Elizabeth, "and that you are likely to involve yourself in some trouble by your generous offers of assistance." Given the limitations of his time and resources and the demands of his position, it was no secret that he leaned on women to help move his projects forward, whether it involved his publications, herbarium, or botanical garden. While he likely compensated the artist Agnes Mitchell, who had designed the woodcuts for the first edition of his *Botanical Text-book*, he also relied on other women to assist him for free. "I receive the most useful help from the ladies," he wrote to Elizabeth. Just as he had enlisted Elizabeth to offer substantive feedback on the book, he had also found a botanist and classics scholar named Mrs. Folsom who volunteered to help proofread an updated edition.[43]

After he was invited to give a lecture series, Asa ran his drafts by Elizabeth hoping that she might offer a second opinion. These lectures paid him the equivalent of his Harvard salary ($1,000), and so were a financial necessity. He likely assumed that Elizabeth frequently attended lectures and would therefore be a knowledgeable test audience. Perhaps he just wanted her to compliment his plan and build his confidence. Asa was not known as a great lecturer, and he fussed over his oratory skills. Even though he was usually happy with the results, Asa confided to Elizabeth that lecturing distracted him from his beloved work in the garden and at his desk writing books. As he prepared for another series, he wrote to Elizabeth, admitting, "I am not *worried* about them, but I assure you they give me no little anxiety." Giving her just a brief list of topics he wanted to cover in twelve lectures, he asked that she use her "lively imagination" to make sense of his outline and suggest images that might prove helpful or "suggest things incidentally connected" that he should cover. He relied on Elizabeth extensively, and she no doubt took pride in that. She often reflected to William that she enjoyed aiding Asa in any way she could, to advance the science of botany.[44]

Countless women attended Asa's public lectures in Boston, and he was able to connect them to his network of specimen collectors. When the British algae expert William Henry Harvey from Trinity College Dublin asked Asa for help collecting seaweed from the North American side of the Atlantic, Asa set women on the task: "Several ladies have offered me to collect for him this summer, and I hope to have some interest excited in our sea-weeds." Elizabeth was happy to contribute specimens and glad for the excuse to comb beaches during her occasional trips to the Jersey Shore and Delaware Bay.[45]

In recognition of their contributions, Asa hosted a picnic for those who attended his Boston lectures to celebrate Linnaeus's birthday in 1846. The event served as both an expression of gratitude and continuous cultivation of his community of prospective collectors who, now educated, would contribute specimens to Asa's garden and herbarium. At the picnic for these female botanists, amid reciting a poem about Linnaeus penned by his proofreader Mrs. Folsom, Asa distributed clusters of forget-me-nots. These, he reminded Elizabeth, who could not be there, were "your Forget-me-nots, from the Pontine marshes—so you must not forget—as we did not—that you thus were part of our enjoyment."[46]

Asa's connections with wealthy women could also prove valuable in other ways. When he wrote expensive, illustrated books, he often needed to get subscribers to underwrite the costs of production by purchasing the book in advance. Asa asked Elizabeth outright if she would subscribe to his books, and she readily complied. Subtly pushing William to join her in her support, she added, "The subscription list should be filled to secure him from loss, for he will be at heavy expenses." Not only did she see this role as a way to sponsor her friend, it was almost an honor to be asked to do as much. Elizabeth often used her own local connections to generate still more subscribers for Asa, encouraging her close friends, neighbors, and botanical contacts. The women who attended lectures and picnics at Harvard would do the same. The illustrations, when the books were eventually published, were beautiful.[47]

Their frequent exchange of specimens meant that both Elizabeth and Asa were always in each other's debt. Asa loved to give gifts. After receiving Elizabeth's thanks for a number of dried specimens he sent for her herbarium, he said he hoped she would be able to take joy in sending the extras on to her other correspondents. "This is really one of the pleasures of Botanists, who while glad to receive are always 'ready to distribute,'" he wrote to her. With an ever-expanding garden and gifts coming in from sources around the world, Asa used this bounty to expand the breadth of his correspondents and reinvest in his stalwarts, offering Elizabeth and others gifts that they were unlikely to encounter elsewhere. After Elizabeth had sent a box of live plants, and Asa had returned an official note of gratitude from the Harvard corporation, he went on to promise more and more gifts that he would send in response. "What you most need is a complete United States Flora," Asa wrote. "Your remarks upon your herbarium give me a desire to become a contributor to it." While he did not have time to press and label native specimens that would add to that collection, he did promise to send some European plants, with a pledge that more would come. At one point he sent her two hundred specimens for her herbarium that had been sent to him from Muzio Tommasini, an Italian botanist. "I am almost demented by the acquisition of such a treasure," Elizabeth admitted to William, offering to share some of the bounty. She would always feel obliged to assist Asa with anything he asked for, given his abundant generosity toward her.[48]

The gifts that Asa sent Elizabeth simultaneously communicated his appreciation for her help while raising her profile in the scientific hierarchy to be closer in stature to himself. Even if she would never be his equal, she still gained esteem and legitimacy just by being his correspondent. Asa valued her and the specimens she gifted him and was able to repay her handsomely with his time. That said, if he had ever offered her cash instead, she would have been mortified and perhaps even insulted. There was a delicate choreography performed between botanists, calibrated based on social class, gender, and other forms of power. Asa had learned the dance moves well, and he kept Elizabeth in his debt through generous gifts and subtle flirting in their correspondence. Their relationship, as well as his professional standing, were built on these gestures of goodwill. Asa would come to pay professional botanical collectors—often working-class or middle-class men—to do this work for him, but for now, with limited funds, he was deeply reliant on talented collectors like Elizabeth and William.[49]

THROUGH THEIR GIFTS, LETTERS, AND FAVORS, WILLIAM Darlington, Asa Gray, and Elizabeth Morris became ever more tightly connected to each other. Their letters show the tangle of messaging threaded throughout. In letters to one friend, they would tuck notes to the other, knowing that the messages would be passed along. Sending a quick letter to William, Asa included a paragraph starting with "If you are writing to our good friend Miss Morris . . ." before beginning his list of requests for Elizabeth. William and Elizabeth often worried about their overburdened friend at Harvard, concerned that Asa was wearing himself out with the stress of balancing all his obligations. "A gentleman who recently returned from Boston told me that he looks pale and thin," Elizabeth confided to William, "and that he works too hard and takes too little exercise." They shared these concerns as they frequently wrote and visited, exchanging snippets of information about Asa's health and the latest projects he had grown overwhelmed with.[50]

Elizabeth also worked to include others by expanding the scientific correspondence network further. Dorothea Dix, better known for her work advocating for reform in mental asylums, was not only a close friend of Margaretta and Elizabeth Morris but also a passionate naturalist. Dorothea had anonymously published *A Garland of Flora* in 1829, a few years after she was introduced by her friend Harriet Hare—wife of Robert Hare—to Margaretta and Elizabeth. The book was a collection of poetry and literary quotations about a number of the "most interesting flowers" in hopes of both charting the complex symbolic language of flowers and inspiring others to start studying botany. Elizabeth kept her copy close, sometimes transcribing selections into her album, which she continued to contribute to all her life. They shared a love of flowers, just as Dorothea and Margaretta shared a love of entomology. Dorothea had also published articles about insects in Benjamin Silliman's *American Journal of Science and Arts* in 1831 and had long been in correspondence with Silliman and the botanist John Torrey about her collections and discoveries. She was, in other words, not only a close friend of the Morrises, but also a scientist in her own right.[51]

Dorothea's primary vocation, though, was in mental health reform. She was interested in improving the conditions of asylums around the country. She became famous for this work, traveling widely. Dorothea regularly stayed with Elizabeth and Margaretta on her travels, occasionally also making stops at William Darlington's home, as she passed through Pennsylvania on her way to political meetings in Washington, DC, or to tour asylums across the country. "She brought to me a few specimens, gathered during her journeys, and pressed in her shawl," Elizabeth wrote to William, "for even in her thorny path of duty to those who have none to befriend them, she finds, and gathers flowers."[52]

Dorothea became connected to this network of naturalists through gifts, and Elizabeth helped facilitate that. Elizabeth passed along a specimen from Dorothea to William, that Dorothea had humbly deemed "not a rarity but a remembrance," along with some archaeological finds for his Cabinet. Dorothea also sent specimens she found directly to Asa Gray, befriending him as well. Given how frequently she traveled—sometimes clocking 6,500 miles in a single trip—Dorothea was something like a one-woman expedition, collecting seeds and plants from all over the country to send to her botanical contacts. Asa eagerly passed a note along through Elizabeth, asking that Dorothea collect water lily seeds for him on the Delaware River and ship them in jars of river water. Knowing how close she was with the Morris sisters and how difficult she was to track down, Asa and William regularly sent their thanks by way of Elizabeth to "our excellent friend Miss Dix," as Elizabeth and Margaretta would see her first or at the very least know her most recent location.[53]

While Elizabeth waited impatiently at her window for spring, when "botanizing will not be confined to parlor windows and dried specimens," as she put it, male botanists were embarking on expeditions across the continent to collect and name plants in what would become New Mexico, Texas, and California. President James K. Polk embraced the idea of Manifest Destiny as part of his party platform and, once elected, had begun the westward expansion of the United States through the Mexican-American War and then through the acquisition of Oregon. Upon hearing that a botanical collector was on his way to Santa Fe, Asa Gray expressed hope that the collections would "turn this wretched war to some good account." Perhaps scientists would get some "new & nice things." In the years that followed, Asa Gray was perpetually busy cataloging and naming the plants and seeds he received from collectors who scoured these territories. Though Elizabeth could only imagine going on such pioneering expeditions, her connection to Gray meant she would eventually receive seeds from all over the American Southwest to plant in her own garden.[54]

Just as Asa sought out patrons to underwrite the costs of his publications, he also enlisted subscribers to pay collectors for their work in the American West. Knowing that Elizabeth, William, and others would be interested in expanding their herbaria to include more North American specimens, he included information in his letters about how they might purchase substantial collections before he sent them to Europe.[55]

American botanists, like their European counterparts, benefited from the colonization efforts of their nation. While Gray ostensibly despised the expansionism that President Polk embraced, he also profited from it. His professional standing teetered on his ability to acquire as many specimens as he could, name as many species as he could, and publish reports about his findings—ultimately to justify the power he was given to shape the future of botany in America. The ability to explore and do so with the protection of the military was central to making that happen. While he did not go on these expeditions himself, he could lean on the collectors doing that work in the American West. Asa and his friends saw all of this collecting, naming, and processing as an immense amount of labor for the sake of science, as they made sense of the beautiful, unknown specimens they had never before seen. Elizabeth, too, delighted in the gifts she received as she saw her herbarium expand to incorporate specimens from the Southwest.

However, this collecting was also a direct act of colonization, as it had been and continued to be for the European explorations abroad.[56]

As Asa's celebrity rose and his workload increased, his letters to Elizabeth became terse. Perhaps he grew so comfortable in their rapport that he did not feel obliged to couch all his requests for plants in niceties. Perhaps he had begun to take Elizabeth and her generosity for granted. Maybe the pleasure of a long correspondence was just another necessary sacrifice as his career bloomed. While Elizabeth's letters with William expanded over time as they shared more personal news about their families, the letters Asa sent grew distant and impersonal, almost a strictly transactional list of requests and gifts. This must have pained Elizabeth, who considered him a friend. Elizabeth served as a distant subordinate, doing her own part to advance science by advancing Asa Gray. She was paid in plants and specimens, boxed impersonally by the various gardeners at Harvard.[57]

If Elizabeth had any thoughts of a romantic relationship with Asa following his earlier flirtations, they were snuffed out in 1847. Asa quickly dropped a note that he was "in the prospective matrimonial way" in an "over-hasty," late-night letter otherwise discussing lichen identification. He had been anxious Elizabeth might learn of this news first from her cousin's wife Harriet Hare, who was acquainted with his fiancée, Jane Loring. Perhaps the increasingly terse letters had been Asa's attempt to set stronger boundaries with his former romantic prospects as he prepared to marry Jane.[58]

Aware of Elizabeth's concern about his being overworked, Asa promised her that marrying Jane would relieve so many burdens, "by diversifying my pursuits and lightening my cares." Even being engaged, Asa assured Elizabeth, "has enabled me to accomplish an unusual amount of work this spring with much better health and spirits than last year." He had shared with others that he needed to marry so that his wife could help organize his work and write letters on his behalf, a role scientific wives often played. He wrote Elizabeth that Jane "claims the privilege of looking over your later letters." Asa sometimes could not answer Elizabeth's questions because he could not get hold of a letter in Jane's possession. Most of Elizabeth's letters would ultimately disappear before they were donated to Harvard's library.[59]

After his marriage, however, Asa still felt overworked and he continued to complain about this to Elizabeth, spending most of his letters listing his various obligations and concerns about his books, trips, and the backlog of specimens he needed to process. He seemed to take her devotion for granted, curtly requesting specimens and plants. "I greatly want fruit in the *Gardenia pubescens.*" When she did not send the specimen immediately, he repeated his instruction in the next letter: "I will thank you for the fruit (one) of *Gardenia pubescens* sent by mail in a letter. Keep one or two, in case this does not suffice. Please add a seed or two separately." Elizabeth was not employed by Asa, nor was she his wife. Her gifts were just that: gifts. However, he had begun making demands as if she were failing to uphold her commitments to him. There were some occasional niceties thrown in the mix, but his tone had changed significantly from the earlier days when he was not nearly as busy or famous, and when he was far more dependent on her collections.[60]

Still, Elizabeth kept sending him plants and sponsoring his books, even when she was seriously ill and struggling to sit up, let alone write letters. For several months beginning at the start of 1849, Elizabeth suffered from "the depressing effects of liver complaints" that had rendered her bedridden and prevented her from even responding to her favorite botanists and friends, much less hiking in search of botanical novelties. She confided in William about her health, as she struggled to write to him six months into her illness. "You have kindly encouraged me to hope for a cure of this depressing disease," she wrote, "and I am always willing to hope, even when the clouds are almost lowering, that brighter hours will come." She struggled to exert energy, whether for "mind or body," and longed for the day when she had interests beyond her medicines.[61]

Mere days after she wrote about her chronic illness to William, Asa sent Elizabeth a letter. Immediately, he jumped from "Dear Miss Morris" into a request for flowering specimens of the *Magnolia grandiflora*. He was preparing a report for the Smithsonian on American trees, and Isaac Sprague, the artist he worked with, needed to see actual specimens for reference material. Writing in June, it was almost too late to get flowers from magnolia trees, but Asa hoped Elizabeth might come to his rescue. He asked that she send the specimens—seeds and blossoms—to him in a tin box by express delivery, whether she procured it herself or found a local gardener who could do the same. Despite her debilitating illness, Elizabeth sent the specimens directly to Asa within three days. Two months later, Asa finally acknowledged the gift that she had "promptly & obligingly procured."[62]

While Elizabeth and Asa's relationship grew more distant, her friendship with William grew stronger and their letters became almost entirely about their personal lives with a bit of botany interspersed. Elizabeth regularly tried to express her appreciation to William for introducing her to Asa, and for all of the mentoring he did. "I have daily cause to repeat my inability to express my feelings in words." She worried that he would not know how deeply she appreciated what he had done for her. "I am indebted for your hospitality, and abiding kindness—for the interest you have taken in my studies, and for all the advantages which you have lavished upon me," she wrote. "I wish I could prove how deep is my sense of your kindness and worth, and my gratitude for past favors!"[63]

THROUGH CORRESPONDENCE, ELIZABETH HAD REBUILT THE intellectual community she had lost when her tutors left Philadelphia so many years ago. She gained mentors and friends, becoming invaluable to them by expanding their network further. The work that she did as part of that network, however, was mostly uncredited. Elizabeth may have been content to remain in the background of America's scientific advancement, collecting hard-to-find specimens and otherwise bolstering the careers of the country's most famous botanists. She had been an important connection and supporter of Asa's as he successfully assembled his international network of collectors and taxonomists, establishing America's botanical reputation and his own professional authority. She might not have minded how credit for her hard work evaporated. The expectations and societal pressures on nineteenth-century women like Elizabeth may have even tempered her expectations about receiving any credit at all.[64]

One thing that did not disappear were her beloved flowers, the forget-me-nots "with love in [their] beautiful blue eyes" that she cared for in her garden. The once rare *Myosotis palustris* (now known as *Myosotis scorpioides*) that Elizabeth gifted in the 1840s to Asa and William, and that she would continue to gift to other botanists she met, spread throughout the United States and beyond. The botanists she sent the forget-me-nots to, Asa especially, then gifted them to dozens of donors, friends, and peers, who in turn distributed them further still. These networks established insiders and outsiders, connected members of the community, and fostered friendship, trust, and credibility among scientists. In some ways, these networks would become less crucial with the start of professional conferences and even the rise of paid collectors later in the century. However, in this moment at midcentury, inspired by the gift giving so central to science, these botanists spread the enchanting tiny blue European flower, showing just how far the botanical networks extended. Forget-me-nots also aided their own dispersal as they tend to seed themselves and are remarkably resilient. If you pull them up, the next spring you might find them sprouting again in unexpected places, to the extent that some American states categorize them as invasive today. They refuse to be forgotten, they refuse to be erased, even if one of their earliest American cultivators has been.

Five

ANONYMOUSLY FIERCE

"I CANNOT EXPRESS THE DREAD, ALMOST HORROR, I FEEL AT having my name brought before the public," Elizabeth wrote to William Darlington. She had opened up the latest issue of the *Pennsylvania Farm Journal*, a monthly magazine edited by William's son, John Lacy Darlington, and found a poem she had suggested for publication with the credit line "Recommended by Miss Elizabeth C. Morris." She recalled that upon seeing it, her "heart fairly jumped!"[1]

There was nothing particularly offensive about the poem itself. Elizabeth had recommended Frances D. Gage's "Home Pictures," which described a rural couple's decision to leave their farm for a brief trip to the state agricultural fair in Ohio. Pennsylvania farmers, she felt, might find it relatable. Gage had become famous for being an outspoken abolitionist and women's rights activist, and so it is possible that Elizabeth was wary of the consequences of their names appearing together. Gage had been targeted for her antislavery speeches with violence and arson, in an attempt to scare her into silence. It may, however, have been simpler still: Elizabeth hated seeing her own name in print.[2]

Elizabeth preferred anonymity to accolades. However, she also loved to publish her thoughts, and she especially loved the praise from readers who made practical use of her advice. Elizabeth was hardly alone in preferring anonymity. Agricultural journals in the mid-nineteenth century were full of anonymous writers, some of whom cloaked themselves in pen names for protection, while others did so to escape accountability for uncivil behaviors. During the Hessian fly debacle, Margaretta had run into the downside of anonymity as writers hiding behind pseudonyms felt free to question her credibility and attack her research on prejudicial grounds.

When authors identified themselves only by initials and town names, it was easy to assume that they were all farmers and that they were all men. However, those pseudonyms hid unknown numbers of women. Elizabeth and Margaretta were two very active authors, and while Margaretta would write just a handful of articles under "the Old Lady," "G. P.," or "M." before abandoning pseudonyms to write under her own name, Elizabeth was anonymously prolific, using the pseudonyms "E. S.," "Americanus," "A Friend to Farmers," "E.," "E. C.," "M.," and perhaps others. In just four years, from 1845 to 1849, she wrote at least seventy-seven articles in the *American Agriculturist* alone.[3]

Anonymity was a protective shield, especially for women at a time when writing under their own names might be perceived as improper or even vulgar. Women who wrote poetry, novels, handbooks, and articles wrestled with how publishing under their own name might negatively impact their lives. A female author with a byline was stepping boldly before the public, opening herself up to criticism, while also requesting credit for her work. There were men and women who saw this as a transgression into a masculine world. This is partly why Margaretta had been so hesitant at first to make her wheat fly work public. When Virginia Woolf wrote, nearly a century later, that she "would venture to guess that Anon, who wrote so many poems without signing them, was often a woman," the same could be said for the women writing public science articles in the nineteenth century. Margaretta and Elizabeth used their pseudonymous personas as a way to be public scientists and purveyors of rural lifestyle advice.[4]

Elizabeth felt her anonymity gave her freedom. While she was carefully reserved and modest in her correspondence with Asa Gray and William Darlington, always deferring to their expertise as etiquette required, in her anonymously penned articles, she was both confident and fierce. "Fear of appearing openly in print would paralyze every thought," Elizabeth admitted. Anonymity enabled her to divorce her social reputation, which she had been raised to protect, from the consequences of expressing her opinions. After witnessing both Margaretta's thrill at sharing her wheat fly discoveries and the vicious backlash that followed, Elizabeth perhaps felt further validated in her publishing preferences.[5]

Cloaked in anonymity, Elizabeth and Margaretta found their voices as science writers. In the 1840s, agricultural journals like the *American Agriculturist* were very popular in the United States, where more than half of the population lived on farms. They served as a type of science magazine, spreading information about new techniques and technologies for farms, as well as different species of fruit, the chemical composition of fertilizers, and how to properly defend crops against a range of insects and diseases. While many of these journals focused on specific regions, A. B. Allen, the editor of the *American Agriculturist*, pitched the journal as being national in scope. The children of farmers, A. B. Allen and his brother R. L. Allen had ventured into agricultural publications after having owned an agricultural supply warehouse. After just a few years, the journal, which was published out of New York City, had a remarkable ten thousand subscribers.[6]

In the 1840s, readers of the *American Agriculturist* were as hungry for scientific knowledge as the wider general public. Demand was high, and that was reflected in the circulation of a number of periodicals, including agricultural journals. Sarah Josepha Hale was weaving science articles and book reviews into *Godey's Lady's Book*, *Scientific American* had just launched in 1845 to cover activities in the US Patent Office, and *Harper's Monthly Magazine* began publication in 1850, incorporating science that its readers were eager to read. Americans, both urban and rural, women and men, old and young, wanted to learn about science, and the more complex and jargon-filled it became over the course of the nineteenth century, the more they needed popular science writers to translate results and theories and explain their significance.[7]

Tensions between "men of science" and popularizers remained high as men of science began to set and assess standards for legitimacy and credibility. Some scientists saw the benefits of cultivating a popular following and continued to write for experts and nonexperts alike. Still, there were others like Yale geologist James Dana who balked at writing to satisfy the "vulgar appetites of the people," fearing that he would have to dilute the rigor of the science he described. As women were increasingly pushed to the periphery of professional science during the nineteenth century, science writing provided an opportunity for them. Taking a cue from the women who wrote educational books—whether textbooks or the conversational books on science intended for home instruction—Margaretta, Elizabeth, and others like them helped to create the genre of science journalism that made scientific breakthroughs accessible while simultaneously sharing tips, experiments, and their own discoveries.[8]

HAVING ALREADY VENTURED INTO THE WORLD OF SCIENCE writing, Margaretta was the first of the sisters to publish in the *American Agriculturist*. In 1845, after the journal's editor

A. B. Allen had a conversation with Margaretta about yet another insect that was devastating wheat and other grains (this time a type of Rove beetle), he asked her to consider writing about it for his three-year-old journal. Margaretta agreed and submitted an article for the March 1845 issue under the androgynous pseudonym "G. P." After the reception of her Hessian fly research in the *Farmers' Cabinet* a few years prior, she was happy to don a mask of anonymity. In the article, Margaretta explained her methodology of carefully collecting the wheat midge larvae, placing them under a bell jar in her house so she could observe them as they matured unmolested by predators, and then—with the help of a strong magnifying glass—comparing the creatures with those in her entomology books in order to properly identify them. The writing was succinct and clear, giving farmers just enough information to identify, if not prevent or eradicate, the pests in their fields. By sharing her methods, she was inviting others to replicate her work, or at least consider their own methods of observation. She described poetically how during the oat harvest the miniscule insects were hard to spot, though "in the slanting rays of the sun they appear like motes in the atmosphere." Seeing Margaretta's success in writing about entomology for non-entomologists, Allen invited her to continue to write for the journal.[9]

Allen also had other ideas about the Morris sisters' involvement. That same year, in 1845, he decided to expand the readership of the *American Agriculturist*. In hopes of including material for the whole family, he introduced a "Ladies' Department" and "Boys' Department." Other agricultural journals had had Ladies' pages before, but Allen hoped his would stand out. While he and his brother felt comfortable writing material for boys about chicken coops and productivity on the farm, he felt "totally inadequate" to author material targeted at women. When he asked Margaretta to write about the insect they had discussed, he seems to have also suggested that she and Elizabeth contribute to the new Ladies' Department. While they would not be paid for their work, the sisters saw an opportunity to reach and educate a large audience.[10]

In the April 1845 issue, a month after Margaretta had made her debut as "G. P.," Elizabeth began writing under the pseudonym "E. S." for the Ladies' Department, in turn outing her gender though not herself. While she would eventually try to sculpt a character for E. S. as a southern woman from the seemingly imaginary town of "Eutawah," she started off simply using initials. Her first article praised the editor for creating a Boys' Department but argued that there should be one for farmers' daughters as well. While girls should feel comfortable doing anything that their brothers do, she argued, they might benefit from information for particularly female activities. She toed a line of simultaneously asking the editor to recognize and make space for young girls while also drawing a boundary around the kinds of activities that would be appropriate for them. Still, she found ways to turn acceptably feminine activities, like preserving eggs or planting flowers, into scientific work. One of her proposed projects for young girls, for instance, was to collect seeds from around the garden and begin the process of labeling and organizing the seeds alphabetically. Elizabeth promised "to teach the girls some of the important properties and uses of plants, as well as their botanical names." While A. B. Allen ultimately did not create a Girls' Department, he encouraged Elizabeth to contribute regularly to both the Boys' and Ladies' pages, where she would continue to intersperse material for girls.[11]

Unlike in her letters with Darlington, Gray, or other botanists, where she barely ever mentioned religion or God, Elizabeth made an argument that girls should learn about botany because it would allow them to better contemplate religion: "The more they study the works of their bountiful Creator, the better they will love Him, and the happier they will be." Margaretta, too, almost never wrote about religion in print or in letters with peers, but found the *American Agriculturist* a good venue to discuss how the close study of insects might be valuable to religious readers. There is "pleasure derived from the ingenious habits and instincts of these little creatures, given to them for their support and protection by their great Creator," she wrote, "and while studying their wonderful forms and curious histories," observers might learn more about how God organized the world. "Whatever God has thought worthy of his creation and care," Margaretta wrote, "is a fit study for us, their fellow creatures." This was the same natural theology that encouraged religious schools to embrace science classes for girls and boys, and sought—through an "argument by design"—to show skeptics evidence of God in nature. By closely observing God's creations, the scientist might get closer to understanding the divine. Elizabeth and Margaretta clearly believed that one of the best ways to entice readers of the journal to take up scientific pursuits was to argue that it would amplify and complement their religious activity. Whether this was a guiding principle in their own lives is unclear, but at the very least, they seemed to think it would be an effective hook for their audience.[12]

Like her sister, Margaretta was eager to convince women that science would improve their lives. In her first article specifically for the Ladies' Department, "What Women May Do," Margaretta suggested that women consider creating social clubs, not unlike what had existed for her grandmothers' generation during the American Revolution. While the older women had come together to discuss topics like sewing, preserving foods, and other domestic arts that aided the boycott of British goods, Margaretta suggested that modern women in the 1840s might gather to read and discuss scientific books about botany, chemistry, and entomology.[13]

She did not stop there. Margaretta argued that a basic understanding of insects and their life cycles could help housewives avoid a host of disasters. Woolen moths, she argued, could be a costly and destructive problem for clothing if knowledgeable women did not intervene. Describing the creatures' habits, Margaretta explained how to best prepare woolen, fur, and cotton clothes for storage by wrapping them in a thick linen surrounded with pepper, camphor, or turpentine. Understanding the infinite variety of household pests, from flies to cockroaches, could help anyone managing a house to develop educated strategies to avoid or handle an infestation. Other writers, like Catharine Beecher, gave readers tips on how to get rid of common pests with various different poisons like arsenic, cobalt, smoke, or even boiling water. Margaretta went into far more detail, educating readers not only about the life and activities of the insects but also about the most effective way to stop their progress.[14]

Margaretta wanted to get women and children outdoors, and she offered plenty of advice for ways they could also learn about and manage insects in their gardens. Orchards—vital to family fruit baskets and income—would only improve if women and children prevented pests from returning year after year. She provided mothers with experiments to do with children, such as putting fallen fruit in a closed cup with dirt to watch plum weevils evolve and emerge. "Many persons believe they have no time to waste on the study of a bug or worm," Margaretta wrote, "but let them think for a moment on the immense amount of produce destroyed by these little creatures in a year, and they will feel that a few minutes of each day devoted to their destruction will be well spent." By focusing her public writing on practical domestic applications, she hoped to convince readers that scientific knowledge was not exclusively the domain of men of science.[15]

While many might have dismissed kitchens as a site of serious science experiments, Margaretta did not see it that way. "Every cook is a practical chemist in a small way, though she may not be aware of it herself," she argued, "and while she is boiling soap, making yeast, or bleaching her linen, she is performing some nice experiment in chemistry which requires knowledge and practice to perform." Margaretta and Elizabeth found science to be applicable in all aspects of their lives, indoors and out. Frustrated by recipes without specific measurements, Margaretta suggested that exact proportions and details about preparation and cooking would allow for more consistent success in the kitchen. The kitchen ought to be a space for science and repeatable experiments. By showing how science was an essential part of domestic life, Margaretta was legitimizing, even elevating, women's work. Generations later, twentieth-century home economists would be making the same kinds of arguments.[16]

Elizabeth embraced her sister's call for more scientific management in households. As her passion for publishing grew, she authored recipe after recipe in the Ladies' Department. She gave detailed instructions for making butter, including recommendations for the appropriate technology. When the US Patent Office released its agricultural report for 1845, they included her anonymous article in the appendix on dairy innovations. Butter, often featured at agricultural fairs in competitions, was a contentious topic and one she regularly returned to. She also published simple recipes for apple pudding, apple butter, doughnuts, potato starch, and mushroom catsup, describing in painstaking detail all the steps—labor-intensive and time-consuming as they were—to achieve what had taken her years to perfect. Elizabeth instructed readers on how best to freeze and defrost meat and even how to design a superior space to smoke meats. Her fans, in turn, wrote to the journal in gratitude, proclaiming that the article about defrosting meat was "worth a year's cost of the *Agriculturist*."[17] In Elizabeth's recipe for ginger syrup, like all her recipes, she made sure to provide specific measurements so that readers could replicate her work:

HOW TO MAKE GINGER SYRUP

SOAK, for twenty four hours, in warm water, one pound of West-India ginger root; rub it well, and boil it in one gallon of water till reduced to three quarts; strain it through a cloth, and to every pint of water put one pound of loaf sugar, and boil it to good syrup, skimming it well. When cold, bottle it for use, and it will keep in a cool place, for any reasonable length of time; and a small quantity, mixed in a tumbler of fresh water, makes one of the most refreshing and healthful beverages that can be drunk during hot weather. It also has the double advantage of being easily made, and fit for immediate use. Another method, even less troublesome, is, to make a rich syrup of water and loaf sugar, and when cool enough to bottle, add to every pint of the syrup, two tea-spoonsful of the best tincture of ginger, which can be purchased at an apothecary's. E. S.

Elizabeth's recipe for a good life did not stop with food. She wrote long articles on how to live a more healthy and productive life. She worried about the chemicals women applied to their faces when they smeared rouge on their cheeks or dusted powders on their faces and instead encouraged young women to exercise to achieve a healthy complexion. They should also put on thick leather shoes, she argued, and walk six to eight miles each day, no matter the weather. Elizabeth gave readers recipes for cold creams, almond pastes, and castile soaps. Good hygiene, daily cold baths, fresh air, and plenty of exercise were her antidotes for just about everything.[18]

Elizabeth's favorite suggestion was that everyone should wake up early. Busy mothers, burdened once their family awoke, could "steal time" in the morning to achieve countless things: they might write entire history books like her now deceased neighbor Deborah Norris Logan, learn a new language, tend to the garden, or take an early walk if they just woke up at four or five in the morning instead of seven. Elizabeth, herself, hated to get her feet wet by walking before the sun dried the morning's dew, so instead washed from head to toe with cold water on a coarse towel to encourage "perfect cleanliness, brightness of complexion, good health, and a complete cure for the lazy fever." She would then throw her windows open, shake out her bedding, straighten her room, and spend time in the greenhouse or garden before joining her family for breakfast.[19]

While many advice manuals of this period argued for the industrious use of time to make money or find space to raise and educate children, Elizabeth argued that it would create time for science. If you woke up early and led a productive day, Elizabeth contended, your evening could be filled with well-deserved leisure hours to join with friends reading about botany and mineralogy. While one friend read aloud or discussed their scientific collections, the others could knit, mend clothing, sew quilts, or even braid straw for sun hats. Margaretta and Elizabeth both prized their own fireside scientific debates, and these evening meetings between like-minded sisters created a dialogue that encouraged creativity and shared knowledge. Elizabeth wanted others to experience something similar.[20]

Regardless of whether Elizabeth regularly achieved this balance of productivity and intellectual pursuits, she described for readers her ideal day and her philosophy for a good life. Aware that she might sound like a bit of a scold for repeatedly suggesting that her readers wake before dawn, Elizabeth poked fun at herself in other articles in a way that might make her seem more relatable. At the very least, she invited readers to laugh with her rather than roll their eyes. She was helping to create the character of "E. S.," one that perhaps revealed a lot of the true Elizabeth Morris, or maybe an idealized version of herself. She certainly would have never written about all of these personal details of her life had she not been protected by a pseudonym.[21]

ELIZABETH'S "E. S." PERSONA ALSO GAVE HER A PLATFORM to be a fierce feminist. While much of her writing ultimately supported women's circumscribed work within households with a gentle push toward intellectual pursuits and scientific experiments, there were moments where she spoke out at injustices, using her pen to combat other writers on the pages of the *American Agriculturist*. When "Solus"—a self-described "lonely and comfortless old bachelor"—wrote in to the Ladies' Department about how he loathed store-bought socks and hoped to marry a woman who would eagerly sit by the fire and knit some for him as he read aloud, Elizabeth could not hold her tongue. Her response, published in the August 1846 issue, wished Solus luck in finding this "ladye love" who might enjoy knitting him long stockings by the fire (unless of course he was a smoker or chewer of tobacco, in which case she retracted her good wishes). And after discussing her own love of knitting, especially mittens, bags, and children's socks, she admitted to hating knitting men's long socks given the tedious size. "I would rather undertake to read Webster's Dictionary regularly through, from A to Z; or count the grains in a sack of flax-seed." One can almost imagine the glint of mischief in her eye as she ribbed the bachelor.[22]

But there was a larger issue at stake—Elizabeth was not just interested in teasing Solus. "Tell me why," she pushed, "the men should never knit for themselves? I *know* it does not *necessarily* make them effeminate . . ." She described how she personally knew two of the "roughest specimens of mankind," who knit their own socks and mittens while listening to someone read aloud. She was exhausted by the implication that knitting was women's work. For the sake of everyone, she argued, parents should teach their sons to knit, just as they taught their daughters. Perhaps Solus could even learn to make his own socks. Weaving in humor alongside jabs, Elizabeth ended her lengthy essay with "I intended to write only a few lines, and I have spun a yarn long enough to knit a pair of stockings for the Irish giant."[23]

This was not the end of the discussion. An author who went by "Reviewer" responded in kind. Reviewer, who identified himself as a man, regularly wrote long reviews of each issue of the journal, giving his two cents on nearly every subject touched upon. He became, along with Elizabeth and the editors, one of the most frequent contributors to the journal, presumably publishing under other pseudonyms as well. It is certainly possible that he was Solus—particularly given that in article after article he argued that farmers' daughters should stop pursuing what he considered fanciful "boarding school" pursuits like painting and music, and instead focus on more practical skills like knitting. Whatever the case, this curmudgeonly, if self-deprecating, anonymous author, after praising Elizabeth's response and suggestion that boys knit, wrote, "if the author is a lady, and single withal, Solus is bound to quit his bachelor's life, and go where he can get his stockings knit *at home*. If he don't I am sure I shall."[24]

Reviewer had tussled with Elizabeth before, and while he often praised both Margaretta's and her articles as the most valuable in the journal in his lengthy reviews, he once implied that she was a bit long-winded. That criticism, however accurate, annoyed Elizabeth enough that when she sat down to respond to his note on knitting, she wrote: "A crusty codger he must be, to wish to deprive a woman of one of the very few privileges even nominally allowed her!" She "laughed off at least two wrinkles" after seeing Reviewer's suggestion that Solus should marry her. Solus's fantasy of domestic comfort was hardly a match for Elizabeth's "cross-looking, tidy, little lady-ship." With all this, she admitted that "I have felt sorry that my incognita would not allow me to let any one share in my mirth." She went on to lay into the Reviewer, suggesting he must not have had sisters, since he so misunderstood women. He must be a "stiff, crusty bachelor, whose far-off cousins sat for his not too attractive portraits of American farmers' daughters." Elizabeth's playful response opened up a new world of anonymous jousting with Reviewer in every issue that followed, and Reviewer became ever more cautious with his responses to her articles in the Ladies' Department, writing: "I recollect how I burnt my fingers in this department."[25]

In 1848, just a few years after Elizabeth battled with Solus and the Reviewer over who ought to knit socks, American women joined together at Seneca Falls in upstate New York for the famed women's rights convention. While Elizabeth would deftly sword fight with men in the pages of the *American Agriculturist*, and Margaretta worked hard to prove herself an equal of her male peers, neither of the sisters joined in the women's rights movement, which captured the hearts of many activists in their generation. They did not leave behind their reasons for their decision, nor did they write much about politics publicly or privately. It is certainly possible that they were so focused on their scientific projects that they found little time for reform work and activism.[26]

Political activism may have also seemed like it could threaten the careers they were trying to build. Some of their contemporaries, including Dorothea Dix, Catharine Beecher, Almira Phelps, and Elizabeth Blackwell, were forging new paths as they published books and articles, jousted with politicians, and pursued their own careers, but were decidedly against women's suffrage. These women—like the Morris sisters—found ways to carve out space for themselves financially and academically within the gendered boundaries of their society. They pushed against the edges, but did not break down the walls, even if they stood to gain significantly from the civil rights for which those like Margaret Fuller, Sojourner Truth, Elizabeth Cady Stanton, and Angelina Grimké advocated. White American women did not all agree on what was needed to lift up their sex or what that might even mean, let alone how they might do so by joining with others across race and class. While the Morris sisters faced prejudice and marginalization because of their gender, their race and wealth meant the system also benefited them and their privileged position may have blinded them to the concerns of others. Elizabeth and Margaretta kept their noses down and continued with their research and writing, not engaging in this larger reform work.[27]

However, Margaretta and Elizabeth were not just concerned with their own work. They also acted behind the scenes to support other women scientists. Elizabeth connected women she both knew and barely knew to William Darlington, giving them access to the kind of mentorship and connections she had found so valuable while opening the door to enter scientific conversations—the first paving stone on the path to legitimacy. She acted as a community builder, connecting women to the other botanists she most admired.[28]

Margaretta, meanwhile, joined the Board of Lady Managers for the Philadelphia School of Design for Women as a way to help women become more financially independent. This school was founded by Sarah Worthington King Peter to train unmarried and widowed women in the industrial arts. Peter saw training in the arts as a revolutionary solution for women because they could use their newfound skills in engraving, lithography, wood carving, scientific illustration, and much more "*at home*, without materially interfering with the routine of domestic duty, which is the peculiar province of women." By preparing women to contribute to the Industrial Revolution, Peter was hoping to benefit not just women but also the nation. One component of the training involved scientific illustration, one of Margaretta's specialties. The Morris sisters might not have been outspoken activists, but their articles in the *American Agriculturist*, coupled with their support for other women in their community, illustrate that they were concerned with women's rights in their own way.[29]

When the activist Lucretia Mott delivered a speech in 1849 calling for women's suffrage and equality, she celebrated the scientific achievements of women worldwide. She asked the audience: "Do we shrink from reading the announcement that Mrs. Somerville is made an honorary member of a scientific association? That Miss Herschel has made some discoveries and is prepared to take her equal part in science? Or, that Miss Mitchell of Nantucket has lately discovered a planet, long looked for?" Mott called on her listeners to consider their implicit biases, asking them to consider which of their beliefs about women's roles in society had been socially constructed. She argued that her listeners should understand women to be intellectual equals of men, and "among the proofs of a higher estimate of woman in society at large" were the articles written for "Ladies' Departments" in periodicals.[30]

Not all agricultural journals focused their "Ladies' Department" articles on science. Most dealt with a combination of housekeeping tips and marriage advice, occasionally veering into women's civil rights in more progressive publications. The *Ohio Cultivator*, for instance, promoted women's rights in large part owing to Frances Gage—the author of the poem Elizabeth had recommended—who served as assistant editor of the journal. Thanks to the Morris sisters and their frequent contributions in the 1840s, the *American Agriculturist* became a source for significant scientific discussion, particularly in the sections devoted to women and children.[31]

Elizabeth clearly took great pleasure in speaking from a position of authority as "E. S." She was proud of her work and she loved the community, anonymous as it was, that she created. While she was simultaneously working behind the scenes to help botanists like Asa Gray succeed in their careers, Elizabeth reclaimed her authority on the pages of the journal in a way that satisfied herself first and foremost. In one of her treasured, leather-bound albums, she copied each article that she submitted. Amid these copies, she also incorporated the feedback she got from readers like Reviewer. After the knitting banter, she and Reviewer continued to throw jabs at each other, though mostly praise, as they established an anonymous friendship on the pages of the *American Agriculturist*.[32]

MARGARETTA ALSO SEEMED TO ENJOY WRITING PUBLIC SCIence articles under pseudonyms. Reviewer regularly praised an author who went by "Old Lady" since she seemed to push back against the growing "refinement" of young women by encouraging them to know the insects in their home and on their farm. The Old Lady was Margaretta. The initial conceit was that an anonymous writer had found the diary of an old lady, and given how useful it had proven for understanding the environment on a farm and within the home, the author decided to publish parts of the diary that seemed relevant to the readers of the *American Agriculturist*. Eventually, this evolved so that Margaretta became the mouthpiece of the character and signed off on the articles as "Old Lady."

After facing a chorus of male scholars putting words in her mouth over the last several years, writing anonymously was a way for Margaretta to find her voice. When she described the Old Lady in her first article under that pseudonym, she might as well have been describing herself. The Old Lady "appears devotedly fond of the contemplation of the operations of nature," Margaretta wrote. She sought out scientific mysteries "in the changing clouds and skies—the still forest—the useful field and garden—or in the homely kitchen and its fireside combinations." Science was everywhere, in all spaces, and offered countless opportunities to explore how the world worked. Margaretta as the "Old Lady" told readers about her driving passion: "Above all, the study of the insect world appears to have been her peculiar delight, and to this she seems to have devoted many of her leisure hours, carefully noting down any interesting fact that has fallen under her notice." While she incorporated cutting-edge entomology into the essays, there was a nostalgia woven through that seemed to assume that Americans were increasingly indoors, losing their connection to nature. Margaretta hoped that sharing the Old Lady's diary entries would not just amuse readers but also encourage them to "find in the book of nature their chief happiness."[33]

Margaretta claimed that the Old Lady "pretends to little scientific information," but her articles were full of details that only a close observer of insects with extensive entomological knowledge would include. Regardless, Margaretta worked to make the story accessible to everyone, avoiding scientific jargon and describing simple experiments that required just a few tools: a paring knife, a box, and a healthy dose of curiosity. The Old Lady had noticed how woodpeckers were especially active hammering at the thick bark on her oldest apple trees. Using the paring knife she always carried with her, the Old Lady peeled back some bark to find cocoons cased in something resembling brown silk. As she carefully removed that silklike layer, she found further layers of white silk embracing a sleeping reddish-brown worm. She collected several of these, put them in a box in a warm room in hopes of simulating springlike temperatures, and the worms emerged from the chrysalises into dust-brown moths. Margaretta was able to convey in a compelling, relatable story that anyone curious about the insect world might be able to watch the awe-inspiring changes insects undergo. She also included enough information so that apple tree owners might be able to combat the common pest.[34]

As she typically did in scientific journals, Margaretta described how a whole system of other creatures could help keep a pest like the Apple Moth at bay. Rather than relay quick fixes like poisonous powders or chemicals, she encouraged readers to foster natural pest control in the form of a garden friendly to birds and other creatures, including "the Woodpecker and his troop of feathered friends, the sparrows, sapsuckers, and wrens with their restless wings and hungry beaks prying into every dark cranny." In other articles, she encouraged girls, specifically, to become champions for birds, in part for their "sweet and merry strains" but mostly for their efficient pest management. She suggested her young readers build birdhouses, get rid of cats prone to hunting hatchlings, and "persuade the boys by kind entreaty and gentle remonstrance to suspend their hostility" and stop shooting birds. Engaging in a long, patient day at her window observing a family of sparrows, Margaretta saw the birds consume 320 worms. If that pace continued, that one family of birds would rid the garden of 10,000 within a month. Given the economic importance of birds for farms and gardens, Margaretta wrote that girls who protected birds would "be of more real service to your country than many a general whose name is written in history."[35]

This holistic view of how different creatures and plants were interconnected was essentially an early understanding of ecology before the word was coined decades later. Margaretta's contemporary, Henry David Thoreau, was writing about similar relationships from his observations at Walden Pond and along his walks through Concord. While Thoreau's descriptions were romantic, Margaretta's were tied to practical, economic uses. Convince children to be kind to feathered friends and the flourishing bird population would keep insects from destroying orchards. She also called on readers to learn to think of moles as "one of our best friends," instead of seeing them as destroyers of gardens, since they "feed on insects, and eat the grubs, and cut-worms, and all those destructive pests that do us so much injury." Close observation of these relationships would save orchard growers and farmers hours of labor, not to mention eliminate the need for pesticides.[36]

Margaretta also incorporated friendly jabs at her sister in her articles. She occasionally featured a character named "Betsey," who shared one of Elizabeth's many nicknames. The character was a stubborn housekeeper who often disagreed with the Old Lady's newfangled ways of handling things. Reading between the lines of her semifictional stories, you can imagine how the two sisters may have debated recipes for soap, proper fertilizers, or even the best way to polish furniture—which they enjoyed rehashing in their articles. These were two strong-minded women with a sense of humor that tumbled onto the pages of the journal as they playfully teased each other.[37]

While some of Margaretta's anonymous articles gave her the space to explore topics beyond insects, most of them were related to her other entomological work. She may have called herself "Old Lady," but she was translating modern scientific knowledge into accessible prose for her readers. She mostly kept her pseudonyms to herself but she unmasked herself several years later to Victor Motschulsky, a Russian entomologist and Imperial Army colonel, during his visit to Philadelphia in 1854. Motschulsky traveled broadly through Europe, Asia, and North America, mainly in search of beetles and other beetle enthusiasts. He had met with several naturalists during his visit but did not want to leave the city without meeting the "distinguished entomologist" Margaretta Morris first. He asked another scientist for an introduction, and the two had a long visit during which Margaretta apparently confided the truth about her authorship of the "Old

Lady" articles. Decades later in 1928, using Motschulsky's notes, German entomologists included all of that material in their extensive bibliography *Index Litteraturae Entomologicae.* That decision and the bibliography that resulted along with remnants of drafts saved her writings from languishing in anonymous obscurity.[38]

Like Margaretta, Elizabeth also found ways to incorporate her scientific expertise into her public writing. Amid all her featured recipes and meditations on the good life, she wrote long articles educating readers on botanical topics like the various species of grasses, their properties, and their uses. "If my very humble efforts in these numbers to awaken interest in the pursuit of natural science among the younger members of the agricultural community are crowned with the least success, . . . I shall have reason to feel peculiarly gratified." Elizabeth did more than educate the journal's readers about grains; she also explored the science of gardening, including how to best time the transplanting of shrubs, how to collect and organize seeds, and how to revive botanical specimens. She made a point of showing how horticulture was both an art and a science.[39]

More than anything else, Elizabeth, like her sister, was eager to share her passion with readers so that they too might find fulfillment and happiness in the garden. Across multiple articles, she confessed how the work of tending a garden soothed her soul. "When I am grieved in spirit, or vexed in temper, by the unavoidable cares of my little world," Elizabeth wrote, "I go out and *work* in my garden; and in the healthful exercise of the body, and the beautiful soul-subduing quiet that pervades the place, and steals like a healing balm over my mind, I soon forget my troubles." She also found gardening amplified her mood when she was happy, as she would "find the flowers a brighter hue, and the birds sing more joyously their welcome to me." By sharing her own connection to the natural world, she was hoping readers might find theirs. While heartsease, or pansies, had become so popular as to be almost invisible in her neighborhood's gardens, she delighted in the "sort of individuality in each flower, a saucy, good-natured confidence," imagining that the flowers had a "quaint way of looking up at one, as if to say, 'I'm laughing at you!' that one cannot choose but to gather, and love them." She understood flowers: she could read them, she could name them, and she nurtured them. Flowers, as she loved to say, were her diamonds.[40]

ELIZABETH AND MARGARETTA NOT ONLY GAVE ADVICE ABOUT seed collection, plant identification, and insect observation, they also gave their female readers tips on how to dress and prepare for scientific adventures. When popular handbooks for botanists and entomologists gave any advice at all, they typically gave clothing recommendations only for men, suggesting things like fustian jackets worn by English hunters that would have been inappropriate for women. Margaretta and Elizabeth worked to correct these omissions. Elizabeth advised potential gardeners to wear short, plain-colored dresses made of a coarse fabric that could withstand brush-ups with thorny shrubs and hide dirt. She recommended gardeners wear a garden bonnet to protect themselves from the sun. Meanwhile, long leather gloves with a broad, stiff cuff would protect them from briars. Women who labored on farms would not have needed advice about not wearing a silk dress out to garden, but fashionable urban women perhaps joining a new husband on a farm or at a summer home might appreciate these tips. In other words, Elizabeth was speaking to women like herself—wealthy, perhaps urban, the female equivalent of a gentleman farmer.[41]

Gum elastic shoes, similar to modern rubber boots, were a particular favorite of both Margaretta and Elizabeth as they traipsed through their garden and embarked on long, muddy adventures along the Wissahickon. Elizabeth was self-conscious about recommending them to her readers, knowing they were not fashionable. After quoting a poem about how women's feet should be like little mice hiding under their petticoats, she conceded that the practical waterproof shoe was embarrassingly ugly: "Why, a lady's feet, cased in high gum shoes, are as ugly as black puddings, or young walruses; and, as to fearing the light, they seem to have such an undue sense of their own importance, that it is almost impossible to hide them." Margaretta cared less about recommending something that might be considered unfashionable. She was grateful for her gum shoes, and how they, along with a thick, hooded coat, enabled her to venture out in the snow without worry. She even furnished advice about how to mend them and optimize their longevity by melting small strips of gum elastic with oil of sassafras for a few days before smearing it over any cracks. While gum elastic shoes were typically recommended for men—particularly soldiers and farmers—they were not advertised for women, least of all in fashionable magazines. The Morris sisters let the readers of the *American Agriculturist* in on their secret for dry and warm feet during outdoor adventures.[42]

There were other ways that women's clothing got in the way of gardening and botanizing excursions. Dresses, for instance, could get tangled up in wheelbarrows. Elizabeth therefore suggested that her readers purchase or build wheelbarrows with long curved handles. Without having to hunch over to haul the load, the gardener could work without worrying about her dress getting caught, putting "her every moment in danger of being *tripped up*, and having her *nose broken*." A few years later, in 1849, writers for the *Water-Cure Journal* would begin recommending bloomers and other radical outfit changes that would make it easier for women to go on adventures and work outdoors. In that journal, Edith Denner asked, "How in the name of common sense is a woman with long, full skirts, ever to become a practical Ornithologist, Geologist, or Botanist with any comfort, or without a great deal of inconvenience, attended by a vast amount of unnecessary labor and fatigue?" Denner described wearing bloomers as she scaled fences, climbed hills, forded streams, and hiked through woods.[43]

Elizabeth and Margaretta likely would have found bloomers or pantaloons to be too radical. Theirs was a feminism that was conservative and constrained, perhaps, but also revolutionary. They were showing women the way out of the parlor and how they might keep their feet warm and dry as they went. The social and cultural restraints on women that prevented them from being comfortable outdoors were just another series of obstacles to be analyzed and overcome with tools and better options, and the Morris sisters were eager to share their findings with their readers.

IN BOTH ELIZABETH'S AND MARGARETTA'S ARTICLES, RURAL life was a constant, whether that involved celebrating nostalgic community events and self-sufficiency or navigating limited access to resources like schools. Given how prolific Elizabeth was, it was inevitable that she would occasionally offend readers. While both Margaretta and Elizabeth imagined themselves as speaking to a rural readership of all classes, there were moments where their own wealth and privilege blinded them to the concerns of average women. Their social circles, after all, were limited to the wealthy, white, and well-educated women of Pennsylvania. As both women described domestic settings, they included descriptions of servants and gardeners whom they often had to scold or educate, details that may have alienated them from some readers.

Elizabeth's elitism was on display when she wrote about the failures of rural schools and suggested small groups of mothers join together to hire governesses to teach their children. A reader, "S. H. R.," who was affiliated with the Albany Normal School in New York, disagreed. Rather than reserving the best teachers for the wealthiest families who could afford private governesses, S. H. R. argued that it would be better to support the growth of schools like hers in Albany to train teachers in every state, so that public schools might be better equipped to educate all students, no matter their wealth. Elizabeth, as E. S., seemed horrified at the implication that she only spoke for the wealthy and carefully contended that anyone might adopt her plan as a stopgap measure until more trained teachers found their ways into public schools across the country. Regardless of her attempts to rationalize her view, this exchange revealed her blind spots.[44]

Elizabeth's greatest success, perhaps, was when instead of preaching, she leveled with readers and admitted to moments of failure. S. H. R. returned to the pages of the *American Agriculturist* to propose that more American women should plant mulberry trees and raise silkworms to both save on foreign imports and make their own dresses. Certain that this would be a more sustainable practice, she asked E. S. to "lend her accomplished pen" in support of the idea. Elizabeth paused. Eager to build a friendship with this former critic, she responded with a list of instructions, including the kinds of mulberry trees necessary, the time commitment required, everything. And while she could have stopped there, she instead proceeded to share how she herself had been swept up in silkworm fever a few years earlier, excited to use her own silk to make purses and gloves to give to friends "as specimens of my success and skill." Things did not go as planned.[45]

Homemade silk had captured many Americans' imaginations in the 1820s and 1830s. They dreamed of an American silk industry that might rival its French, Turkish, and Chinese counterparts. Farmers in New England believed raising silkworms was a way to keep their daughters from moving to mill towns in search of work. Investors saw a chance to get rich quick through purchasing large numbers of mulberry trees that would feed the worms. Still others embraced nostalgia for an imagined homespun American culture when women and men had the power to make what they needed with their hands before they became dependent on faraway markets. Elizabeth wanted to warn her readers away from getting caught up in it all.[46]

While Elizabeth regularly championed DIY efforts, she had experienced the difficult realities of raising silkworms. Knowing a great deal about entomology thanks to her sister and always fond of pets, she grew overconfident. She purchased several thousand silkworm eggs and placed them in her writing desk drawer—her first mistake. They hatched prematurely, enjoying the warmth of her sitting room, and Elizabeth had nothing to feed them. She gave them lettuce as she waited for local mulberry trees to leaf out. She then took whatever worms survived the poor diet up to the airy garret rooms where she and Margaretta conducted experiments and maintained their collections. As the silkworms grew, Elizabeth scrambled to create more space for them, laying out boards, chairs, and parts of an old bedstead. The work became relentless and all-consuming, as she traipsed across town to gather leaves from a neighbor, and constantly cleaned up after the worms.[47]

Elizabeth did not love insects the way her sister did, but she persisted. "I cannot think, even now, without a shudder," she admitted, "of the disgust, almost amounting to horror, when any of the heavy, cold worms fell upon my hands, or crawled over my dress." Exhausted and overwhelmed, Elizabeth decided to hire someone to attend to her cocoonery. All she had to show for it was a single skein of silk "to be kept as a talisman against future temptations to misspend my time." Elizabeth, long a voice for rural self-sufficiency in the pages of the *American Agriculturist*—even if it meant spending sixteen hours stirring apple butter over an open fire—finally came out in support of just buying fabric. Silkworms were just too much. Margaretta enjoyed "a good deal of mirth" at her expense, and likely the teasing at home helped make it possible for her to share the story with the public. "I am willing to let others be amused," she wrote, "though they may not be much benefitted by my experience."[48]

Both Elizabeth and Margaretta savored the rural, somewhat self-sufficient lifestyle, where they might dabble in raising silk-worms but also more successfully cultivate an orchard and ride horseback over fields. A large portion of their articles celebrated rural life, just at the moment when Germantown was becoming increasingly urban. The railroad connected their neighborhood to Philadelphia, and once-seasonal residents were making German-town their year-round home, giving it more of a bustling, urban feel than it had ever had before. Industrialization was also reaching Germantown, with a number of new mills and factories opening up around the Wissahickon Creek. In light of the encroaching tide of urbanization, many of the sisters' articles celebrated nostalgic activities like cornhusking bees, apple-paring bees, and other community gatherings to support new families and help with the labor of harvests. Even Margaretta's "Old Lady" moniker reached back to simpler, less refined, less urban times. America was transforming, and the sisters captured and championed the agricultural life, skills, and crafts of early antebellum America that the country seemed on the precipice of abandoning.[49]

The sisters also defended local environmental knowledge at a moment when they believed the city and its built environment threatened to encroach upon the diversity of creatures and plants they cherished at home. When Margaretta stayed for a stretch in her cousin's row house in Philadelphia, she longed for her semi-rural home in Germantown with its "grain fields and sweet flow-ers" instead of the "unnatural place" in the city. She wrote that she spent hours trying to "cheat myself into a dream of the country, by sitting upstairs and looking directly into a large tree that shadows my window and listening to a little dunce of a bird that has built its nest there, instead of the country, where it had wings and might have chosen better." She saw herself as a country woman, even as her home became increasingly suburban or even urban. "That there should be such perversion of taste in man we need but wonder, but in a *bird*," she continued. "I love and am grateful to the little creature notwithstanding, and am not sorry its views are not more elevated." Despite their anti-urban proclivities, the women benefited from their proximity to Philadelphia where they could easily spend a day collecting the latest books and scientific journals at Carey's bookstore or attending lectures at the various institutions around town. They were connected to this community of scientists *because* of their proximity to the city.[50]

BOTH MARGARETTA AND ELIZABETH USED THE PLATFORM OF agricultural journals to reach a large, popular audience. They imagined their readership to be a broad swath of rural Americans who embraced a simpler sustainable existence that was harmonious with nature. They translated complex botany and entomology into practical, relatable articles intended not only to help solve problems on farms and in gardens, but also to seed the country with many more scientists and enthusiasts. Both women, in their own way, conveyed a love of science and scientific observation in their writing, encouraging others to see science in their kitchens and backyards. They wanted society to take the domestic work that women performed in the kitchen and garden seriously. The Morris sisters may not have left records about their stance on suffrage, but they left their marks on the pages of the *American Agriculturist*.

When Elizabeth reflected on this period of her life, she admitted to William how she "was often gratified by the praises bestowed upon my anonymous articles." Public science suited her. Elizabeth was able to command a wide, devoted readership, with the freedom of doing so anonymously. But that very anonymity, protective as it may have been for women like Elizabeth who were worried about disrupting social norms, also erased them from the historical record. One of the reasons why Elizabeth Morris is so unknown today is because her widely read articles were never attributed to her. She preferred it that way. She was not overly concerned with renown or celebrity, let alone being credited for her work.[51]

Margaretta donned the mantle of the Old Lady to invoke a romanticized nostalgia for early America while encouraging the public to redefine their relationship with insects. While anonymity did not truly suit her, she cut her teeth writing for an audience beyond the men at the American Philosophical Society and the Academy of Natural Sciences—a practice that would prove useful once she made her next big discovery.

Six

HIDDEN AT THE ROOT

ON THE HUMID MORNING OF AUGUST 21, 1850, MEN FILED into the College Chapel at Yale, shuffling into the pews and greeting each other as they removed their hats and settled into their seats. Scientists from all over the country had traveled by train, steamboat, and stagecoach to be there for a week of sharing discoveries and making connections at the fourth meeting of the American Association for the Advancement of Science. Modeled after its British counterpart, the AAAS was only two years old and was the first of its kind in the country to bring together men of science from all disciplines in an effort to establish science as a true profession in the United States. Throngs of onlookers clamored to catch sight of American scientific celebrities "whose personal history is almost a record of the birth and progress of the sciences," as the *New Haven Journal* put it—men like Joseph Henry, the head of the new Smithsonian Institution, and Benjamin Silliman Sr., Yale University's first science professor. A reporter from the *New-York Tribune* shuffled his papers and readied himself to take copious notes on the proceedings and presentations for several days of front-page coverage.[1]

The goal of the Association was not only to advance scientific research by bringing investigators from around the country together, but also to stimulate broad interest in the sciences. There were more than just men of science in the pews that day. Local New Haven residents, many of whom were women, joined the crowds for presentations on everything from the speed of currents through telegraph wires to the analysis of what was believed to be a fossilized elephant's tooth found in rural New York. The members of the AAAS would later officially thank the ladies for having "adorned" the meetings. This was New Haven's first chance to host the conference, and it was a way to showcase Yale's new Scientific School, which was founded just a few years earlier. More than simply celebrating the host city, founders like Louis Agassiz hoped these meetings would raise the self-confidence of American men of science. The popular enthusiasm and news coverage certainly helped.[2]

From the front of the brick chapel, Alexander Dallas Bache, the amply bearded president of the AAAS, called the meeting to order at nine. The business that morning on the third day of the conference concerned the election of new members to the Association. The Standing Committee—a learned group of eleven scientific men—gave Bache a list of ninety-two names that he read aloud. It was a long list, so long that many in the audience, journalists included, missed a revolutionary moment: the election of the first two women, Margaretta Hare Morris and the astronomer Maria Mitchell. Margaretta missed it too since she was 165 miles away in Germantown. Though she would later describe how she felt "a deep sense of the honour conferred" on her, the decision to permit her entrance into professional science was made in absentia. She was only later informed of the outcome first by returning friends and later with an official letter reluctantly written by her old rival, Edward Claudius Herrick, the secretary of the AAAS.[3]

Present or not, Margaretta's election to the Association certified the significance of her contributions to American science on the national stage. This membership was a credential that would stand to elevate her in the eyes of even her most skeptical peers, a shorthand for her credibility. Far from just a lucky break, the events that would alter the trajectory of her career and catapult her to professional heights typically unknown for her gender were thanks in large part to her strategic decisions that began on a bright and chilly day four years earlier.

IN MARCH 1846, MARGARETTA HAD BEEN STANDING BESIDE a gardener instructing him about what she wanted as he dug a trench around a failing pear tree. After the shovel's blade thumped against the deeper roots, the evidence she hoped for was flung out of the trench and into the accumulating pile of dirt. The recent heavy rains had melted the remnants of snow that had blanketed the garden, making this experiment possible but also making the earth decidedly muddy. Eager to confirm her suspicions, Margaretta knelt down in the mess to sift through the pile and collect the countless seventeen-year cicada larvae whose long, pointed proboscises were embedded in the severed roots of the tree. While the gardener kept digging, Margaretta guided a handful of the cicadas into a glass jar filled partly with soil, keeping others on a segment of root, and gathering still hundreds more to pin, measure, study, and share with other scientists.[4]

Since Margaretta was unearthing the cicadas several years before they were scheduled to emerge in 1851, she was able to make observations that had never been logged before. From one thin segment of root about three feet long, she counted twenty-three larvae with "their suckers piercing the bark," clinging desperately to their food source. Up until this point, it was generally believed that cicadas found sustenance in the decaying vegetable matter around tree roots during their long subterranean existence, having little effect on the trees themselves. However, these cicadas were so deeply and firmly embedded in the root where they harvested fluids from the xylem that their impact had to be more than superficial. To Margaretta it seemed apparent this was more than just correlation; the cicadas were damaging the trees. They remained attached to the roots for half an hour after being lifted from the earth. As she scraped the bark off of the abandoned stretch of root, she saw how deeply their punctures had gone.[5]

Margaretta also noticed something curious. The cicada grubs, though ostensibly all the same age, having hatched and burrowed

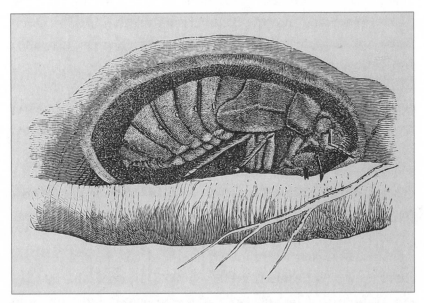

When Louis Agassiz supervised the posthumous 1863 edition of Thaddeus William Harris's *A Treatise on Some Insects Injurious to Vegetation*, he incorporated an image of the *Cicada septendecim* larva puncturing a root for sustenance, based on Margaretta Morris's findings.

underground in 1834, varied significantly in size. Some were as small as a quarter of an inch, while the larger ones were closer to a full inch. No matter their size, they were side by side on the same roots. This was a mystery to solve later. For now, Margaretta hurried to collect them all, small and large alike.[6]

No matter how novel the findings, she knew that her peers might not immediately accept what she presented. She tempered her expectations, given Edward Claudius Herrick's and Thaddeus William Harris's public contradiction of her wheat fly research five years prior in 1841. If she were to successfully present this new discovery to the world, she would need to maintain control of her message from the outset. She would also have to strategically reinforce her legitimacy by finding allies willing to endorse her findings.

THE LEAFLESS TREE WHOSE BRANCHES QUIETLY OVER-looked the back garden of Morris Hall was the cause of all this activity. This pear tree seemed to be dying. Like several other pear and apple trees in the Morris family garden, it had gradually ceased producing fruit, the leaves had grown increasingly sparse over the years, and its bark was now covered in a moss that kept spreading. Pomology was a major obsession of many ambitious orchard owners and even casual gardeners and they all had a stake in protecting their trees from harm. Agricultural and horticultural journals were filled with articles devoted to learning about, caring for, and propagating new fruit trees in American soil. With the expansion of railroad networks in the 1840s, fruits—especially apples—were quickly becoming a popular export for farmers, where they could sell dried apple slices around the country, and even fresh apples in London. During this time, horticulturists struggled to understand why some varieties of fruit trees only produced fruit for a generation or two before the trees began failing when there was no visible insect infestation or disease. Some believed that the trees had a set lifespan, that the soil grew exhausted, that the climate was wrong, or that there had been some kind of neglect.[7]

Margaretta had another theory. Back in November 1812, when she was fourteen, her older brother Thomas had been digging beneath an apple tree and found cicada grubs a few feet below the surface. In the years afterward when the cicadas emerged—1817 and 1834—Margaretta had witnessed the creatures claw their way up the branches of the fruit trees in her yard and molt, leaving behind the crunchy shell of their nymphal skin. She watched closely as the males joined together in a noisy droning chorus to attract mates and the females then went about scratching the bark of branches with their strong pincers, laying ten to twenty eggs in each slit before moving on to find another spot. After five or six weeks, the little cicadas began hatching and—in a move that always baffled their human observers—leapt to the ground, faithful that this was their life's purpose and that they would survive the harrowing fall. Margaretta then lay on the ground with her magnifying glass to watch the miniscule cicadas burrow down below the trees, preparing for seventeen years underground. She could not help but wonder if those creatures might pose a threat to the trees themselves.[8]

So, when Margaretta, Elizabeth, and their mother, Ann, contemplated the future of their failing fruit trees as the spring planting season neared in 1846 all those years later, Margaretta saw an opportunity to test her long-held hypothesis. Elizabeth had been watching the trees closely for years, noting the dwindling leaves, the loss of verdure, and the growth of moss on limbs that seemed to spell doom. Elizabeth and Margaretta conferred on a method, and instructed the gardener to remove all of the dirt and roots from a circular trench three feet from the tree, digging two feet down. The gardener applied this treatment to another pear tree as well, and once Margaretta had finished collecting the grubs with surgical precision, he filled both trenches with a combination of fresh soil and manure. There was significant risk that the failing trees would continue to decline, particularly with a number of their roots severed, but at least the fertile soil would give them a chance.[9]

For almost seven months, Margaretta studied these creatures in glass jars filled with dirt and roots at her desk in the library, adjacent to others that held wheat flies. If any of these experiments failed, or she needed to test new theories, she knew that she could return to those trees, and perhaps even others, to unearth more larvae. After all, those pear trees were not the only ones suffering—other fruit trees seemed to be failing, and, if Margaretta was correct, perhaps they too hosted many an immature cicada, ravenously sucking fluids from the roots of sickly trees. Her garden, as always, was her lab.[10]

Years later, when Elizabeth would encounter the cicadas in the garden mere months before their 1851 emergence, the general public had caught up with Margaretta in their cyclical excitement about cicadas. The insect-averse would worry needlessly about being stung once the muddy brown larvae molted into their mature form with bulbous red eyes and membranous wings marked by black veins. Others were concerned that they were like the biblical "locusts" bent on widespread environmental destruction. The American public often interchanged "cicada" and "locust" even though the creatures were scientifically distinct. Regardless of their formal classification, when an insect dominates the soundscape and landscape like clockwork every seventeen years, it can be difficult to ignore. However, in 1846, five years before the emergence, few besides Margaretta Morris were thinking about the creatures and whether they might be quietly impacting the trees they latched on to.

Her theory, if proven correct, would help orchardists decide when to plant saplings based on predictable cicada years, saving the most vulnerable trees from harm. This knowledge might also help orchardists salvage older, suffering trees by applying the method Margaretta and her gardener had used—and with demand for apples booming both domestically and overseas, it would have significant ramifications for this growing sector of the economy.

MARGARETTA HOPED TO DO MORE EXPERIMENTS WITH HER trees the following spring before publicly announcing her findings, but the November 1846 issue of the *Horticulturist* caused her to act. Sitting down to read the latest journal, as she and Elizabeth did each month, Margaretta came across an article by someone named "JBW" describing an experimental approach to resuscitating fruit trees. The author had a number of pear trees that seemed to have reached the end of their lives, and instead of uprooting and replacing them, JBW followed the personal advice given to them by famed landscape architect and editor of the *Horticulturist* Andrew Jackson Downing. He or she dug a trench around two trees, pulling out roots and replenishing the soil with a rich compost mixed with charcoal, iron shavings, and potash. Over the course of three years, the author found that the trees had regained their green leafy youth, producing plenty of large pears. Both the author and Downing were convinced that this experiment proved how to extend the lives of dying fruit trees—an important discovery for anyone maintaining commercial orchards or even humble backyard fruit trees.[11]

Margaretta's heart must have been racing as she read this. Perhaps she first assumed that she had been scooped, that someone else had made the same discovery. Her eyes likely sped across the article, pausing on the diagram of the trench around the tree, so similar to the one she had instructed her gardener to dig, before she slowed herself down to read the article from the start in a deliberate way. JBW and Downing had stumbled onto the problem of dying fruit trees, but fixated on poor soil as the cause.[12]

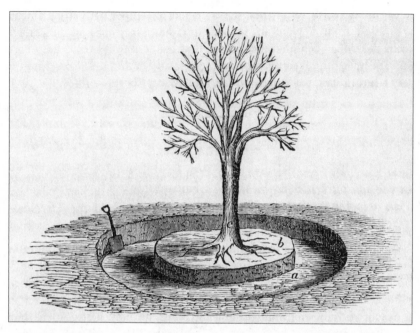

Margaretta's attention was drawn to the illustration JBW provided in the November 1846 *Horticulturist and Journal of Rural Art and Rural Taste* article, as it resembled her own technique when she discovered cicadas at the roots of her pear trees earlier that spring.

Almost immediately, Margaretta began to correct the record by drafting a letter to Walter Rogers Johnson, corresponding secretary for the Academy of Natural Sciences, while preparing specimens for shipment. Johnson—an educator, chemist, geologist, and one of the first Americans to dabble with daguerreotypes— was a trusted old friend of the Morrises. He had initially met them more than two decades earlier when he was hired as the principal for the Germantown Academy, where Margaretta, Elizabeth, and their friends regularly attended evening science lectures and used the telescope. He had even married one of their friend's sisters. Margaretta was confident, therefore, that her specimens and research would be taken care of by him. At the December 15, 1846, meeting, he presented the specimens Margaretta had donated to the museum—some of them preserved, others alive—and read her remarks to the men of the Academy present.[13]

Margaretta was never one to mince words, and she began her letter by asserting the significance of her findings and stressing their implications for orchards around the country: "I have, for a number of years, believed that the failure of fruit on trees over twenty years old, was mainly owing to the ravages of the larvae of the *Cicada septendecim*, though Entomologists have heretofore considered them harmless." She wove a rebuttal to the *Horticulturist* article through her report. Engaging with recent literature, particularly published by names with a certain amount of renown like Downing's, was an evolving component of scientific legitimacy. Challenging Downing in such a public way was also, simply, a power move on Margaretta's part.[14]

Though there was more research to be done, Margaretta included the discrepancy in size between the different larvae in her report. While she had first thought that the tiny cicada larvae might be starved relatives of the larger ones, she was increasingly convinced that the species were distinct. Her mind was open to such possibilities, given her experience with the wheat flies. "I am inclined to believe that there are two species, differing sufficiently in size," she announced to the members of the Academy. She added that in her observations of the cicadas in 1817 and 1834, she found that "the note of the smaller variety, or species, is much shriller than that of the larger." Since she had so many specimens of both sizes, she very well could have proposed to name the smaller cicadas then, but she may have decided to wait for confirmation from the members first or at least until she could collect adult specimens in 1851. However, since she was not a member of the Academy, she could not be present while her work was shared, nor could she take part in the conversation about whether it was time to name the new species. Although the Academy had

recently elected their first woman as a corresponding member—
Lucy Say, Thomas Say's widow, thanks to her donation of her
husband's collections to the Academy—a corresponding membership was a gesture, not an invitation for involvement. In other
words, women were still not welcome at these meetings. Margaretta would have to wait for Walter Johnson or others to report
back on how it all went.[15]

Margaretta did not rest after submitting the report to the
Academy. She immediately began drafting an article to send to
the *American Agriculturist*. Margaretta and Elizabeth now had
other avenues to speak to the public at their disposal, and they
used that to their advantage. While Margaretta drafted, Elizabeth schemed. Elizabeth submitted an article under a new
pseudonym—"An Amateur"—mischievously describing herself
as a "very humble-minded man" interested in the cultivation of
fruit trees. The character she portrayed seemed to be fumbling his
way toward destroying his orchard, having taken too much advice
from those who professed to understand the subject (likely a jab
at Andrew Jackson Downing and his *Horticulturist*). Elizabeth
went on to ask nine questions about tending trees that she likely
knew the answers to, one of which laid considerable groundwork
for Margaretta's article: "I have heard it asserted," she wrote, "that
when an apple or pear tree does not bear well it can be made to
do so by digging around it, two or three feet from the trunk of
the tree, and cutting off the ends of all the horizontal roots. Is
this a well ascertained fact?" She must have felt a wry sense of
satisfaction as she passed the baton to her sister once the article
was published that January 1847.[16]

Margaretta's article answered the question Elizabeth had
seeded directly in its title: "Apple and Pear Trees Destroyed by the
Locusts." Writing for a broad audience, Margaretta transformed
the way she discussed her findings, by using "locusts" in the title, knowing that, though entomologically incorrect, it would be

familiar to her readers. While she did not condescend, she incorporated more contextual information than she had in her scientific report, aware that not all readers would be as versed in the cicada life cycle as she and her peers were. Aware that readers would want practical advice, Margaretta not only shared her technique about digging the trench, but also suggested that readers could feed the cicadas to hogs, ducks, chickens, or even moles. She deliberately omitted her theory that there were two species of cicadas as it would have complicated her primary message that the cicadas were a threat to orchards.[17]

Margaretta's decision to submit the article to the *American Agriculturist* herself was a tactical move. She published the article under her own name, shortened to the androgynous "M. H. Morris" rather than using her "Old Lady" pen name. By writing for herself rather than being represented by one of the Academy members, she avoided being referred to in the third person as "Miss Morris"—which had seemed to trigger the ire of her critics during the Hessian fly controversy. As Virginia Woolf would later put it, "'Miss' transmits sex and sex may carry with it an aroma." Submitting the article herself also meant avoiding the pitfalls of others, however well-meaning, mischaracterizing her work. Her multipronged approach, writing for both a scientific audience and a popular agricultural audience while also directing the way her research reached both, showed her hard-won wisdom from the controversy five years prior. As a woman trying to navigate both worlds, she had to control her message, and therefore her name. For the sake of the popular agricultural audience, she would remain "M. H. Morris" for now.[18]

As winter began to warm into the first traces of spring, Margaretta decided to perform her experiment on seven more of the garden's failing pear trees. On March 5, 1847, a year since the first successful experiment, Margaretta set out with a gardener once again. While late-breaking stories sped over telegraph lines from the battlefront of the Mexican-American War and onto the pages of newspapers, Margaretta's thoughts were still with cicadas. The recent thaw had made it possible to wedge a shovel down into the mud-clumped roots of her orchard's trees. As the gardener dug the familiar trench around the first tree, Margaretta gathered somewhere between four hundred and five hundred grubs. Among the other trees she, Elizabeth, and the gardener examined, there were variations. One neglected tree that had been clearly suffering had been the recipient of piles of compost and dirt, and that unintentional gift had triggered it to grow new, healthy roots close to the surface of the earth that prevented the gardener from reaching the cicada-ravaged roots below without damaging what was thriving.[19]

Another tree, in the far back corner of the garden, was a different case entirely: it had stopped producing fruit years ago, and so the Morrises had given it up as a lost cause. But in 1845, new, vigorous shoots had burst forth from its old dried branches. Curious whether there was a cicada-related explanation, Margaretta targeted this tree because of its unexplained revival. As she knelt down to brush away leaves and examine the soil around the tree, she found mole tracks extending in every direction. The moles had tunneled around the roots of the trees, devouring the cicadas, and alleviating the strain on the tree itself. The same was true for four more trees that had exhibited similar patterns of growth in that corner of the garden. Moles—so often scowled at by gardeners for their disruptive tunnels and hills—had rescued that part of the orchard.[20]

As Margaretta once again packed a box for Walter Johnson at the Academy of Natural Sciences later that day, she included not only a note with her observations of each tree and the various ways cicada larvae had affected them, but also a slice of damaged root with the cicadas still attached—stronger evidence than the loose cicadas of her previous samples. "This experiment was in every way satisfactory," she proudly declared, "and proved beyond a doubt the correctness of my former observations."[21]

Keenly aware of how so many had doubted her scientific skills when she shared her wheat fly findings, Margaretta decided on a new strategy for this discovery. In order to build a network of allies and strengthen her credibility, she needed to restage her discovery so other scientists could vouch for her observations and methods. Along with the specimens, she invited Johnson to come and see it all for himself. Margaretta was practically giddy that so much of her orchard was being decimated by cicada larvae when she wrote: "All this it will give me great pleasure to show you at your earliest convenience, having reserved several trees, that from [their] sad looks I judge have a glorious company of Cicada at their roots." While Johnson lived in central Philadelphia, he was a regular presence in Germantown and likely visited soon after.[22]

While Walter Johnson may have been one of the first guests to witness Margaretta's discovery himself, he would hardly be the last. When Margaretta and Elizabeth got word from their cousin Robert Hare that Louis Agassiz—the Swiss-born naturalist and newly appointed Harvard professor of zoology—was going to be in Philadelphia for several months in the winter and spring of 1849, they invited him to visit. In the two years since Johnson's visit, Margaretta had been busy defending herself against Asa Fitch's slights, writing articles as the Old Lady, and sending many more specimens to the Academy. Agassiz planned to give a series of lectures on animal embryology in and around the city. While a local critic in Germantown found that Agassiz's struggle with English detracted from the lectures, many were actually charmed by his accent. With a sweep of dark hair and an air of gentility, the tall, red-faced Agassiz made a point to make complicated scientific topics accessible to a wide audience, as Asa Gray put it, "so that ladies might attend." He even choreographed pauses in his lectures for laughter. Ralph Waldo Emerson was perhaps less than enchanted, describing Agassiz as a "broad-featured, unctuous man, fat and plenteous as some successful politician." Regardless, he remained popular with the educated public as well as students and established scientists, all of whom were willing to pay three to five dollars for a ticket each night.[23]

For years, Elizabeth had heard about the charismatic Agassiz from Asa Gray, who raved about the Swiss scientist in his letters. "I wish you could hear Mr. Agassiz lecture," Asa wrote her in December 1846, soon after Agassiz's arrival in the country. His public and private lectures were well attended, and Asa reported to Elizabeth that his friends at Harvard were "all charmed with him." Asa was rarely so effusive in his letters about fellow scientists, so Elizabeth must have been curious to meet Agassiz for herself and see whether he lived up to his reputation.[24]

Between lectures, Agassiz spent much of this two-and-a-half-month visit at the Academy of Natural Sciences, using the collections and library assembled by local naturalists. One of Agassiz's particular interests involved the origins of the races of men, and he had seized on the proposals put forth by Philadelphia's Samuel Morton, one of the Academy's members, who had recently published *Crania Americana* and *Crania Aegyptiaca*. In these books, Morton essentially argued that racial hierarchy was prescribed by nature, as evidenced by the shapes of skulls and corresponding sizes of brains. Agassiz made long visits to Morton's home during this trip, continuing conversations they had begun a few years prior amid Morton's eerie collections of human skulls from around the world. Agassiz, who declared to his mother that Morton's "collection alone [was] worth a journey to America," proved to be a devoted and charismatic mouthpiece for the scientific racism Morton espoused.[25]

When Agassiz made his way out to the Morrises' home in Germantown with Robert Hare, Margaretta and Elizabeth were happy to greet him. In some ways his visit signaled a kind of acceptance of the Morris sisters into the male scientific community. Margaretta's bylines, Asa Gray's endorsement of the sisters, and their relation to Robert Hare made this all possible. Their tireless scientific work combined with their social connections to make them worthy of such recognition in Agassiz's eyes. Knowing how much could ride on this visit, the sisters made sure everything was just so. Aware of her visitor's expertise in marine biology, Elizabeth likely brought out the preserved sea anemone she planned to donate to the Academy of Natural Sciences later that spring. While Margaretta and Elizabeth had never expressed much interest in human anatomy like Morton's skulls, the women would have had much to discuss with their guest. They would have talked about his lectures, their connections to many of the scientists Agassiz had been meeting with in Philadelphia, as well as the botanical and entomological experiments they were engaged with.[26]

The real purpose of the visit, though, was waiting in the garden. Summoning her gardener, Margaretta had him once again dig a deep hole so that she could remove a segment of the tree's roots. While Margaretta omitted her gardener's labor in her reports, erasing him grammatically with the passive voice ("I gathered [the specimens] while the earth was being removed from a trench four feet wide and two deep"), he played an important role during these visits. If Margaretta had dug the hole herself, assuming she had been strong enough, Agassiz would have likely judged her as unladylike. Growing up around their mother and great-aunts, the sisters knew that propriety and social reputation could make or break professional success. And her gender was, professionally, a liability—and not one that could be disguised face-to-face. She had to walk a fine line between asserting her professional capability as a scientist and behaving according to the unspoken rules of etiquette between genders. She succeeded.[27]

It is impossible to recover who the gardener was as Margaretta never gave him credit. Perhaps it was Massy Tango, an Irish immigrant and servant who lived at the Morrises' home or perhaps someone else they hired who lived in the neighborhood. Margaretta, Elizabeth, and their mother were often hiring gardeners from the local community, some of whom were immigrants briefly staying in Germantown before making their way west to start farms or seek gold. Margaretta's erasure of her gardener's labor was not unique. Other scientists typically did something similar, rarely giving credit to those working behind the scenes, just as several better-positioned scientists erased Margaretta's and Elizabeth's contributions in books and journal articles. In stepping into the actively forming power structure of professional science, Margaretta intuited that she was expected to act like her peers in order to gain standing in their community. This meant knowing who was and was not supposed to get credit.[28]

Margaretta's carefully choreographed reenactment of her experiment worked. Agassiz witnessed the feeding behavior of the cicada larvae, as well as the variance in size Margaretta had noticed between insects. He was intrigued by the mystifying life cycle of this North American creature and eager for specimens. Robert Hare would later report that "Mr. Agassiz was perfectly satisfied of the existence of the grub at the roots of the trees." This was a significant celebrity endorsement, and one that all but certified Margaretta's discovery.[29]

Robert Hare also continued to promote his cousin's work whenever he could. In August 1849, a few months after Agassiz's visit, many of the country's leading scientists had gathered in Cambridge, Massachusetts, at the height of a cholera epidemic for the second meeting of the American Association for the Advancement of Science. The growth of this national community, which would not have been possible without the expanded railroad system that brought all of the scientists together in Cambridge, helped to build a professional network of university professors, engineers, and doctors, as well as amateur enthusiasts. A year before the meeting that would take place in New Haven, it was already becoming clear that the AAAS, coupled with Harvard's and Yale's development of scientific programs and the founding of the Smithsonian Institution, had kickstarted the professionalization of American science.[30]

As the president of the AAAS opened the weeklong conference, he instructed the men assembled in the brick-clad Harvard Hall to keep their presentations of "new truths" as brief as possible so that they could spend time mingling between sessions. During some of that downtime, Hare struck up a conversation with a group of scientists, including Thaddeus William Harris, Harvard's librarian and one of Margaretta's chief critics during the wheat fly controversy; Charles Pickering, a Philadelphia naturalist who had famously gone on the Wilkes Expedition a few years prior; and Alfred Langdon Elwyn, yet another Pennsylvanian and the treasurer of the AAAS. The conversation flowed swiftly from topic to topic but centered mostly on insects. Hare, along with others, began discussing Margaretta's work with cicadas. Harris, who had once been so dismissive of Morris, was taken aback by how much respect these men held for her and her research.[31]

Harris, no doubt trying to change the subject, turned to Charles Pickering. He had been reading Pickering's private account of his time on the Wilkes Expedition, and he had questions about some drop worms from Philadelphia that Pickering had mentioned. Pickering's response, though, must have thrown Harris: "ask Dr. Hare's niece, she doubtless will know these '*drop-worms*,' & will get them for you." Even if they mischaracterized her relationship with her cousin, Pickering and the other men spoke about Margaretta Morris and her research seriously, and readily recommended her expertise—which was as much a social instruction for Harris as it was a suggestion about how to obtain specimens. Louis Agassiz, who was currently taking the AAAS meeting by storm, presenting twenty-seven papers in just five days, also endorsed Margaretta and her findings, further confounding Harris. He was beginning to reconsider his dismissal of her.[32]

That evening, Harris went back to his desk and began to write to Morris. Over six years before, Margaretta had sent him a letter defending her wheat fly discoveries and he had never written back. Their relationship was icy, but he tried to call on her goodwill and smooth things over. After attempting to build a friendly connection by mentioning her cousin's name as well as the others who had been singing her praises, he began a list of favors: asking for the drop worms he had been discussing with Pickering, asking for *Cecidomyia* samples, asking for any and all butterflies and moths from Pennsylvania, asking for a slug caterpillar to replace a specimen destroyed by parasites. He had a lot of requests and hoped to transform Margaretta Morris—someone he once expressed such disdain for—into a collector of specimens.[33]

Yet Harris was simultaneously careful to keep Margaretta in her place. He highlighted the hierarchy he wished to reinforce when, after begging for butterfly and moth specimens, he wrote: "Should your own taste lead you to take an interest in these things, you would confer a great favor on me, & might thereby enable me to advance science, if you would occasionally forward to me such objects as these." However much Harris might have liked to diminish her role, Margaretta was more than just an assistant—she was doing her own part to advance science. Harris was not sure how to categorize her, and may have looked to the example of his colleague Asa Gray and other botanists who regularly worked with women scientists. Harris, like many men of science, decided that Margaretta could be a trusted supplier of specimens, invaluable to the community, but not herself a professional. Elizabeth was content with that role for herself, but that was not enough for Margaretta, particularly at this stage in her career. While she was happy to exchange gifts of specimens, she saw herself as Harris's equal, not his subordinate.[34]

Regardless, Harris, in his own limited way, had to show a certain respect for Morris. His colleagues at the conference had made it clear that it would no longer be socially acceptable to ostracize her. When he sat down in the coming years to write the second edition of his encyclopedic book on New England's agricultural pests, Harris referenced Morris regularly, introducing new paragraphs to his old text that highlighted her discoveries with cicadas and other creatures. While he credited her with the discovery that cicadas consumed sap from the roots of fruit trees, he neglected to mention her having discovered another cicada species, though she had written about it in at least one letter to him.[35]

MEANWHILE, AS HARRIS'S DISDAIN SUBSIDED INTO GRUDGing acceptance, Margaretta Morris was about to name her first insect. Samuel Stehman Haldeman—a good-natured Pennsylvania naturalist who had championed Margaretta's work—encouraged her to officially name the new species of wheat fly she had discovered. She decided to name it *Cecidomyia culmicola* (Morris), a reference to how they lived and ate at the culm of the wheat plant. This was the first time her name was connected to an insect and she was uncomfortable with the prospect. "You deserve to have your name attached," Haldeman assured her, reminding her that her careful observations and persistent collecting had led to the discovery. She had waited as long as she had—nine years—because she was hoping a new infestation might allow her to collect a full set of the insects to include with the naming. While wheat farmers celebrated that there had been no infestations since 1843, Margaretta grew frustrated. Still, Samuel Haldeman encouraged her to submit a name to the Academy of Natural Sciences along with a description, despite the lack of new specimens. On August 21, 1849, the Academy accepted the new name of the wheat fly species she had discovered. Margaretta was not invited to attend that meeting.[36]

Margaretta's scientific fame was growing to new heights. In August 1849 alone, she was hosting several scientists, including the British naturalist Richard Chandler Alexander, whom she took on a walk to discuss her long history of entomological discoveries. She also was called upon by John Wilkinson, the new director of the Mount Airy Agricultural Institute, to help decipher what insect was destroying the potato crops at the agricultural school. Not only was Margaretta becoming a noted regional expert, but Morris Hall was now a regular stop for many scientists as they traveled to Philadelphia, the city at the heart of American science.[37]

When Henry Goadby, the English microscopist and entomologist, arrived in Philadelphia in April 1850 to give a series of lectures, Margaretta and Elizabeth not only attended the lectures twice a week but also befriended the man, inviting him to call at their home and introducing him to their friends. Goadby's lectures were well attended and inspired many in Philadelphia to become armchair entomologists. Goadby also brought his compound microscopes to Morris Hall to show specimens to the Morris sisters and their visitors that he did not have time to exhibit in the lectures. His artwork and microscopic skills were particularly astonishing—Elizabeth was left almost speechless by what she witnessed. "To see the two hundred and forty teeth of the gizzard of a cricket! the beating of the heart of an aquatic insect so small as to be visible only when in motion, the whole nervous system of a cockroach," she exclaimed. "It is impossible to give the faintest idea of the beauty of these most curious & interesting preparations."[38]

Goadby's visits gave Margaretta yet another wonderful opportunity for a respected entomologist to witness and vet her cicada discovery. As she had done so many times in the preceding four years, Margaretta directed her gardener to dig down to the roots of one of the declining fruit trees where the cicadas were allowed to feast unchecked. In Goadby's words, the withering tree "gave sufficient evidence to her practiced eye of the attacks of this vegetable vampire." Goadby was immediately convinced, but Margaretta continued, taking him to yet another tree. In all, they found thirty-five cicada larvae during that visit. Goadby returned home eager to endorse her work.[39]

Later that summer in August 1850, Robert Hare continued to publicize his cousin's findings. As a member of the Pennsylvania Horticultural Society, he hoped to inspire a committee to make a similar visit to the Morris Hall garden to witness what so many entomologists had already observed. Given the scale of their possible threat to fruit trees, cicadas caught the attention of the horticulturists. When the committee arrived in Germantown, however, things did not go as planned. Elizabeth likely chatted with the visitors about the fruit trees as the gardener got started, relaying that she was friends with William Darlington, who was also a member of the Society. As the gardener lifted the shovel heavy with roots and earth to show Margaretta, Elizabeth, and the men, the cicadas were very active and their rapid movements caused them to immediately disengage from the roots. Sixteen years into their seventeen-year transformation, the larvae were lively and relatively close to maturity. The visitors were disappointed to not see what they had hoped, and Margaretta was likely also frustrated by this failed presentation.[40]

Annoyed that the bugs did not perform as expected and showcase Margaretta's theory, Robert Hare tried to quiet the doubtful horticulturists by sharing Agassiz's and Goadby's observations, which seemed to appease them. Margaretta's investment in cultivating prominent endorsements continued to pay dividends. Months later, Hare circulated an abbreviated version of this report for publication in a handful of newspapers. Once again, and this time mere months before the cicadas around Pennsylvania and the surrounding area were set to emerge from the ground in a much-anticipated event, Margaretta Morris's name appeared repeatedly in newspapers and in agricultural journals, linking her to her discoveries.[41]

Not everyone was keen to accept her findings. Unwilling to accept that cicadas could damage trees, a reader named E. Nichols wrote a letter to the *Ohio Cultivator*, arguing that the creatures were a benevolent creation of God "because they sing and fly elate with joy, during their stay with us, and their plump appearance on their arrival is significant of past comfort." The reader was not willing to believe "those who are often croaking about the evil natural arrangements of the world." Later, "An Observer of Nature" wrote to *The Friend: A Religious and Literary Journal* to explain that they had a hard time believing that cicadas were at the root of the trees' problems. Even though Morris's research had been "corroborated by the learned Professor, and Doctor, together with the celebrated Naturalist," the writer, sounding like the cicadas' defense attorney, declared: "I would free the Locust from the charge of occasioning the injury, until some further proof of their guilt be established." Ultimately, the critiques that Morris received were mild, far-fetched, and unsubstantiated by science. Notably, though, unlike during the Hessian fly debate, these critics did not cast aspersions on her gender.[42]

The true testament to Margaretta's research could be found in the trees. Those that had been successfully cut from the cicada-infested roots, were flourishing. Goadby championed this as nature's ultimate endorsement:

> The lady, to whose intelligence and persevering industry the discovery of the habit and instincts of this insect is solely due, Miss Morris, has set the example of relieving her trees from these persistent enemies and she has succeeded, by this means

alone, in saving some trees from threatened dissolution, and restored them all to that robust health by which they have been enabled to bear very large crops.

Indeed, except for the few struggling trees she kept for study, her orchard had made a full recovery. Margaretta and Elizabeth began to see their trees decked out in "healthy verdure" and "finer fruit" than they had enjoyed in years. Whether these trees revived because their cicada-infested roots were severed or because the nutrients in the soil were replenished with the fresh soil, or some combination of the two, remains unclear—though it was likely some combination. To Margaretta and Elizabeth, the experiment was a success.[43]

Nearly two centuries later, there is still no consensus about the ways cicadas impact trees. One twentieth-century entomologist found that infested apple trees had limited wood growth compared to cicada-free trees. Others found that trees like oaks ultimately were little affected by the creatures sucking at the roots. Still others have found that fertilization from decomposing cicada bodies at the end of the emergence year replenishes and exceeds the nutrients they extract during their long stay underground. Nowadays when orchard owners turn to agricultural school extension centers, such as Penn State Extension, they are advised that the cicada presence underground "probably affects the tree productivity, though this has never been fully documented." The entomologists suggest that orchard owners wrap their trees in nets during a cicada season to keep the cicadas from laying eggs there, or use pesticides to keep them at bay. In short, it seems that orchard owners still believe there's a connection between cicada larvae and their harvests.[44]

In the nineteenth century, Margaretta Morris's peers accepted her findings. It was due to these discoveries that Louis Agassiz endorsed Margaretta's election to the AAAS that summer in 1850 at the conference in New Haven. Agassiz, who had recently married Elizabeth Cary, who herself would become an advocate for women's education, was not only a member of the AAAS and its standing committee but also its newly elected president. Any member had the prerogative to nominate candidates who did not otherwise qualify through membership in local scientific organizations, by holding positions at universities or colleges, or through careers in engineering or other fields. As most of these organizations, institutions, and careers were traditionally populated with men, it is no wonder that few women were a part of the AAAS initially. The Association's constitution did not explicitly prevent women from joining, but the membership qualifications were such that they privileged men and made it possible for women to join only if they were nominated by an existing member and seconded by a standing committee member—a much higher hoop to jump through.[45]

While the AAAS had from the start opened its meetings to men, women, and members of the general public, there were subtle ways that they made women feel like less than equal members. Repeatedly, when discussing the purpose of the Association, the leadership referred to its being founded to bring together "scientific men." Nine years later, when the next woman, Almira Hart Lincoln Phelps, was elected, the president of the Association capitalized on that moment to call out the organization's inclusiveness, remarking that "the door stood as wide open to ladies as gentlemen—that there was nothing in the Constitution to hinder their nomination and election." Puzzlingly, he continued by warning women against membership, "lest it should be deemed a challenge to the public to admire their scientific acquirements." While many scientific women preferred to remain anonymous and out of the public limelight—Elizabeth Morris included—it is difficult to comprehend how the public might better admire the "scientific acquirements" of women if they were not honored for their achievements and work, let alone given equal access to the platform of the AAAS or other societies.[46]

Margaretta Morris and Maria Mitchell, then, were even more notable for having achieved that status, and fortunate to have the connections through the scientific community to marshal support from Agassiz and others. Mitchell, though twenty years younger, was similar to Morris. Born into families that valued education for girls, unmarried and balancing a life of scientific exploration while caring for aging parents, Morris and Mitchell both carried themselves in similar ways, with a stoic seriousness. Mitchell, who had worked as her astronomer father's assistant, had discovered a comet in 1847 that became known as "Miss Mitchell's Comet," winning her international accolades, including Louis Agassiz's sponsorship to the AAAS. Mitchell, ultimately, would become more of a public figure as the decades went on, giving public lectures and winning a professorship at Vassar College in 1865.[47]

If women had been welcomed into the AAAS in larger numbers and with fewer obstacles, on equal footing with their male counterparts, perhaps there would have been a broader community of women scientists sharing strategies and histories as a way to lift each other up and amplify voices. As it was, there was barely lukewarm support. After the Civil War, more women would join, causing the Association's leadership to reconsider how to categorize its members, ultimately deciding in 1873 to create a category for "fellows" who were "professionally engaged in science," with very few women qualifying. All of this meant there was nearly no national community for women in science initially, beyond the direct friendships and connections cultivated by the women themselves.[48]

At that same conference where Margaretta was elected, Agassiz also personally presented her work—a badge of honor. He read her report on cicadas at the conference and took questions on her behalf, though he neglected to submit the report to be printed in the conference proceedings, as was the case for many of the reports he read for others. While Robert Hare was excited that Agassiz might publish something of his own on cicadas, he never did, instead incorporating what he had learned from Margaretta into his book *The Natural History of the United States of America* several years later in 1857. While musing about the unequal lifespans of various plants and animals, he briefly lingered on the curious life cycle of the seventeen-year cicada "so fully traced by Miss M. H. Morris."[49]

AGASSIZ'S ENDORSEMENT AND THE RECOGNITION OF THE AAAS vaulted Margaretta's career to a level of scientific authority that had previously been unattainable for women. While she was never fully welcome in the innermost circles of its fraternal ranks, she was now part of the community in an official way. Editors and scientists recognized and deferred to her expertise on Pennsylvania's insects, particularly those that troubled farms and orchards, and they called on her repeatedly for advice and identifications.

After her election to the AAAS, Asa Fitch—the entomologist from New York State who had been so quick to dismiss Margaretta's findings on the wheat fly in his writings for the *American Journal of Agriculture and Science* in 1847—finally responded to Margaretta's public call to take her evidence seriously or prove her wrong. Though Margaretta had written that letter to the journal in April 1847, and though the editor, Ebeneezer Emmons, had promised a response from Fitch in the next issue, it instead took him until 1851 to finally address the topic.[50]

In an article in the *Cultivator*—yet another agricultural journal—a reader asked Fitch to identify a wheat-eating joint-worm that was plaguing Virginia farms that year. In the midst of answering the reader's question, however, Fitch went off on a tangent and started writing about Margaretta's findings. Evidently this dispute had been weighing on him. He argued that he had initially ignored her because her observations contradicted his own, which he knew to be right. Seemingly unwilling to trust that a Pennsylvania woman had seen a fly different from any species he had personally witnessed locally in upstate New York, he had discounted her report outright.[51]

Perhaps, over the years, Fitch had come to realize his mistake. Or perhaps he had been humbled by the fact that the AAAS—of which he too was a member—had elected Morris.[52] Regardless of what drove Fitch to apologize, he now recognized that Morris had observed a completely different form of *Cecidomyia*. In what must have been a particularly satisfying line for Margaretta, Fitch wrote: "I may remark that much credit is due to Miss Morris for having detected discrepancies and made discoveries which none but a close investigator would have been apt to notice." He did, however, defend his choices by pointing out how she had initially believed her new species of *Cecidomyia* was the well-known Hessian fly. Fitch added, "Mistakes more gross than this have in repeated instances been made by the most acute and experienced observers." Perhaps he meant to soften the reminder of her mistake with a concession. Perhaps he meant to reassure her that he did not mean to undercut her authority. But by distinguishing her from "the most acute and experienced observers," he managed in one stroke to apologize while reinforcing the distance between her world of entomology and his. Though they studied similar creatures and similar problems in neighboring states, and could have benefited from collaboration and correspondence, Fitch and Morris never built those connections. Fitch's apology was ultimately too little, too late.[53]

Margaretta and Elizabeth, however, were always happy to reciprocate the kindness shown to them by more supportive scientists. Feeling indebted to Agassiz and eager to continue their friendship, Margaretta—with the help of Elizabeth—collected hundreds of cicada larvae to send to his collection at Harvard. Agassiz clearly had as much to gain from the Morris sisters as they did from him. In the years to come, Agassiz would continue to call on Margaretta to help him collect regional fish, a favor she was pleased to grant.[54]

IN EARLY JUNE 1851, THE CICADAS THAT HAD LURED SO MANY scientists to the Morris sisters' garden emerged from the ground in "countless numbers." Those that Margaretta, Elizabeth, and their gardener had not unearthed loosened themselves from the roots and tunneled upwards, navigating through topsoil and tangled roots until they surfaced into the warm summer air. Margaretta had been monitoring their progress since March, writing her observations in her notebook as the insects prepared to make their way aboveground. Leaving behind their tunnels, the creatures scrambled across the ground as they adjusted to their new environs, to the delight of chickens. Those that escaped the fowl climbed tree trunks in preparation for molting. Margaretta watched as they instinctively chose the trees whose roots they had been feeding on, passing over more convenient shrubs.[55]

All around Philadelphia and the surrounding region where this specific brood of seventeen-year cicadas resided, children and adults alike watched the emergence unfold. Newspapers reported on sightings, spreading word about what readers might look for as the sound of an "incessant, monotonous, somewhat melancholy multitudinous roar" reverberated from the trees. Margaretta preferred to describe their drone as sounding like "Faaaa <u>ROO</u>." It was the sound of summer, and a nostalgic sound for those remembering back to a June seventeen years prior. While some locals grumbled about the nuisance of their clumsy flights and the noisiness of their hums, those like Margaretta savored a rare moment to watch their life cycle progress aboveground.[56]

As the cicada season concluded in the summer of 1851, newspapers reflected upon the quiet absence of "the song of the locusts." Margaretta, whose repeated demonstrations of her discoveries had been so closely tied to this emergence, would undoubtedly relate to the solemn emotions that accompanied the cicadas' return underground. It would not be unlike Margaretta, who was now fifty-three, to measure her lifespan in terms of cicada years perhaps wondering if she would live to see the next.[57]

Though her professional peers had accepted her findings about how cicadas subsisted underground during their long subterranean existence, few had acknowledged her claim to have discovered a new, smaller species. Amid the "Faaaa <u>ROO</u>" sound she heard in the chorus of cicadas, she had also heard another cicada song, something "sharp and shrill, like the noise made by the loom of a stocking weaver." Germantown was known for its woolen mills, and Margaretta, like her neighbors, would easily be able to identify the rattling clicks and shrill buzzing of a stocking weaver. The creatures singing this distinct song were also much smaller in size and "extremely active, springing backwards with a sudden motion," which made it difficult for Margaretta to catch them. She had noticed these differences in size, song, and habits among the cicadas as early as 1817, when she was nineteen years old—the first cicada year when she would have been old enough to take notes and carefully observe her subjects. She had then noticed the two different types of cicadas seventeen years later in 1834, when she was thirty-six. By 1851, she was certain that there were "two distinct species in the same swarm." She had been arguing her case since her first report on cicadas to the Academy of Natural Sciences five years earlier in 1846, as well as in subsequent letters and reports to scientists.[58]

It must have come as a shock then, when Margaretta read in the latest issue of the *Proceedings of the Academy of Natural Sciences*, that two of its members—James Coggswell Fisher and John

Cassin—were declaring that they themselves had discovered a new species of cicada that was smaller, darker, and had "a note *entirely different* from that of the larger Cicada septendecim." This sounded awfully familiar. By the time the cicadas had stopped buzzing in 1851, the two men prepared separate descriptions for the Academy of Natural Sciences, and a name that jointly celebrated them both: *Cicada Cassinii* (Fisher), commonly referred to as Cassin's cicada. Given the frequency with which Margaretta Morris had sent reports and specimens to the Academy, and her close connections with several members of the Academy, it is unlikely that Fisher and Cassin would have been unaware of her claim that there was a distinct species of cicada that had previously been conflated with the *C. septendecim*. Perhaps her notes on the small cicada species inspired them to look closely once the cicada emerged, but regardless, they gave her no credit in their reports. John Cassin was much more interested in ornithology than entomology and though he named several creatures, they were almost all birds. Fisher's interests were in geology and mineralogy. Yet these two men, who certainly benefited from Margaretta's close observations from a period of over three decades, have achieved a sort of scientific immortality as their names are repeated by cicada researchers and enthusiasts more than a century and a half later.[59]

Perhaps Margaretta regretted not putting forth a name herself earlier. If she had been a member of the Academy in 1851, and welcome to attend the meetings alongside her male colleagues, she would have been able to contest their claim in person. Margaretta, however, had never been particularly concerned with the scientific fame that came from naming creatures. She knew that what she could do with science was far more important than what it could do for her. There was so much yet to be understood about the ecological landscape of North America, and Margaretta was driven by curiosity, wonder, and a deep sense of connection to the natural world, not fame. In fact, she continued to send reports of what she had been learning about the different cicadas to scientific institutions, like the Boston Society of Natural History, seemingly unfazed by professional slights like Cassin and Fisher's claim of discovery.[60]

Despite all of this, Margaretta had reached the pinnacle of her career as an entomologist. Her expertise was recognized locally and nationally by scientists, horticulturists, and farmers. Some of her harshest critics in years prior, including Thaddeus William Harris and Asa Fitch, came to embrace her findings about the impact of cicadas on tree roots, incorporating her work into their own studies, even if they were still wary of counting her as an equal. Nearly every achievement seemed to be tempered in this way. Though elected into the AAAS, a sign of her inclusion among professional scientists, she was still marginalized and treated as a second-class citizen. The boundaries of the scientific community and its institutions were constantly shifting, but Margaretta continued to find ways to leverage insights into the inner workings of this quickly coalescing community to her advantage. Her strategy of inviting esteemed male naturalists to her garden where she could stage her work had been successful in building a system of allies ready to defend and promote her discoveries. These carefully cultivated relationships had served her well. Regardless of the exclusions and restrictions that kept her from being an equal among the scientists, Margaretta had found a way to push entomology forward.[61]

Seven

LITTLE TIME TO CALL MY OWN

As the fall approached in 1850, Margaretta longed to be wandering the Wissahickon. "The cares and anxieties of a close attendance on an aged mother has engrossed all my time, though not my thoughts," she reflected. "They are free, and will haunt the leafy forest, where dwells my little friends the Insects, and there I have longed to be." Margaretta and Elizabeth's mother Ann was eighty-three and it seemed like her illnesses came and went like the tides, sometimes receding but never fully gone. As Margaretta tended to her mother, bringing her broths to keep her hydrated, administering medicines, changing bed linens, and reading books and newspapers aloud from the chair at her bedside, her mind was elsewhere. She daydreamed about searching the woods for Luna moths, those majestic pale green, leaflike creatures with feathery antennae that fly at night and rest on forest trees during the day. Given how Margaretta was recently elected to the American Association for the Advancement of Science, many were reaching out to her with entomological puzzles they hoped she might help them with. With signs that her mother's health might be improving, Margaretta anticipated future days filled with experiments and "more pleasant hours in the woods," but for now she remained in Morris Hall.[1]

Margaretta was not alone. Elizabeth also struggled to balance her scientific work with the responsibilities of caring for their mother and managing the household. She explained to William Darlington that it was difficult for her to complete any task, even copying down a lecture or book passage, given the persistent need to help. "My dear mother's very precarious state of health leaves me no *uninterrupted* time," Elizabeth wrote, "and when I sit down to write, it is always in momentous expectation of being called away." Her ability to collect specimens, attend agricultural fairs, or even write letters became increasingly limited as her mother's health worsened. She tried to schedule trips to visit William and her other friends, but found she could not leave home even for short visits while caring for her mother.[2]

Unmarried daughters bore the brunt of being caregivers for their aging parents, as well as other family members. Margaretta and Elizabeth had been balancing these roles for decades, collecting specimens and writing articles while tending to nieces and nephews, aunts and uncles, cousins, siblings, and especially their mother. While they certainly had more freedom to pursue vocations than their married sisters, Elizabeth and Margaretta were responsible for the upkeep of Morris Hall. Particularly as their mother aged, they took over all of the household management, with the assistance of hired help. In the early nineteenth century, as the number of unmarried women rose—for reasons ranging from personal decisions to demographic changes like the large numbers of single men moving west in search of gold and farms—a number of women like the Morris sisters struggled with the conflicting demands of their domestic obligations and their vocations. Many of these women submerged themselves in caring for family for long stretches so that they could feel less guilt devoting time to activism, creative projects, studies, or paid jobs. Margaretta and Elizabeth felt this push and pull, complaining of the interruptions to their beloved scientific work, while still finding some joy in their domestic work as they nursed relatives to health. As was clear in their anonymous articles for the *American Agriculturist*, science permeated their household responsibilities as well, as they tried to make it as efficient as possible. They were proud of the recipes and methods they had perfected.[3]

Margaretta did not particularly excel in her role as a nurse, though she accepted its necessity. Her mother may have also been a demanding patient. Ann had once complained to neighbors that under Margaretta's care, she could not be nursed as she would have preferred. "Whatever M. desires must be submitted to," whether she liked it or not. Margaretta was known for her efficiency, but not so much for her patience. Elizabeth did not seem to receive the same complaints, but both sisters similarly felt the strain of their caregiving responsibilities increase as their mother aged. For women whose work required them to be outdoors in the forest, field, and garden, their roles as caregivers kept them cloistered inside.[4]

MARGARETTA'S CLOSE INVOLVEMENT IN HER MOTHER'S HEALTH issues began to change the way she thought about the diseases affecting peach trees and potato vines. Frustratingly, the scientific demands on Margaretta's time had increased exponentially just as her freedom to focus on entomology contracted. Still, the combination of these different types of work had its benefits. In this period before the germ theory of disease and our contemporary understandings of bacteria, when doctors debated theories about how cholera and tuberculosis might spread, there were also plenty of unanswered questions about how plants contracted diseases.

Earlier that year, in May 1850, Margaretta had published an article in the *Horticulturist* about the peach bark beetle (*Tomicus liminaris*). She argued that this tiny beetle that burrowed inconspicuously under bark was spreading a disease called the "Yellows" to peach trees. The disease caused fruit to prematurely ripen, rendering it bitter and inedible, and consequently of no value to the orchard owners. The yellowed leaves that hung from sickly branches gave the disease its name. Horticulturists had long debated the cause, speculating about everything from weather conditions to bad pruning.[5]

As the disease radiated out from the Philadelphia region across the country, orchard after orchard fell victim. Margaretta's theory about the hard-to-spot bark beetle spread quickly, bolstered by *Horticulturist* editor Andrew Jackson Downing's cautious praise: "The foregoing is, we think, entirely new; and from the reputation of MISS MORRIS, as an entomologist, is entitled to attention." Editors of agricultural and horticultural journals from around the country reprinted this article over the next few months. Having lost many of her own peach trees, Margaretta had been intent on diagnosing the origin of the destructive disease so she could help others treat and prevent it.[6]

The praise she received for her article was extensive. The naturalist Samuel Stehman Haldeman, having first learned of Margaretta's discovery in letters they exchanged, announced to his peers at the Academy of Natural Sciences that "Economic Entomology has been enriched by the discovery of Miss Morris." The anonymous reader "Agricola" declared in the *American Agriculturist* that they hoped "that we shall often hear from Miss Morris, as it is only by close investigation, that we can ascertain the causes of many of the mysterious diseases of fruit trees, and vegetables, and every laborer in the cause, increases the prospect of success, especially when the investigator brings to her aid, the assistance of a good knowledge of entomology." The anonymous "Reviewer" added his own praise to the pages of the *American Agriculturist*: "Here the labor of a scientific mind has been applied to make a most important discovery for the benefit of every one engaged in agricultural pursuits." But while Andrew Jackson Downing expressed his deep respect for Margaretta in his editorial comments at the end of her article, he did not fully accept her findings. Downing had published *The Fruits and Fruit Trees of America* in 1845, a popular handbook that had already been reprinted several times by 1850. In that book, he theorized about the causes of the Peach Yellows, arguing that the disease was constitutional rather than transmitted by insects. He blamed exhausted soil and poor cultivation primarily, though he recognized that the definitive cause was still unknown.[7]

Margaretta's article presented the evidence for her argument and outlined her method but, most importantly for peach growers, also included practical recommendations for treating an infected orchard: cut down the afflicted tree and burn everything, even the roots. That was the most effective way to eradicate the Yellows and keep it from spreading throughout the orchard or to neighboring properties. Margaretta also found that dumping chamber pots full of human urine at the base of peach trees helped the ailing trees, but she chose not to include that in her report, likely to avoid ridicule. Instead, she anonymously submitted that piece of advice to the readers of the *American Agriculturist* using the initial "M." Perhaps some of the more humbling work of caring for her mother had led to this particular discovery.[8]

Plant pathology was only just finding roots in the United States in the 1850s, and botanists, horticulturists, entomologists, and orchard owners were all eager to combine their observations to try to make sense of what was happening to the trees. Margaretta's theory about the bark beetle—whether it weakened the trees or spread the ailment—advanced the conversation and drew renewed scientific attention to the problem. In his notes at the bottom of Margaretta's article, Downing shared that he found that the Yellows spread from tree to tree after he used the same paring knife on both, perhaps hinting at the possibility of contagion. The beetles, meanwhile, were found on some trees with Yellows but not all. As authors and editors took part in a messy peer review, bouncing theories back and forth in the pages of agricultural journals, they edged toward collaborative answers, though with a lot of chaos and confusion mixed in. It would take another fifty years for plant pathologists to determine that microorganisms now known as phytoplasma caused the Yellows, transmitted by plum leafhoppers, insects that flew from tree to tree.[9]

INVESTIGATING THE ORIGIN OF PEACH YELLOWS WAS NOT Margaretta's only professional focus as her mother grew sicker. While she consulted with doctors about her mother's symptoms, she was also trying to determine what was ailing American potatoes. The same rot that was famously plaguing potatoes in Ireland and across Europe had been affecting American potato yields since 1843. When it reached Ireland two years later, the result was disastrous, leading to at least a million deaths from starvation, and pushing a million more Irish people to emigrate within the span of a few years. This humanitarian and agricultural crisis was unprecedented, and it spurred a frantic effort to make sense of the tragedy. While Americans were not as uniformly dependent on potatoes for their diet, the question of what caused this agricultural disaster was a transatlantic obsession. Scientists, farmers, even armchair agriculturists all had theories about the blight and fiercely fought over them in the pages of journals, magazines, and newspapers. Margaretta never shied away from contentious agricultural mysteries. Armed with considerable entomological expertise, she decided to enter the fray.[10]

When John Wilkinson, the director of the new Mount Airy Agricultural Institute in Germantown, found the school's potato crops withering thanks to tiny grubs inside the troubled stalks in 1850, he called on Margaretta's expertise to help identify the pest, bringing the infested potato stalks to her at Morris Hall. Her election to the AAAS had only cemented her status as one of the country's leading agricultural entomologists, and editors and others sought her out for advice. Wilkinson had set up the school not only to educate future farmers but also to have the property serve as an experimental farm where agricultural scientists like Margaretta could work alongside students to distill and devise best practices for farmers. He carefully noted the types of fertilizer they applied to the potatoes—pine coal and ashes from the local railroad companies combined with animal manure from the farm—and inspected the stalks carefully for blight and other issues. Baffled by the beetles he found, he delivered the infested stalks to the local expert—Margaretta—in hopes that she could offer an explanation for the disease. After placing some of the stalks under bell jars in her library, Margaretta took notes on the insects, identifying them as *Baridius trinotatus*, a type of beetle. "I had the gratification of following them through their various transformations," Margaretta reported. Observing the insect in all its forms was the only way to be confident in the identification and truly understand its impact.[11]

Margaretta wondered whether this tiny beetle could bear some responsibility for the terrible potato blight. In 1849 and 1850, as Margaretta examined potato stalks and the beetle, scientists like Margaretta understood that a fungus was involved but debated whether the fungus was the cause or the consequence of the blight. Theories about the cause proliferated and included everything from wet summers and volcanic exhalations to industrial pollution and airborne miasmas. The weather theory, which held that wet and cold conditions led to the fungus, was the most popular, promoted by leading British botanists on the pages of agricultural journals. Margaretta's theory that a beetle might be at the heart of the potato pandemic was not a far-fetched or outlandish explanation within this context. Still, she was careful not to jump to conclusions. "Whether this prove the origin of the potato rot or not," she wrote, "it is an evil of great magnitude and demands the attention of entomologists and farmers."[12]

Margaretta decided that the best way to find out whether the potato curculio (one of the beetle's common names) was causing blight worldwide was to consult her ever-growing network of colleagues. Given that the blight was present in North America two years prior to its European introduction, American agricultural scientists had insight to offer their European colleagues. This was an early instance in which American science emerged on the international stage, shaking off some past insecurities to collaborate in order to make sense of this crisis. After a week of observing a large number of specimens, Margaretta sent sets to the entomologists Thaddeus William Harris in Massachusetts, S. S. Haldeman in Philadelphia, and T. C. Westwood in England. While Haldeman praised her work, both Harris and Westwood were skeptical that the insect could be at fault for the global blight. Westwood confirmed that the *Baridius trinotatus* was a North American insect not found in Britain, as far as he knew, making it unlikely to be the source of the Irish crisis. Margaretta also sent notes to a number of other friends and naturalists, accompanied by explicit instructions that explained how to look for the grubs in their own potato fields. From their responses, she determined that the "small, dusky black beetle" could be found wreaking havoc throughout North America from Maine to Mexico.[13]

Regardless of whether the potato stalk weevil (yet another of its many common names) was at fault for the worldwide blight, it caused North American vines to decay prematurely, preventing the tubers from growing to their full size. Studying this little-known beetle, Margaretta argued, could prevent future scourges. In the article that she published in the *American Agriculturist*, which was then reprinted in the *Valley Farmer* later that year, Margaretta drew connections between the potato rot and the cholera pandemic that was on everyone's minds. "Like the cholera," she proposed, "may it not be checked and restrained by timely care, and the removal of exciting causes?" She suggested that the beetle might be making plants more susceptible to rot, just as doctors believed that certain behaviors or conditions made humans more vulnerable to cholera. The blending of Margaretta's immersion in her mother's health and her professional interest in plant health enabled her to draw constructive connections between otherwise disparate areas of knowledge. At a moment when neither the pathology of potato rot nor the source of cholera was fully understood, Margaretta's was a viable theory.[14]

In the midst of these evolving debates about the potato blight, Margaretta's gender once again became a topic of public conversation. "Reviewer" praised her work as "a most valuable and learned article . . . which shows that it flows from a well-informed mind." A supporter of scientific agriculture, Reviewer argued that Margaretta's articles were exactly what agricultural journals ought to include. Hoping to shock subscribers, he continued: "No doubt from the pen of some professor, or scientific gentleman, says the reader, perhaps. And yet, I say, if I am not mistaken, this valuable article is from the pen of a lady." While Margaretta had continued to use "M. H. Morris" to avoid unwanted criticism that would distract from the scientific content of her articles, she was becoming prominent enough that those like Reviewer reveled in outing her to the general public. How remarkable, he seemed to be saying, how novel, that a woman could be so smart and such an accomplished entomologist. It was meant as a compliment, but it came at the expense of the rest of her sex. One way or another, Margaretta's gender remained a persistent presence in every professional conversation.[15]

While he never explicitly disparaged Margaretta on the basis of her gender, Thaddeus William Harris made it clear that he did not accept her potato weevil theory. After reading an article in the *New England Farmer* by another author separately claiming that the potato weevil was at the root of the rot making the plants susceptible to fungus, Harris wrote a response discussing his exchange with Margaretta in which he outlined the logic he used to refute her claim. Curiously, he chose not to use her name. He did, however, refer to John Wilkinson, the director of Mount Airy, by name, crediting him for initially spotting the grubs. When he discussed Margaretta and her theories, however, he just referred to her as "a correspondent." Perhaps he chose to do this so as not to insult her publicly. Alternatively, omitting her name may have been its own kind of slight, the tactic his friend Edward Herrick had used a decade earlier during the wheat fly debate. One way or another, he ended the article by trumpeting his own theory that potato rot was "propagated by a kind of *malaria*, or atmospheric poison, capable of extensive diffusion." Harris wanted Margaretta and everyone else to stop blaming insects.[16]

Margaretta continued to study the beetle from the potato stalks, finding that they accompanied the rot in every case that she studied, but that the rot did not always accompany the beetles. She shared her findings with Harris, noting that the beetles deserved more attention. "I wish to keep my mind free from *theory*, that stumbling block to truth," Margaretta jabbed. And though Harris worked Morris's information into his new edition of his *Treatise on Some of the Insects of New England*, he remained skeptical of her findings. She, however, continued to hold her ground, puzzling over the connections between the beetle and the rot, gathering specimens from across the country. When Harris charged that he still was not finding the beetles in New England, Morris responded with a polite but pointed suggestion that he try harder: "They may be found by millions in September and the beginning of October in the potato fields here, yet everyone says they cannot find them unless I go to show them where to look." Open your eyes, Margaretta implied, and you will see them everywhere.[17]

As Margaretta had done in her previous clashes with Harris, she refused to throw out her conclusions without indisputable evidence that she was wrong. "While I am careful not to give up an opinion hastily, I am equally desirous of keeping my mind clear from prejudice," she wrote. "And when I am fully convinced that the potato vines have been well looked into in New England, as they have here, I will decide at once that the rot is the potato cholera," referencing Harris's belief that the rot was an airborne disease. She did not trust that everyone was looking closely enough at their potato vines to rule out the beetle connection. It was not improbable, given her past experiences, that Harris or even Westwood was overlooking her suggestions, assuming she was far more of a novice than she was, and therefore not seeing the beetles and the threat that they posed. If Harris had found her unnecessarily stubborn, Morris explained why she refused to abandon her conclusions. "If in this matter I appear pertinacious, I again beg indulgence," she wrote, "and ask if I have not some reason to be slow in yielding to anything but proven facts, when we remember how long it was before the world would listen to the fact that the cecidomyia that destroys the wheat in this neighborhood, fed in the center of the straw, or that the Cicadas killed one's fruit trees." This was not her first rodeo, after all.[18]

Harris exploded. As she had promised, Margaretta had sent him a new set of infested potato stalks, as well as a number of other specimens, including a glass bottle that she labeled "potato grubs and pupae." Claiming that she had mislabeled the bottle, Harris seemed to delight in her mistake: "You have surely fallen into an extraordinary error." He then spent a long paragraph detailing how dangerous these kinds of mistakes could be in terms of the future education of farmers and entomologists. "Scientific men, who understand scientific descriptions, would attribute the injury in question to the Lepidopterous borer," Harris chided her, "while you and your neighbors would have in view only the Coleopterous insect." In a single sentence, he invoked their power dynamic, drawing a line between "Scientific men" and her parochial Germantown community. Lest she think him arrogant, Harris assured her that "the study of larvae & transformations of insects dates with me from my boyhood, and I am now 57 years old. It would, indeed, be very strange, if I could not speak with some confidence on the subject now." Of course, Margaretta, just two years younger, had been studying insects since she too was a child.[19]

Ultimately, despite his confidence, Harris was incorrect about the origin of the potato rot. So was Morris, for that matter. By 1861, the German botanist Anton de Bary, building on the work of other researchers, published his account proving that the spores of a fungus caused the disease, naming the fungus *Phytophthora infestans*. The feverish transatlantic search for the source of the potato blight ultimately gave birth to the scientific field of plant pathology, opening possibilities for new ways of understanding how plants could become vulnerable to diseases.[20]

While Margaretta was ultimately wrong about the blight, she had uncovered the fact that American potatoes were simultaneously suffering from a different but still substantial threat—the *Baridius trinotatus*. This beetle, whose influence was not widely understood until Margaretta began writing about it, would continue to ravage North American potatoes far into the twentieth century. She had been the first to truly describe the scale of its impact.[21]

THE POTATO BLIGHT INCIDENT ILLUSTRATED HOW THADDEUS William Harris and Margaretta had grown to have a strained, if sometimes amicable, relationship. Amid his swagger over his superior observational skills and training in his letters to

Margaretta, he was also forthright about some of his failures. Ever since the letter he wrote Margaretta in 1849 after meeting Robert Hare at the American Association for the Advancement of Science meeting, he had been asking her for information and specimens of the drop worms that infested Philadelphia's trees. He wanted to study them, identify them, draw them, and write about them. Margaretta sent specimens, as did one of Harris's other contacts in Philadelphia, but he confessed that he had difficulty getting the worms to emerge from their cocoons into their full, "perfect" adult forms. "The drop worms too, disappointed me," Margaretta wrote, as she had also been unsuccessfully trying to watch their transformation in the Morris Hall library, "but the future is before us, and I never give up." A tree in her neighborhood was covered in them, and she promised to send him more. While Margaretta kept her cocoons on a cedar branch under a bell jar, Harris took it one step further and tied the cocoons to his beloved arbor vitae tree in his garden, in an effort to see if they would thrive outdoors when they struggled inside.[22]

Harris loved that tree. He called his fifteen-foot-tall tree "a magnificent specimen" and described it as "a dense, graceful pyramid from bottom to top of deep green foliage." He had planted it just outside his parlor windows so that his family and their guests could sit and admire its trim evergreen boughs. As the pace of his work at the Harvard Library picked up, he completely forgot about the twenty cocoons he had tied to it. Next thing he knew, months had passed and the tree was completely covered, particularly at the top where the drop worms "had begun the work of destruction in earnest." Multiplying rapidly, the caterpillars not only conquered the arbor vitae, they began to spread with the wind, attaching themselves to apple trees hundreds of feet away. Harris put his youngest son to work collecting them with the tallest ladder in their possession. But the worms kept at their destruction, and Harris resorted to spending at least two hours every day trying to save his trees from ruin. "There must have been millions of the drop-worms on the tree," he wrote, exhausted. Only the cooling temperatures and pounding rains of early autumn halted their progress. Still, though his drop worm story would not become the stuff of entomological legend, like the accidental introduction of the gypsy moth (now known as the spongy moth) to North America by an enthusiastic but careless naturalist nineteen years later—Harris very likely introduced a new pest into the trees of Cambridge. He was proud, though, to be able to draw and preserve some of the adults, which he identified as *Oiketicus*, a genus of moths. While his beloved tree may have survived the infestation, its bedraggled appearance forever reminded Harris of his folly.[23]

Through their correspondence, Margaretta determined that they shared a mutual friend—or something close to it. Margaretta sent Harris an entomology paper that the perpetually traveling reformer Dorothea Dix had shared with her, only to learn in his response that Dorothea was actually his cousin. Harris and Dix despised each other. Margaretta tried to sidestep the family disputes by remaining cordial in response to Harris and whatever snide remark he made (lost, like the letter, to history): "I was not a little amused by your remarks," she wrote. But then she came to her friend's defense: "She is indeed the prisoners' friend, and very proud am I to say she is mine too. Our house is one of her homes, when she visits Philadelphia and when absent, we are constantly gratified by her tokens of remembrance. She is now in Washington, and gives us hopes that she will soon be amongst us once more. Need I say how joyfully we will welcome her." Margaretta was carefully cultivating a professional relationship with Harris as one of her few peers in the small world of agricultural entomology, but she quickly established boundaries and made it clear that she would not privilege that connection over her relationship with Dorothea. Perhaps Harris bristled at the slight.[24]

WHILE MARGARETTA GOT CAUGHT UP IN THESE DEBATES over potato rot and Peach Yellows, drop worms and Dorothea Dix, she could not separate herself from the health concerns in her own family. Between writing articles and letters and checking in on the insects living in her library enclosures, Margaretta had to relieve her sister of duties caring for their mother. The women could not escape to the Wissahickon as often as they would have liked, but they could bring the scientific world into their home where they could still be close enough to their mother to check in on her. They would try to do it all at once.

To fill their need for community when they felt so isolated, the sisters created what Elizabeth called their "little social circle." This group of intellectuals—their neighbors J. Jay Smith, D. B. Smith, Ann Haines, and Lloyd Mifflin, as well as some occasional professors from the University of Pennsylvania—met regularly to discuss literature and natural history around a fire. Ann Haines, who had joined them on hikes along the Wissahickon when they were teenagers, and helped encourage Margaretta to fight back when Asa Fitch had publicly discredited her, continued to be part of their scholarly conversations. This social circle, or salon, became the highlight of their homebound weeks. "We have *books*, and *work*, and *talk* in abundance, all of which we find very pleasant—no ceremony to hear the most unreserved conversation," Elizabeth told William Darlington, hoping to convince him to visit and join the circle. The group met every two weeks or so. This was the intellectual community Margaretta and Elizabeth had craved since the disbanding of their tutors for New Harmony, and which they regularly promoted anonymously in the pages of the *American Agriculturist* as the ideal capstone for a productive workday. They had finally taken their own advice.[25]

Parlors had been the site of science discussions for decades, not only in revolutionary Paris but also across Philadelphia. The Wistar Party, convened by Caspar Wistar in his Philadelphia parlor with a select group of men from the American Philosophical Society, became a famous site for the scientific in-crowd, welcoming visiting naturalists into their conversations and demonstrations. The Morris Hall "social circle" was not nearly so exclusive, but the sisters had mixed socializing and science since they were young, and it came naturally to them. Similar groups in early America fostered their own communal libraries and specimen collections, while others welcomed children and young adults as a way to nurture their scientific curiosity. Margaretta and Elizabeth's group of men and women seems to have debated recent scholarship and shared readings, occasionally welcoming visiting friends. With the sisters' minds so consumed with their mother's well-being, it was comforting to have her in earshot while they spoke about discoveries.[26]

MARGARETTA AND ELIZABETH'S SCIENTIFIC WORK WAS A welcome escape from the stark reality of family tragedy. Looking back on this period years later, Elizabeth would identify the string of unfortunate events that dramatically changed their lives. Just as she was about to take the train to West Chester for a long-awaited visit with the Darlingtons, her brother's health took a turn for the worse. Thomas Morris—widowed for almost a decade and living in Maryland—had come to visit his childhood home in February 1852 so he could consult with Philadelphia doctors about a worrisome cough. The doctors diagnosed it as consumption, or tuberculosis. "Consumption is a new disease in our family," Elizabeth agonized, wondering whether they had missed symptoms or early warnings. By the time he packed up to return to Maryland, his health was in rapid decline. He was weak and his cough distressing. Though his mother and sisters begged him to stay at Morris Hall, he insisted on returning home to settle his affairs. Margaretta followed him out to Maryland to help care for him and his children, the youngest of whom was in her mid-twenties. Back in Germantown, Elizabeth and her mother were filled with anxiety and dread each time the mail arrived, anticipating the worst. The cruelest fate, Elizabeth felt, was feeling not only that nothing could be done to save him, but also that she could not leave Germantown and her elderly mother to tend to her beloved brother in Maryland. "I feel at times as if my head would break." Though they hoped that he would regain enough strength to return to Morris Hall, he had left his childhood home for the last time. Thomas died in Maryland in May 1852.[27]

Elizabeth and Margaretta were devastated. The death of her only brother, Elizabeth grieved, was "the severest affliction and greatest calamity combined." As they buried Thomas, the trees above dropped flowers into his open grave. Throughout his life, he had maintained a passion for natural history, just like his sisters, and he had regularly sent them specimens and questions. As a teenager he had dug up cicadas in the garden with Margaretta, he had hiked through the Wissahickon with Elizabeth. The women adored their older brother. Elizabeth considered him a lifelong "advisor, friend, everything!"[28]

Then Ann Willing Morris, whose own health was precarious at best, fell deeply and alarmingly ill. Margaretta and Elizabeth often worried that their mother's health was affected by bad news, such as the deaths of childhood friends, and now she had outlived another child. For months, her health improved briefly, only to be set back again by an overnight attack or a case of influenza.[29]

As their mother's illnesses lingered, Elizabeth and Margaretta found it hard to manage everything. Despite letters from the botanical community piling high on her desk, Elizabeth once again found herself frequently interrupted and completely absorbed by caregiving duties and anxiety over her mother's well-being. Margaretta, too, found herself so preoccupied with domestic responsibilities that it was difficult to focus on anything else, including the entomological work that she loved. "My time is so filled with household cares and attendance on my aged parent," she wrote, "that I have little time to call my own." She had promised to create a catalog of Pennsylvania insects for William Darlington but kept having to postpone the project. Wiping dust from the worktables, she gazed longingly at the insects pinned in her collections, remembering each location where she captured them and wishing she could be outside finding more. "I have so often been disappointed," she sulked, "that I begin to feel as if it were not my vocation, and that the humble track of Pennsylvania housewife was my true path." She savored her scientific work, and while it did not carry any substantial income, it was her life's calling. "I can't give up yet," she wrote Darlington. Trapped between professional aspirations and domestic demands, Margaretta and Elizabeth struggled.[30]

One highlight during this dark time was Dorothea Dix's regular visits to Morris Hall, including a long visit for several weeks around Thanksgiving. Margaretta and Elizabeth adored Dorothea, Margaretta most of all. As Margaretta struggled with the death of her brother and the deterioration of her mother's health, Dorothea seems to have tried to comfort her by suggesting she listen to her grief and learn from it. She shared a poem that began:

> *Do not cheat thy Heart and tell her*
> *"Grief will pass away,*
> *Hope for fairer times in future,*
> *And forget to-day."—*
> *Tell her if you will, that sorrow*
> *Need not come in vain;*
> *Tell her that the lesson taught her*
> *Far outweighs the pain.*

Rather than look for distraction, Dorothea encouraged Margaretta to "nurse her caged sorrow / Till the captive sings." As they aged, poetry remained a central way that Margaretta, Elizabeth, and their friends and relatives processed and shared their emotions and honored those they lost.[31]

Morris Hall was filled with the various gifts that Dorothea gave the sisters, one of which was a small wooden cross bearing both Dorothea's and Margaretta's initials, perhaps a sign of romance, but at the very least a testament to their intimate friendship and kinship. Most of their letters have vanished, perhaps destroyed by the women or their descendants. In the few remaining gifts and letters, though, their deep connection and familiarity, even love, are palpable. Following her visit, Margaretta wrote to Dorothea while making a pudding for dinner, reminding her how she thought of her constantly. "I snatch my pen to assure you that you are not forgotten now or ever, the pleasant remembrances of

This wooden cross, bearing Margaretta's and Dorothea Dix's initials on either side, may be evidence of a romantic relationship, or at the very least of a deep and affectionate friendship.

(*Special Collections, University of Delaware Library*)

you, we scattered too thickly over the house to allow of your being forgotten for a day, and so in fancy you are always with us, making the *past*, present always." As she concluded the letter, filling the envelope with some pressed flowers from the bush in front of her house, she returned to the dish she was in the midst of preparing: "Now for the pudding. I wish you were to eat some of it." These notes and gifts show a closeness and intimacy that was cherished. Dorothea's presence at Morris Hall during some of the darkest days of their lives must have soothed the sisters.[32]

In this house in Germantown filled with illness and grief, the holidays burst to life with light and activity, crowding the Morris Hall parlor with beloved nieces and nephews, sisters and brothers-in-law, who came not only to celebrate Christmas and ring in the New Year with Elizabeth and Margaretta, but also to surround Ann in what they suspected might be her final holiday season. Christmas and New Year's was traditionally a boisterous time for the larger Morris family, who all returned to their Germantown home to celebrate, though this year without their beloved brother, Thomas.[33]

It would prove, however, to be the last Christmas of its kind. In early January, after most of their guests had left, Ann once again fell ill, this time so severely that she knew she would not recover. After three days of intense illness, she spent her final day in prayer, resting peacefully before dying. Margaretta and Elizabeth were distraught with grief.[34]

Having spent the last several years devoted to caring for their elderly mother and maintaining the house, Margaretta and Elizabeth had grown accustomed to complaining about the tediousness of domestic work. It seemed as if they were trapped inside. But when these obligations lifted, they felt unmoored. "My weeping is all selfish & sinful," Elizabeth wrote to William Darlington, feeling she should not grieve as much since her mother died such a peaceful death. "But, after having had but one duty for so many years, I now feel that I have nothing to do." Margaretta, too, mourned her mother deeply, confiding in her cousin that her sustained grief had sunk her into depression. "The *realities* of life are now all before me and for the first time I feel that there is nothing here that is worth hoping for." This was a turning point for Margaretta and Elizabeth, not unlike the one they faced when they reckoned with the loss of their tutors and prospects for marriage in the 1820s.[35]

The two women, now in their mid-fifties, had never lived apart from their mother. Years before her death, Ann wrote a will that would protect the sisters, by granting them ownership of her property, belongings, and wealth. "I am anxious," Ann wrote, "that my two unmarried daughters—Elizabeth and Margaretta should continue and maintain, by amicably residing together, on this place wherein I now live." After divvying up heirlooms among her children and larger family, Ann dedicated most of the will to ensuring that whatever remained of her wealth would be jointly managed by her son-in-law John Stockton Littell and Margaretta to keep the sisters afloat.[36]

Elizabeth struggled to return to botany as grief washed over her. She confessed these feelings to Asa Gray and William Darlington, knowing they would understand. Asa encouraged her to get outdoors: "You will find the care of your garden a pleasant and placid occupation." Having struggled with his own grief, Asa shared how he coped with his father's death by returning to nature. Years prior, Elizabeth had anonymously written in the *American Agriculturist* about how gardening served as a balm for her when she felt low, something Asa would not have known. His suggestion, in other words, suited her.[37]

Margaretta, meanwhile, came to face her own mortality after having witnessed the death of her mother. "So much of Life has passed," she wrote to her cousin, "and so short a journey before me." This realization proved surprisingly empowering—she was determined to create meaning from loss. When William Darlington asked her if he could publish something she had casually written in a letter to him, she did not hesitate. "I have arrived at a time of life when *fastidiousness* appears wrong when *good may be done*," she replied to him. "I often shock the more sensitive feelings of those who have not been so much accustomed to do for others." Years prior she had wrestled with drafts and hesitated for long stretches before sharing her findings. Clear-eyed about the finite nature of her life, she discarded the time-consuming anxieties about gendered conventions and propriety that had crippled her in the past.[38]

Elizabeth, too, became more aggressive about following her passion. She decided that her first priority going forward should be her friends. Her friendships and connections with botanists like William Darlington, Asa Gray, Ann Haines, and others brought her deep joy, and she realized in the depths of her grief just how central they were to her happiness. After William Darlington mailed her a colored lithograph of the *Darlingtonia californica*, or cobra lily, Elizabeth knew what she wanted. William Breckinridge had found this pitcher plant in California on the Wilkes Expedition, and the New York botanist John Torrey in turn had described and named the carnivorous plant that rose out of bogs for his beloved friend Darlington. Like John Torrey, Elizabeth was always eager to honor her mentor and when she saw the lithograph, she offered to color a collection for him, saying it would be a "labor of love." "I have long wanted a good excuse to resume my pencil," she wrote, "and this is a rare chance of enjoying a pleasure, under the pretext of obliging a friend." She was characteristically self-deprecating when she returned the results of her monthslong work to William and some of their mutual friends, wringing her hands over what they would think. The project, however, had given her a renewed purpose.[39]

The *Darlingtonia californica* was the first of many such subjects. Next came the *Nepenthes rafflesiana*, another carnivorous
plant that Elizabeth would draw for the *Philadelphia Florist and
Horticultural Journal.* The editors credited Elizabeth as the artist of
the drawing that inspired the woodcut, likely revealing her name
without permission. While the *Darlingtonia* was from the newly
acquired American states of California and Oregon, the *rafflesiana*
was from Southeast Asia. Elizabeth was developing a specialized
expertise in illustrating and coloring these curious carnivorous
pitcher plants that thrived in bogs. In the article that she illustrated, the editor of the *Philadelphia Florist* articulated how odd
plants like these had the power to captivate even those otherwise
uninterested in botany, arresting "the admiration and attention
of the most thoughtless of mankind." While most of these exotic
pitcher plants required careful cultivation in humid greenhouses,

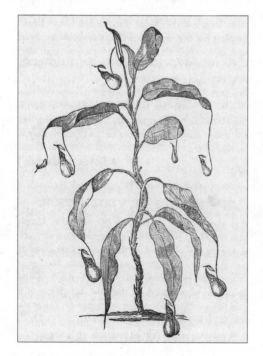

Elizabeth Morris gained
expertise in coloring
and illustrating images
of carnivorous plants,
such as her woodcut of
a *Nepenthes rafflesiana*
seen here from an 1853
*Philadelphia Florist and
Horticultural Journal.*

Elizabeth was able to bring them to life on the pages of the magazine for those who could not get access to rare specimens.[40]

Margaretta similarly busied herself with a variety of new projects, such as helping to coordinate the collection of fish around Philadelphia for Louis Agassiz and continuing her work compiling a list of native insects for William Darlington. When her cousin Robert Hare gifted her a large collection of minerals, she took the opportunity to reorganize Morris Hall's library in order to properly display it. Robert always looked after his younger cousin, and perhaps the collection was meant to refocus and reawaken her scientific curiosity after the deaths of her brother and mother.[41]

Robert Hare, now seventy-two, was himself struggling with mortality and grappling with what happened after death. He had previously been skeptical of the budding spiritualism movement that was taking root in the United States, in which audiences believed they could communicate with spirits through mediums, connecting the earthly world to the afterlife. Hundreds of thousands of Americans embraced the idea in the mid-nineteenth century, including a number of high-profile celebrities and politicians, like Abraham Lincoln, Mary Todd Lincoln, Harriet Beecher Stowe, Jenny Lind, and Horace Greeley, among others. Invited to a séance and unable to easily dismiss it as a sham, Hare was determined to use the scientific method to determine whether the messages originated with spirits or their mediums. After a long and successful career studying electricity, matter, and magnetism, Robert began to pivot in his retirement, devoting increasing amounts of his time to designing a variety of Spiritoscopes (essentially complicated Ouija boards) and writing a 460-page conversion narrative explaining the scientific basis for the spirit world. There was comfort in spiritual readings such as when Robert's long-dead sister communicated that after his own death, Robert's "freed spirit would rove the endless fields of immortality with those loved friends who have gone a little while before."[42]

Robert Hare's scientific peers were aghast. When he tried to share his research at the 1854 meeting of the American Association for the Advancement of Science, his colleagues allowed him to speak only out of respect for his long career and service to his profession. The next year, however, when he stood up to announce the time and location of a meeting of spiritualists at the very same conference, he was heckled off the stage. Robert's old friend Benjamin Silliman—who had once shared a boardinghouse and makeshift basement laboratory with him in Philadelphia as a younger man—wrote to him, saying he regretted the direction Hare had taken. "It would have been happy if your public career ended with science," Silliman opined. Robert's peers saw his new evangelism for spiritualism as a break from evidence-based scientific work and perhaps even a sign of senility.[43]

Hare, however, did not consider his new beliefs to be a fundamental departure from his earlier career. "We live in a wonder-working universe, which becomes more and more wonderful as we learn more of it, instead of being brought more within our comprehension." He called on his peers to keep their minds open. The more he learned, the more humble he had grown about the limits of his own knowledge. There was still so much more to know, so much wonder to behold. To Robert, the spirit world was simply a new frontier for natural science.[44]

Margaretta and Elizabeth's greatest spokesman and advocate was being openly discredited. What they made of Robert Hare's public conversion is unknown. Their cousin had encouraged Margaretta to share her research, presented her work to scientific societies on her behalf, defended her roundly from criticism, and connected both the Morris sisters to his network of friends in the scientific world. He was their ally, but his name had lost the currency it once carried in the community. Regardless of what they thought of spiritualism, for Margaretta and Elizabeth family ties ran deep, and they stuck with their cousin, no matter his professional fall from grace.

As Margaretta and Elizabeth adjusted to an emptier house in Germantown after the deaths of their brother and mother in 1853, Robert Hare was only just beginning to fixate on the next world—an obsession that would consume his last few years. Margaretta and Elizabeth mourned many of the changes in their lives. The holidays, once so boisterous and full of visiting family members, had become quieter now that the family matriarch, who had pulled so many children and grandchildren close, was gone. Even the social circle that Elizabeth and Margaretta organized, filled with professors and naturalists who discussed the latest scientific discoveries, no longer met as regularly as it once had. "Trouble and afflictions in various forms have visited almost every member of it," Elizabeth recounted.[45]

IN THE BACKGROUND, WHILE THE MORRIS SISTERS CELE-brated personal triumphs and suffered from profound losses, Germantown was changing. The Wissahickon, their beloved forest and laboratory for all kinds of scientific explorations, was becoming increasingly industrial. Philadelphians could still hike and explore its trails, as Margaretta and Elizabeth had wistfully longed to do in the last years of Ann's life, but the sounds of the mills had grown louder. Dams obstructed both fish and the kinds of small boats Margaretta and Elizabeth had once rowed with friends. One map from around 1848 labeled the area as "Paper Mill Run," eliminating the name Wissahickon entirely. Germantown had more factories, more mills, and more people. It was hardly the sleepy rural village that Margaretta and Elizabeth had known as children; it was far more lively and diverse.[46]

The railroad had connected Germantown to the center of Philadelphia for decades, shaping it into a vibrant suburb for its residents who worked in the center of the city, and enabling the mills' and factories' products to reach larger markets. Philadelphia itself was only two square miles, but it was surrounded by dozens of interconnected townships like Germantown. In February 1854—a year after Ann's death—Germantown and its surrounding townships and districts were incorporated into the city of Philadelphia, doubling the city's population and expanding its footprint to become the largest American city by area.[47]

The Consolidation Act, as the legislation came to be known, was driven by the belief that a united city would be better equipped to manage the problems facing the smaller townships. Inefficient volunteer fire companies, corrupt ward bosses, and local gangs governed these communities, and politicians promised that consolidation would lead to a better-regulated city. The expanded government would also have centralized authority to invest in public goods like parks and coordinated infrastructure. Those developments, however, would take months, even years, to become reality.[48]

The same winter that Philadelphia's politicians argued about consolidation, arsonists terrorized Germantown. Barn after barn and stable after stable lit up in flames each week. One of those destroyed barns was a celebrated landmark—the artist Gilbert Stuart had used it as his studio when he painted his famous portrait of George Washington that is now on the dollar bill. The newspapers published a flurry of reports as the cases began to mount. The editor of the *Germantown Telegraph* found it difficult to pin the crimes on anyone specific but did not hesitate to speculate. "All these fires are doubtless caused by boys—indeed the fact is notorious," the editor opined, "and yet our police officers never know who they are."[49]

Margaretta and Elizabeth had been keeping up with the reports, wondering, like their neighbors, who the culprit was and whether they would be targeted. So it may not have come as a complete surprise when on Friday, March 31, 1854, they both heard the telltale shatter of broken glass. Their newly built stone barn was soon ablaze, ignited by a fiery object hurled through the second-floor window. After a flurry of activity and a seemingly interminable wait, the local fire companies eventually arrived, connected a hose to the nearby cistern, and suppressed the flames.[50]

As they picked through the rubble and assessed the extent of damage in the days that followed, the Morris sisters believed that their barn was just the latest casualty in a string of random misfortune. Then the arsonist returned. Exactly two weeks later, nearly to the hour, someone set fire to Margaretta and Elizabeth's stable. The women moved quickly enough to rescue their cow, but the stable—adjacent to the ruins of the stone barn—was wooden and full of dry hay. It burned quickly.[51]

Rumors began to swirl because none of the other victims of arson had been targeted twice. Soon after the second fire, police arrested a teenager named Andy. He admitted to setting several fires, and while the *Germantown Telegraph* reported that he admitted to those at Morris Hall as well as others, Elizabeth told William Darlington that he had not actually confessed to theirs. The *Germantown Telegraph* suggested that since Andy was himself a volunteer firefighter, he enjoyed setting the fires so he could rush back with his fellow volunteers to fight the flames. "He was very fond of running with the engines," they wrote, "and was at all the fires an active assistant!" It may have been that widespread property destruction and the resulting publicity made him feel powerful. Philadelphia's rowdy volunteer fire companies often raced to arrive at fires first, sometimes battling in the streets to the dismay and frustration of those whose property was at stake. The arsonist would have been giving his company an edge in the competition.[52]

Margaretta and Elizabeth, however, had other ideas. Given that they had been targeted twice, the attack felt personal. Undoubtedly, this belief was fostered by years of professional attacks and fueled by their unrelenting grief. Elizabeth confided to William Darlington that she suspected that someone had bribed Andy to attack their property to send the women some sort of message. Perhaps he had not admitted to their fires because he was protecting whoever paid him; perhaps it was someone else altogether hiding their violence in the string of random attacks. The person behind it all, Elizabeth told William, "we honestly believe to be quite capable of such an act, and who is, as far as we can conjecture, the only human being who has threatened or wishes ill to such quite old maids as Sister Margaretta and I."[53]

Exactly who Elizabeth was alluding to is a mystery, vanished behind intentionally vague references in her letter and murmurs of long-lost conversations. The police seem to have considered the case closed with the arsonist's confessions, incomplete as they were since he had not admitted to them all. What is revealing, however, is that Margaretta and Elizabeth felt that someone was out to get them. They read the attack as an attempt to silence them, to put them in their place.

The sisters could have succumbed to their fear and retreated from public life. They had certainly lost a lot in the fires—a feeling of security, maybe even specimens and experiments half complete, certainly some of the plants that grew in the garden along the edges of the charred structures. It was spring, though, and even among the ashes, some of the plants began to regrow on their own, like the forget-me-nots and even the curly dock. Perhaps some unexpected new plant sprouted as well, given that sometimes only fire can germinate a seed.

Since Margaretta had long managed their finances and the maintenance of their property, she got to work right away. She hired men to rebuild their structures almost immediately, combining the barn and stable into a single two-story stone stable with a separate stone wash-house. They completed the work in just two months, and the new outbuildings were as fireproof and durable as was possible. Elizabeth admitted to William that focusing on this project had cost them time they would have otherwise spent on their scientific projects. "This has thrown everything back," she wrote, "and the trouble and expense of rebuilding are by no means trifling." They had built the stable just two years before and they certainly did not anticipate having to rebuild it so soon.[54]

Within a week of the first fire, Margaretta had reached out to the Franklin Fire Insurance Company to purchase a policy for Morris Hall. They would not be caught vulnerable again. She hovered while the inspector counted windowpanes and noted the pine floors, checked the Franklin stoves in the bedroom fireplaces, counted all the bathrooms, measuring doorways and room dimensions as he went. Showing the sun-filled garret rooms to the inspector, perhaps she was able to imagine it from his perspective, seeing the microscope and bell jars filled with insects, dirt, and plants lining the tables, the plant presses and cabinets filled with collections thoughtfully organized, the shelves full of books. Perhaps she explained some of her active projects as he took notes on the cabinets, the dormer windows, the short closets tucked in the eaves filled with specimens and tools.[55]

HAVING FACED INCONCEIVABLE LOSSES OVER THE LAST FEW years, Margaretta and Elizabeth examined their lives with a clarity that comes from these kinds of ruptures. While the health issues in their family had inspired Margaretta to think creatively about how insects affected plant health, much of the sisters' scientific work and even their place in the larger scientific community was restricted by their need to remain within the walls of Morris Hall. Confronting their own mortality, they took stock of what felt most important now that they could return more fully to the work they loved. Elizabeth had found that her greatest happiness came from her usefulness to friends. Margaretta decided that life was too short to fret over perfection. Over the years she had spent so much time reinforcing her arguments to withstand all kinds of attacks, chasing impossible standards, and waiting for a mythical right moment before publishing, and she was done with all of that. So when Charles Darwin, an up-and-coming naturalist in England, asked to hear more about her water beetle research knowing that it might inform his current project on species, Margaretta did not hesitate to respond.

Eight

A LIFE OF EXPERIENCE

IN 1855 WHEN CHARLES DARWIN SAT DOWN TO READ MARgaretta Hare Morris's letter, he did not intend to skim it. Rather he took his pencil and underlined the details he found most important, including the name of the species (*Dytiscus marginalis*, a type of water beetle) and the locations where she found them. He scribbled some notes in the margins, and then crossed out the rest of her words with a swift vertical pencil line.[1]

Margaretta had not sent the letter to Darwin, though she likely knew it would end up with him. She did not know much about why Darwin, a rising figure in British scientific circles, was interested in her water beetles. He was incredibly secretive when it came to his species project. Still, Margaretta had vowed to stop overthinking before sharing her work with others. She therefore sent this letter to her acquaintance, Richard Chandler Alexander, who had reached out to her on Darwin's behalf. Alexander had hoped Margaretta might recount for Darwin the details she had shared with him about the discoveries she had made in the 1840s.[2]

Given the popularity of science in the nineteenth century, some men of science were edging closer to a kind of celebrity based on their publications, lectures, and press coverage. Darwin's book chronicling his time on the HMS *Beagle* had earned him some fame, and with that came the power to lift peers up and leave others in obscurity. America's leading scientific authorities were courting a similar kind of influence and enjoyed the media attention given to them at events like the AAAS meetings. In order to gain worldwide recognition, though, the upstart and still-coalescing American scientific community needed to build connections with and earn the respect of European scientists. This determined who could advance into the upper echelons of scientific celebrity or even simply who could gain recognition for their contributions.

Margaretta had arguably reached the peak of her career. She had skillfully navigated the sexist culture of professional science and earned respect and membership in their institutions. She had developed strategies to marshal support for her work from within the community before advancing it, and learned to carefully control the narrative when her scientific discoveries were made public. Thanks to the acclaim she had won through this work, she was closer to the culture of academic celebrity than ever before. The one thing she could not control was how her work was received. Gender, race, class, even proximity to more influential scientists, whether through connections or geography, all played a role in which discoveries were respected and which were dismissed. Even an accomplished scientist like Margaretta could not escape how her peers saw her sex as a detriment. No allies or institutional laurels could overcome that basic fact about her. To her male peers, she was fundamentally a different species of scientist.

IN 1849, SEVERAL YEARS BEFORE CHARLES DARWIN WOULD read her water beetle account, Margaretta had walked alongside

Richard Alexander—likely hiking the forest paths beside the Wissahickon Creek—discussing some of the more curious insect behaviors they had observed in their careers. Alexander was visiting from London that August and had stopped in Germantown to meet Margaretta and Elizabeth before heading north to attend the American Association for the Advancement of Science meeting. A wealthy physician whose poor health caused him to retire young, he spent his time studying literature and botany, while traveling often. Alexander was a fellow of the famous Linnaean Society of London, and American scientists were happy to have the chance to connect with him.[3]

As their conversation meandered from insect to insect, Margaretta's thoughts drifted back to a visit she had made a few years earlier in 1846 to her friend Sarah Miller Walker's home in northeastern Pennsylvania. Sarah lived with her brother and father at the Woodbourne estate outside of Montrose, 175 miles north of Philadelphia. Margaretta and Sarah, who as young women had linked arms as they attended scientific lectures together, had stayed in touch, regularly making the several-day trip to visit each other's homes.[4]

Margaretta told Alexander about how one discovery she made on that visit had stayed with her all these years later. As she and her hosts sat inside, a large, muddy-brown beetle had flown through an open window, attracted by the light of a lamp. Never one to swat away an opportunity to study a new insect, Margaretta began examining it in detail. Based on what she knew of the British species from her copy of William Spence and William Kirby's *Introduction to Entomology*, she identified it as a water beetle, or *Dytiscus marginalis*, though she was likely observing a similar North American *Dytiscid* relative.[5]

Margaretta was eager to examine the visitor because she had read an account of someone finding eggs attached to the beetle's hairy, oarlike legs. Try as she might, though, she could not find

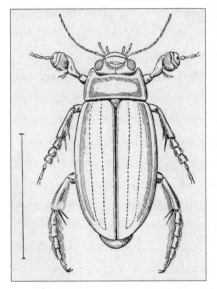 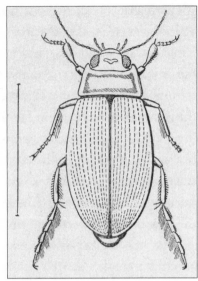

The water beetle that Margaretta found in northern Pennsylvania was a relative of the *Dytiscus marginalis*, which transferred fish eggs from lake to lake thanks to the sticky fluid on its legs. The male and female are seen here in L. C. Miall's *The Natural History of Aquatic Insects* (1895).

anything on the legs of this particular beetle. Perhaps it was a female *Dytiscid* that did not exude the sticky fluid from its legs that would have held eggs. This made her all the more determined, however, to visit the nearby mountain lakes where she might have a chance of observing more beetles. Margaretta and Sarah galloped over three miles on horseback through forests of hemlock, birch, sugar maple, and white ash trees to visit the lakes nestled in the sunny hills and meadows of northeastern Pennsylvania. As they waded through the muck at the water's edge, the women found salamanders and frogs, and, most importantly, plenty of water beetles whose legs were covered with roe. To Margaretta, this left no doubt that the beetles, capable of flying far distances at night, had the ability to populate unconnected lakes with the fish.[6]

While Margaretta never published her observations of these beetles or their egg-covered legs, her story—as she relayed it to Richard Alexander on their walk—traveled far. Years after he returned to London, Alexander told his acquaintance and fellow Linnean Society member Joseph Dalton Hooker about the beetles that transported fish from lake to lake. Hooker, in turn, mentioned it to his close friend, Charles Darwin. It was like a nineteenth-century game of telephone. With each retelling, the new storyteller was endorsing it as believable, or at least worthy of being repeated. Hooker would have never shared it with Darwin if he thought it was too far-fetched.[7]

Darwin had recently finished a major project on barnacles in 1854 and had begun devoting himself completely to expanding his "species theory." Wrestling with this theory for decades, he worried about the backlash he might face from both his peers and the general public. Frequently sick with digestive issues and preferring the quiet of his countryside home at Down in England, Darwin depended dearly on the favors of correspondents near and far, some of whom he barely knew, to help him aggregate data about the behavior of species from around the world. This information flowed in one direction—Darwin's growing fame enabled him to use international scientific correspondence networks to his advantage without the expectation of reciprocation. As his biographer Janet Browne has put it, "Like countless other well-established figures of the period, Darwin regarded his correspondence primarily as a supply system, designed to answer his own wants." He spent enormous sums on postage and paper, and even angled a mirror in his study so that he could see exactly when the postman arrived. Information poured in from naturalists, pigeon fanciers, farmers, civil servants, really anyone who might have collected observations or specimens that could help him answer various questions he had as he privately worked through the holes in his theory. Darwin and his friends were well-known and the culture of sharing natural history specimens and observations was robust enough that it was not hard to get this kind of assistance. This immense correspondence network, much larger than Asa Gray's or William Darlington's, allowed Darwin to refine his theories on an international scale.[8]

In order to assemble a unified theory about species, how they evolved, and how they came to exist in different locations, Darwin needed to rely on regional experts, including collectors and hobbyists, who were fonts of local environmental knowledge. The kinds of observations people like Margaretta and Elizabeth had been making in their gardens, along the Wissahickon, and as they visited friends throughout the mid-Atlantic states were integral to validating a unifying theory. This was true for Darwin and for anyone else working on that kind of scale if they wanted their theories to be grounded in specifics and data. Without the transatlantic scientific culture that had inspired so many to attend lectures, devour science articles and books, and start their own collections and experiments, projects like Darwin's would have been impossible.

In April 1855, as Darwin worked feverishly to gather information on species' transformations and distribution, Margaretta's water beetle story was brought to his attention by his close friend and confidante, Joseph Hooker. Joseph was the son of William Hooker—the director of Kew Gardens—and was a serious botanist in his own right. He had spent long hours with Darwin debating the distribution of species, and knew the story would be relevant. But a story many times retold was hard to make sense of, let alone trust, particularly if it had transformed during the many retellings and lost essential details along the way. Darwin was eager to trace it back to its source. He had asked Hooker to get Richard Alexander to elaborate and connect him with the entomologist. At the end of a letter to Hooker about experiments he was performing with seeds and salt water, Darwin tacked on:

"Remember to ask about my *distinct* case of 'a lady in N. America' who saw fishes' spawn adhering to a Ditiscus." He did not even have Margaretta's name.[9]

Darwin knew he would focus a section of his new project on the distribution of species, and both his saltwater experiments and Margaretta Morris's observations of water beetles were linked to that line of inquiry. He pondered whether seeds could be whisked across oceans and survive to germinate, prompting him to soak seeds in salt water for varied lengths of time before planting them in soil. In the case of Margaretta's beetles, he was looking to explain how freshwater fish might come to populate unconnected ponds and lakes. He had his theories, particularly about waterfowl carrying eggs on their feet, but this "*distinct* case" mentioned by Margaretta Morris, the "lady in N. America," might explain things further.[10]

Darwin would not be the first to imply that species evolved over time, challenging ideas of divine creation and even the static taxonomic categories that naturalists relied upon, like genus and species. If plants and animals evolved, then the categories themselves would need to be elastic, too. Decades earlier, the French naturalist Jean-Baptiste Lamarck and even Darwin's grandfather, the popular scientific poet Erasmus Darwin, pushed forward the radical idea of transmutation, which described a gradual development toward more complex organisms. More recently still, in 1844, the Scottish publisher Robert Chambers anonymously penned *Vestiges of the Natural History of Creation*, arguing that the progressive transformation of living things was governed by natural laws rather than miracles. The wildly popular book captured the imaginations of many of the social and intellectual elite, including Elizabeth Morris, who was eager to share its provocative ideas with William Darlington after she first read it.[11]

Vestiges received not only praise but also a hearty amount of criticism, kindling a debate about natural law. Critics were angry that the anonymous author of *Vestiges* both erased God's direct creation of nature and described it as viciously uncaring. Among the many suspected authors were Harriet Martineau and Ada Lovelace, along with many other British women involved in the sciences. As women tended to write anonymously and *Vestiges* was a synthetic book with popular appeal—characteristics thought to be unique to female talents—they seemed to be prime candidates. Critics also used the presumption of female authorship as grounds to discredit the book and its controversial theories. Darwin had been tinkering with an essay about the evolution of species, but the condemnation of *Vestiges* made him realize that he would need to mount a watertight, research-driven defense for his theory of evolution before he could present it to the world. He was anxious to accumulate more evidence.[12]

Darwin sought more details about how the same species could be found in vastly different locations because the phenomenon was at the crux of his evolutionary ideas. Naturalists were engaged in debates over whether there were single or multiple sites of creation, or if species distribution had happened due to fluctuating land and sea levels or climate changes. Some naturalists theorized that long-lost land bridges had once connected continents and distant islands. Others suggested that seeds and creatures might also be transported by wind, water currents, shifting glaciers, or in the fur or stomachs of animals. There were so many variables that might explain how even a single species came to reside in a given location. Darwin described the geographic distribution of species not only as "a grand game of chess with the world for a Board," but a "key-stone of the laws of creation." The more alternative explanations for species distribution he could supply, such as with the salt-water-soaked seeds and the water beetles, the weaker his potential critics' position became. If what Margaretta Morris observed was true, the presence of the same fish species in distant ponds and lakes might not have been the work of a Creator, let alone changing water levels, but rather the work of a far-flying bug.[13]

Even before Darwin received further information from Margaretta about her observations, he was not inclined to trust her. In his next letter to Joseph Hooker after he had reminded him to obtain more information about the case, he wrote, "If Dr. Alexander can give no more information, the case even for *credulous* me is worthless." He proceeded to explain his skepticism: "I have constantly observed eggs of some parasite adhering to the bodies of Nepa, a *water-bug*, if the Lady uses the terms correctly, it is perhaps these ova, & not of Fish." Having already reduced Margaretta to two descriptors based on her gender and location ("a lady in N. America")—both of which raised questions about the credibility of her information—Darwin was perhaps less than open-minded about the reliability of her observations. It was more than just sexism and some vague postcolonial superiority, though. Darwin was also an expert on beetles, having obsessed over them for decades. His confidence in the comprehensive and superior nature of his expertise, bolstered as it was by his scientific celebrity, would have made it challenging to accept another's authority when it ran counter to his own. And yet, the *Nepidae* that Darwin described witnessing were far different from the beetles Margaretta encountered.[14]

Part of Darwin's skepticism may have also stemmed from Margaretta's connection to Richard Alexander. Darwin did not know Alexander directly, and even Hooker could only call him an acquaintance. Years prior, Hooker had described Alexander to Darwin as a "monied dabbler in European Botany, who travels much." This was no great endorsement. The fact that this unreliable source was delivering Margaretta's research could not have helped. British and American scientific networks were tenuously linked, but geographic distance made it harder for Darwin to judge Margaretta's reliability without easy access to the kinds of endorsers she had in the United States.[15]

Richard Alexander, meanwhile, was under pressure from Darwin and Hooker to get Margaretta's story in writing. Growing impatient while waiting for her answer, he began brainstorming ways to get more details. He reached out to Asa Gray, knowing Margaretta was somehow connected to him, to see if he knew any details about the beetles or if he could perhaps pull some strings with Margaretta for a quicker response. "Darwin attaches great importance to it if it could be authenticated," Alexander wrote Asa, anxious to help confirm the story. Darwin's budding celebrity was enough to create urgency and compel action.[16]

While Asa did not know the details about the water beetles himself, he decided to include a mention of Margaretta in his own letter to Charles Darwin. Gray had spent some time with Darwin at Kew Gardens during a trip to England in 1851, but Darwin had only just sent a letter to Gray trying to reconnect in April 1855. "I hope you will remember that I had the pleasure of being introduced to you at Kew," Darwin wrote as he sought answers about the distribution of North American plant species. Though they would soon become good friends and confidants, Gray and Darwin were still following a careful choreography of trying to impress each other and establish trust—a game that Asa Gray knew well. The English and American scientific circles operated separately, for the most part, but connected in these highly personal ways through visits, letters, and gifts, though with travel and shipping so expensive, it was the rare few who could make these kinds of connections. As Gray's and Darwin's letters passed across the Atlantic that spring, Asa vouched that Margaretta *could* be trusted. He knew that Darwin had to sift through the observations of so many unknown naturalists. "This Miss Morris is a good observer," Asa assured him. He pledged to write to Margaretta's sister "who I know very well" to encourage an expeditious answer. Elizabeth's carefully cultivated relationship with Asa was benefiting her sister. Asa Gray's confidence in Margaretta's scientific acumen might have reassured Darwin, but their relationship was too new and untested for the endorsement to mean much.[17]

As such, when Darwin finally received Margaretta's letter and sat down to read it several months later in the summer of 1855, he was not especially generous. As he annotated the letter, he assessed the significant details as he went along. When he reached the third page, though, he encountered her claim about the beetles and the fish eggs. Margaretta wrote: "Those feeding on the margin of the lake were covered with [roe]—leaving no doubt in my mind as to the fact that they thus carried the eggs, from lake to lake and peopled them with fish that had no other means of being transported." He put down his pencil and grabbed his dark orange crayon to highlight that passage with two lines in the margin. With his pencil he scribbled, "This is the most interesting statement in Letter." Then he thought better of that, crossed out "most" and added a snarky "only" in its place: "This is the ~~most~~ only interesting statement in Letter." With that, Darwin dismissed the "N. American lady" and her water beetles.[18]

In writing *On the Origin of Species*, Darwin was making a significant claim to universal scientific authority. In order to do this, he had to be able to lean on his correspondents from around the world, many of whom he barely knew. Determining whom he could trust and what information was reliable came down to endorsements from people he knew and perhaps to other categories and assumptions that might make them seem more knowledgeable and trustworthy than others. He had to make a high volume of quick decisions, which invariably meant that he did not always make the right calls about people, and he recognized that. In his introduction to *On the Origin of Species*, published a few years later in 1859, he warned readers, "No doubt errors will have crept in, though I hope I have always been cautious in trusting to good authorities alone." Darwin had to rely on instinct, and that instinct was reinforced by cultural and community prejudice against women and Americans.[19]

American men had been concerned about how their scientific work, training, and intelligence were being judged abroad for decades. The creation of professional organizations and publications often echoed British models, to prove, in part, that Americans were peers, not inferiors. It was the kind of thing Asa Gray worked doubly hard to counter, proving his value by becoming a trusted supplier of specimens to European botanists—thanks to talented collectors like Elizabeth. To some extent, Margaretta was marked by this postcolonial demerit, this assumption that she was too far from the seat of scientific power to have the training or connections to know what she was talking about.[20]

A year after dismissing Margaretta's account, Darwin received a letter in 1856 from his close friend and mentor, the geologist Charles Lyell. Darwin had devoured Lyell's *Principles of Geology* while a young man on the *Beagle*, and it greatly influenced how he made sense of the world he encountered during his trip. The two had formed a close friendship and correspondence in the decades since Darwin's return. Like Hooker, Lyell was one of the few friends in whom Darwin had confided about his work-in-progress. In that 1856 letter, Lyell shared accounts with Darwin that sounded awfully familiar. He described how he had learned thirdhand about a man who had caught a large water beetle (*Hydrobius piceus*) with a mollusk (*Ancylus fluviatilis*) adhering to it. "Here is a new light as to the way by which these sedentary mollusks may get transported from one river basin to another," Lyell wrote. He followed this with yet another thirdhand account, this time from John Curtis, a nearly blind entomologist who had recently been in an accident that broke his shoulder. The recuperating Curtis told Lyell that a French entomologist had once brought him a *Dytiscus marginalis*—the same beetle Margaretta had found—"with a small bag of eggs of a water spider under his wings, evidently put there by a parent." Lyell added, "It was not a parasitic insect which had done this." He was contesting Darwin's reigning theory that only parasites placed eggs on water beetles—one of the ideas that had prejudiced him against Margaretta's story. However, Curtis could not look up his notes about this case to provide further details to Lyell due to his ailments.[21]

As Darwin eventually set about writing the section of *On the Origin of Species* that dealt with species distribution, he decided that even though he had been so incredulous of Margaretta Morris's observations, he found the thirdhand stories passed along by Lyell to be much more believable. He had not gotten verification from Lyell's contacts about their observations—but Darwin trusted Lyell, and that was enough. Darwin respected Lyell as a mentor, as someone above him in the scientific hierarchy, and so he could take him at his word. In *On the Origin of Species*, after discussing how eggs adhere to the feet of ducks that flew from lake to lake, he added a summation of what he had learned about beetles, bungling which kind of beetle had which kind of stowaway: "Sir Charles Lyell informs me that a Dytiscus has been caught with an Ancylus (a fresh-water shell like a limpet) firmly adhering to it; and a water-beetle of the same family flew on board the 'Beagle,' when forty-five miles distant from the nearest land: how much farther it might have flown with a favouring gale no one can tell." He made no mention of Margaretta Morris and her Pennsylvania *Dytiscus*, though.[22]

Margaretta's goal in recounting the story of the water beetles she observed to Darwin was simply to share her findings in the hope that they might help advance scientific knowledge. "If this meagre account can throw any light on the subject," she wrote, "be assured that it will give me much pleasure." Margaretta promised to help with further investigations into the beetle should he require her services. "It will make me most happy to communicate with you on this subject or any other question in natural history that may have fallen under my observation." Darwin had crossed all of this out.[23]

Margaretta, however, had been correct. The beetles did transport the fish eggs. Marine biologist Daniel Pauly describes in *Darwin's Fishes* how *Dysticus marginalis* larvae not only prey on fish eggs but the males of that species have little suckers on their front legs that produce a sticky fluid that retains small objects like eggs. Water beetles and fish are involved in cross-predation, where the beetles prey on fish eggs, and then the fish prey on larval and adult beetles. "The existence of such feeding loops," Pauly writes, "may actually confer evolutionary benefits on water beetles that bring fish eggs along when they colonize newly formed water bodies." In short, "Margaretta 1: Charles 0," as Pauly puts it.[24]

Though women ultimately made up just 5 percent of his correspondents, Charles Darwin came to rely on them as contributors of data and editors for many of his projects, particularly for the books and pamphlets he wrote in the 1870s and beyond. He was not as dismissive of other women's observations later in his career as he had been of Margaretta's. In part, this came from his own confidence following the success of *On the Origin of Species*. Beyond that, connections and proximity still played a big role. By the time Asa Gray introduced the American naturalist Mary Treat to Darwin by letter in 1871, Gray and Darwin's friendship was well established. Gray's endorsement meant much more than it had in 1855, eliminating whatever pause Darwin may have had with this other lady from North America. Proximity also helped some women gain Darwin's respect faster. At the same time that he was dismissing Margaretta's theory, he was welcoming information about barnacles, silkworms, and other subjects coming in from Charles Lyell's connections, whether it was his wife, sister-in-law, or even his neighbor.[25]

Despite his dependence on many intelligent women, including those in his immediate family, Darwin's ultimate estimation of the sex was low. Sixteen years after he read Margaretta's account of the water beetles, he wrote in his 1871 publication, *Descent of Man*:

> The chief distinction in the intellectual powers of the two sexes is shewn by man attaining to a higher eminence, in whatever he takes up, than woman can attain—whether requiring deep thought, reason, or imagination, or merely the use of the senses and hands. If two lists were made of the most eminent men and women in poetry, painting, sculpture, music,—comprising composition and performance, history, science, and philosophy, with half-a-dozen names under each subject, the two lists would not bear comparison.

Never accounting for the social and cultural restrictions of Victorian women's education and roles, Darwin ultimately concluded that the "average standard of mental power in man must be above that of women." Perhaps these prejudices partially influenced how Darwin assessed the merit of the submissions he reviewed. He certainly was not alone among his peers when it came to these views.[26]

WHILE DARWIN'S TREATMENT OF MARGARETTA MIGHT BE attributed to distance and tenuous social connections, she also continued to face sexism close to home from scientists who knew her and her work personally. In the decades to come, scientists—Darwin included—would debate the essential differences between men and women, oftentimes using their findings to justify differing social roles. Even at this early moment, though, biological determinism and gender essentialism impacted the way scientists categorized and treated Margaretta. There was little she could do to control this, and it was often happening in spaces where she could not defend herself. In 1855, around the same time that Darwin was reviewing her water beetle observations, Ezra Otis Kendall sent a brief note to John Lawrence LeConte. Kendall, who had interests in mathematics and astronomy, was then the principal of Philadelphia's Central High School. He wrote to his friend LeConte alerting him that "Miss Morris is extremely anxious to see you in reference to some of her June *Bugs*." The two men, both a generation younger than Margaretta, made arrangements to meet at the railroad depot and dine together, after which they would "call on the fair Entomologiste." Who knows what the men said, or whether Margaretta picked up on the mixture of mockery and dismissiveness when they joined her after their meal. Regardless, Margaretta certainly recognized that she was treated differently by her peers.[27]

Margaretta was not alone. Elizabeth Blackwell, a generation younger than Margaretta, was the first woman to earn a medical degree in the United States. Though by no means a supporter of the women's rights movement, Blackwell had grown frustrated by the mistreatment of accomplished women like herself. She wrote an article in the *Philadelphia Press* in 1857, outlining the difficulties ambitious women faced: "An occasional artist, physician, merchant, astronomer, is an eccentricity to be stared and wondered at, rather than approved and helped forward." Blackwell argued that American society ought to educate girls and support women, so that their paths might get easier in future generations. "When a woman has won herself an honorable position in any unusual line of life," Blackwell continued, "she is still excluded from the companionship and privileges of the class to which she should belong, because her course is unusual." Experiencing something similar,

Margaretta would forever be categorized as the "lady entomologist" or the "fair Entomologiste" rather than simply an "entomologist" like her peers. That outsider status meant her observations were more likely to be dismissed. Blackwell called on her Philadelphia readers to change this. "Let us learn to regard women as human beings as well as women."[28]

Margaretta Morris and Elizabeth Blackwell were far from alone in the way they were treated, and it was not unique to the United States. There were women known primarily for being the wives of their scientist husbands—like botanist Orra White Hitchcock, conchologist Lucy Say, bryologist Eliza Sullivant, and botanist Isabella Batchelder James—despite being accomplished scientists themselves. They were forever discounted as simply part of their husband's team, no matter their contributions or training. Many of these women served as illustrators for their husbands' books, articles, and lectures, and scientific illustration was often undervalued because of its ties to the expertise of women. There were women, like the British entomologist Eleanor Anne Ormerod, the American astronomer Maria Mitchell, and still others, who pursued careers on their own but found themselves stymied by unequal pay and blocked opportunities. The fact that they were always tagged as *women* scientists meant that if they were acknowledged at all, they were seen as different. They were exceptions to the rules, but the rules remained the same.[29]

Even Margaretta's accolades were generally awarded with an asterisk. While she was deeply honored by her election to the AAAS, she did not attend the meetings. Perhaps she did not feel safe traveling alone or did not want to put out another scientist by asking him to accompany her by train or stagecoach from Philadelphia. Perhaps, attuned to the ways she was treated at even casual meetings such as the one with Kendall and LeConte, she did not seek out further uncomfortable interactions. Perhaps she stayed at home because of responsibilities to her family. Whatever the case, after the 1856 meeting of the AAAS, Asa Gray wrote to Elizabeth: "Your good sister should have been at the Albany meeting with Miss Mitchell & many ladies—all having a good time." Good-natured as his remark was, inclusive as he was trying to be, he was still segregating Margaretta with the women scientists.[30]

AMID THIS MIX OF ACCLAIM AND EXCLUSION, MARGARETTA kept her head down and persisted with her work. The sexism she faced may have kept her from achieving a wider influence, but it did not prevent her from continuing to pursue the work she loved. As she neared sixty years old, she studied and wrote more extensively about the insects that threatened fruit trees. J. Jay Smith—the Morris sisters' Germantown neighbor and a regular attendant at their evening "social circle"—became the editor of the widely read *Horticulturist*. Andrew Jackson Downing, the famed landscape architect and editor who had at times sparred with Margaretta, had tragically died in a steamboat accident in 1852, and the magazine had been run temporarily by another editor before landing with J. Jay Smith. Smith held a deep respect for the Morris sisters and encouraged them to contribute. He also mentioned their research in passing in his own articles. In a brief note about how to save pear trees from cicada larvae in November 1855, Smith summarized Margaretta's findings, calling her "an enthusiastic and discriminating naturalist," and suggesting that orchard owners use her discoveries to change their practices.[31]

Margaretta focused her articles for the *Horticulturist* on the creatures that threatened pear, plum, cherry, and peach trees, often illustrating the magazine's frontispieces to match her articles' content. After Thaddeus William Harris's death in 1856, she grew more comfortable gently criticizing his work in her articles, pointing out moments where his text was muddled and contradictory. "In Dr. Harris's Treatise on Insects," she wrote, "I find the statements of observers so conflicting, that I shall pass all by, and give only the history of my bell-glass and its inhabitants, with the dates as I find them in my note-book." Her confidence, while never in short supply even in her earliest writings, grew stronger still as she asserted claims not only about the impact of various insects, but also about the strength of her own methods. "Though most of the isolated facts may have come under the notice of observers," she wrote in an article about the plum weevil, "I believe no one has hitherto put them to so careful a test as I have been able to do in the last two years, trusting to nothing but ocular demonstration and tangible proof." She also called on other experts, including William Hammond, the US Army's "eminent microscopist," to further support her claims, particularly when she contradicted other entomologists. She had come to rely less on Harris's book and other texts and more on what she could see with her own eyes.[32]

She grew bolder still with her byline. Shedding the androgynous "M. H. Morris" that she carefully had chosen for earlier articles, she instead published under "Miss Margaretta Morris" and "Margaretta Hare Morris." She was emboldened after decades of carefully skirting discrimination. She tried on different bylines in different articles, including "Miss M. H. Morris" in the *Gardener's Monthly*, but each was identifiably female now. Margaretta no longer seemed to care about shielding herself from the kind of gendered criticism that plagued her early career.[33]

Partly this confidence came from age, but it also grew from Margaretta's rising status among scientists and the constant praise publicly heaped on her by editors. As they introduced her work or tacked on their own comments to the end of her articles, editors referred to her as a "distinguished entomologist," "the well-known entomologist," and as one "who has long been distinguished for proficiency in Entomological research." Even when an editor of the *Valley Farmer* retracted his support of her conclusions about the cause of Black Knot in fruit trees, he did so with deference and respect for her skills. Following one of her articles on peach trees and insects, Thomas Meehan, the editor of *Gardener's Monthly*, went further still with his celebration of Morris. "We insert this communication with great pleasure," Meehan crowed, "as probably no one in the world has devoted more time and careful study to the subject than the distinguished authoress." Meehan, who had recently settled in Germantown, also made sure to point out that her fruit trees were the best around, so she must have known what she was talking about.[34]

MARGARETTA MAY HAVE ESTABLISHED HERSELF AS AN AUTHORity within the male scientific structures of antebellum America, but the sexism woven tightly through nineteenth-century culture meant that young entomologists like LeConte and Kendall could joke about the "fair Entomologiste" and Darwin would quickly dismiss her account of egg-covered beetles. There was little that Margaretta could have done to prevent this, even if she knew it was occurring. While she could control her message, she could not control its reception. She could not change who she was. Darwin had to make quick decisions about the reliability of his correspondents' accounts of foreign species, and whether he was conscious of his biases, they ultimately impacted what was immortalized in the scientific canon. In terms of her legacy, it did not matter that she was correct about the water beetles if Charles Darwin had decided she was not.

Margaretta had lived through virulent, published critiques against her wheat fly observations, had learned a careful choreography of publishing anonymously or under androgynous bylines for years to have her research taken seriously, had built complex networks of supporters, and had won accolades and honors. Even with all of this, she was still having her discoveries dismissed. Despite being a pathbreaker, for centuries Margaretta has remained mostly unknown. World historical recognition requires the endorsement and support of celebrities on the level of Darwin. Though she had allies among America's leading scientific figures, that ultimately was not enough. Her lack of access to that kind of celebrity herself, and the prejudices of figures who came to possess it, cemented her obscurity. She was not the only one whom the glass ceiling trapped beneath its oppressive weight. Collective scientific knowledge was simultaneously limited when environmental experts were silenced because of their sex, race, class, or distance from power. As Margaretta put it, her observations had been drawn "from a life of experience." She certainly had experience.[35]

Nine

PLANTING AND PRESERVING

IN A LETTER TO HER "VERY DEAR OLD FRIEND" WILLIAM Darlington, Elizabeth Morris lamented that she had not convinced any of her nieces or nephews to study botany. At seventy-seven and sixty-four, William and Elizabeth were facing their own mortality, and the two aging botanists had an urge to bequeath their passions, like their property, to younger family members. Elizabeth occasionally suffered from what she described as "liver complaints" and realizing the limitations that age and poor health might bring, she increasingly gave her attention to botanists of the next generation. "All of the Morris tribe inherit a love for the cultivation of plants," Elizabeth wrote, drawing a distinction between an appreciation for gardening and a devotion to science. However, she was frustrated that she had not been able to "inoculate" any of them with "the desire to know more of flowers than their common names & their outward beauties."[1]

Elizabeth saw her favorite science threatened by the massive urbanization taking place in the United States. Philadelphia, like many American cities, was expanding rapidly with an influx of rural and overseas migrants. The industrial jobs that drew so many were located in poorly regulated factories scattered throughout the city on railroad lines, rivers, and creeks, like the Wissahickon. Germantown was not immune to change even in just the six years since it had officially become a part of Philadelphia, and Elizabeth worried that its budding scientists were looking for "studies that can be pursued with less personal fatigue, and in the smoke & dust of the city." Like many contemporaries concerned with urban pollution, she argued that everyone would benefit from the "fine country air" and the exercise required during botanizing adventures. If Americans were growing sickly in their smoky cities, botany "would ensure vigor to mind and body."[2]

Botany *was* changing. However, it was not because of urbanization, even if it was becoming harder to spot wildflowers among the dense buildings. The decades of identification and classification work that botanists like Asa Gray, William Darlington, and Elizabeth Morris had undertaken—gathering specimens, determining and cataloging scientific names, defining the boundaries between varieties and species, observing life cycles and transformations—enabled celebrated scientists with extensive correspondence networks to make far-reaching statements about universal patterns in the natural world. Elizabeth was sensing that the kind of botanical fieldwork that she cherished would not be as central as it had been during her lifetime. In the decades to come, botanists—or biologists, as they would come to be known—would perform the majority of their experiments in laboratories at universities and natural history museums, using expensive equipment like specialized microscopes, dissecting tools, and staining chemicals that were out of reach for most individuals. To some extent, fieldwork would be tagged as a feminized recreational space, or a space for working-class collectors.[3]

Even if her nieces and nephews were ambivalent about botanical science, Elizabeth had hope for the future of the profession.

"Some of the rising generation seem to take to it," she reassured William Darlington, "even in spite of obstacles thrown their way." As Elizabeth settled into her sixties, she found joy in mentoring her successors, exchanging specimens with them and connecting them to her ever-expanding scientific network. After her mother's death, she had embraced this role for herself as a behind-the-scenes supporter of the community of botanists. "I hope I shall never lose my love for young people," she wrote. While word of Margaretta's water beetle discoveries traveled across the Atlantic (albeit with mixed success), Elizabeth's young friends set out on their own international adventures in search of specimens, many of which they shared with her.[4]

WELL INTO HER SIXTIES, ELIZABETH WAS STILL HUNGRY TO explore, study, and tinker. She embraced her spinster status with humor as she aged, describing herself as "*a real old maid* with as queer looks, and as many queer fancies as are usually found in the species." Her so-called queer fancies included her long-held preoccupation with ferns, studying their unique attributes and perfecting their cultivation. While she had always studied grasses and had begun to dabble in mosses, she was enamored with ferns. "The ferns are my special favorites," she admitted, "and as far as I have been able, have added to them until I have now a valuable, though not a huge collection."[5]

So fervent was Elizabeth's passion for ferns that she purchased Wardian cases—essentially enclosed glass terrariums—where she carefully controlled the humidity and other conditions necessary to cultivate them year-round. These terrariums had been invented by Nathaniel Bagshaw Ward, a London doctor and horticulturist who had struggled to keep his ferns alive in the smoky, polluted air of the coal-fueled city. His invention revolutionized the distribution of species worldwide, as the traveling version of the cases made it possible for plants to survive long boat rides. Suddenly

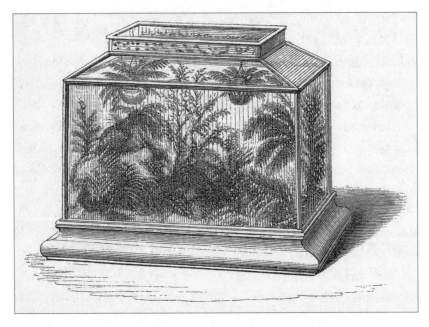

Wardian cases made it possible for botanists like Elizabeth to cultivate ferns at home under controlled conditions. In sturdier forms, they also made it possible to ship live plants globally. (N. B. Ward, *On the Growth of Plants in Closely Glazed Cases*, 1852)

orchids from Florida and bloodworts from Australia could make the long journeys across the globe to join the collections of illustrious international botanical gardens and nurseries. While seeds, like Elizabeth's forget-me-nots, could travel far, they were not failproof. They could die if they dried out, or become moldy if kept too moist. The cases made it possible for plants to persist without the need for the ship's crew to water them while they made their way. In turn, collecting expeditions and the market for live plants expanded rapidly, transforming botany and environments around the world.[6]

Elizabeth was not alone in her passion for ferns. During the middle of the nineteenth century, a fern craze—or "Pteridomania" as it was called in 1855—blossomed in Britain and to a lesser extent in the United States. This was thanks to a number of factors, including the discovery that ferns reproduce through spores, the public's passion for botany, and the growing popularity of Wardian cases. Publishers put out illustrated guides, and it became popular to "hunt" ferns, sometimes to near extinction. Ornamental Wardian cases also became centerpieces in many parlors in the United States and England. Elizabeth had two, a large one in the parlor where she kept her exotic ferns, and a smaller one in her bedroom where she tended to rare North American species. In his review of Ward's illustrated book on the cases, Asa Gray unreservedly endorsed them, declaring that any botanist who studied ferns and clubmosses would not be content without one. Botanists like Elizabeth who hoped to study live ferns and mosses rather than the pressed specimens in her herbarium finally stood a chance at watching her more finicky subjects thrive close to home.[7]

It was through her affection for ferns that Elizabeth became connected to the eccentric Prussian botanist Augustus Fendler. Having dropped out of college as a young man, Fendler eventually emigrated to the United States in 1835, working in tanneries and lamp factories before living as a hermit on an island otherwise inhabited only by turkeys outside of St. Louis. After a narrow escape from the rising Mississippi River, Fendler decided to try his hand at botanical collecting. With the expansion of institutional, university, and private botanical collections, Fendler saw an opportunity to capitalize on the burgeoning market for rare specimens. After consulting with a Prussian scientist, he returned to St. Louis eager to get his start collecting whatever he could, bringing them to George Engelmann, a well-connected botanist and physician who often funded and supported botanical collectors as they headed west. Learning of Fendler's skills from an impressed Engelmann, Asa Gray connected Fendler with the secretary of war during the Mexican-American War, so that he might accompany the army into the Southwest to collect specimens for Harvard.[8]

Asa obtained many new plants thanks to Fendler, but he also found the botanical collector decidedly odd. "He is full of sentiment, and has a fine, delicate taste," Asa wrote to Elizabeth, describing Fendler in ways that questioned his masculinity and sometimes even his sanity. Jane Gray, more restrained and subtle in her written descriptions than her husband, pointedly referred to Fendler as "a very interesting man." After struggling to make ends meet following his collecting trip in the Southwest, Fendler decided to move to Venezuela in hopes of finding work. "I hope I have plenty of leisure to make botanical collections besides," Fendler wrote to Asa. Collectors like Fendler relied on buyers—whether individuals or institutions—to purchase collections of specimens. "Do you think 15 sets, each containing from 1000 to 1500 species of plants collected about Caracas could meet with a ready sale?" he asked Asa, eager for a valuation. Asa Gray and George Engelmann sold Fendler's collections for him, since Fendler was often on the move and unable to handle the sales himself. Asa Gray was able to leverage his extensive network to facilitate sales and encourage his friends to purchase what he had on hand, something Fendler would not have been able to do on his own.[9]

Increasingly, Asa Gray and other powerful botanists relied on a growing class of professional collectors whom they could pay to collect new specimens for them. Asa was able to afford this by soliciting subscriptions from connections like Elizabeth and others, which funded collectors' work in the field. The collectors may have struggled to scrape by with these meager payments, but the system was beneficial for Asa and his ilk. Money gave Asa the power to be brusque. He was able to make more specific and frequent demands than when he relied on favors, gifts, and goodwill alone—and while professional collectors could at times be high-maintenance and demanding, these relationships did not require the same

choreography of mannered obligations that gift-giving relationships did. This meant an increasing reach for Asa, but it also meant
a diminished role for those like Elizabeth who privileged the social connections and specimen exchanges that the gift economy
afforded. With professionalization, the traditional botanical community was giving way to more market-driven relationships. Gift
giving would continue and collectors remained diverse and idiosyncratic, but professional collectors offered new avenues for acquiring specimens for professional botanists.[10]

Fendler settled on a small farm in Colonia Tovar, a German
immigrant community in Venezuela west of Caracas, in 1854. "A
visitor who never before lived in a valley like this finds here many
peculiarities of vegetation, surface, and climate which make him
feel that he is not far off the land of perpetual peace; for he has
entered the happy region of the ferns." Fendler loved ferns. He
kept regular meteorological notes about the delightfully temperate climate in Venezuela, sending them back to Joseph Henry at
the Smithsonian. He made very little money, though, complaining to Asa Gray that his entire income during the first two years
in Venezuela did "not amount to one Dollar." Fendler's desperation meant he pressed Asa even more to secure subscribers for his
botanical specimens.[11]

While the scattered, footloose collector might have seemed
to have had little in common with the well-heeled anonymous
science writer Elizabeth Morris, the two shared a love of ferns,
flowers, gossip, and Asa Gray. Ever generous with her network,
Elizabeth was happy to connect him to her childhood friends
from Germantown who lived in Caracas once she learned of
Fendler's isolated existence in Venezuela, forwarding him a letter
of introduction via Asa. Fendler was stunned by this small act of
kindness. He recounted to Asa that his "fair friend Miss Elizabeth C. Morris" had managed to cheer his spirits by lifting him
out of his solitude. "I hold companionship with almost nothing

but trees, and shrubs and flowers, and where the exciting din and roar of an uninterrupted line of cascades is the only music that I can listen to, and the infinitely variegated fall of the rushing and leaping mountain stream my most exalting spectacle." Elizabeth was a deeply empathetic person, and their shared experiences on the edges of the botanical community, solitary as that could feel, helped her see Fendler as a kindred spirit. She worried he was lonely in the forests of Venezuela and cherished how overjoyed he was to receive any kind of note. "How isolated he is in that distant Paradise!"[12]

Fendler returned briefly to the United States to deliver his collections to Asa Gray, restock supplies, and collect his badly needed income. As he made his way from Boston down to Philadelphia to catch his ship back to Venezuela, the "happy land of perpetual spring," he made a point to pass through Germantown to call on Elizabeth. As she welcomed the collector into her home, Elizabeth was enthralled by his stories. "He is a simple hearted, enthusiastic man, longing to be again among the flowers and ferns of the tropics," Elizabeth relayed to William Darlington. She was captivated by what she described as "his glowing descriptions of the climate and vegetation of the elevated region in which the colony is located, almost making one sigh to feel chained in our miscalled Temperate zone." The conversation must have turned to business, as Elizabeth expressed concern about Fendler's waning commercial prospects and resulting financial challenges, which were not insignificant. "But like all botanists he is very hopeful," she wrote, "and looks forward to a rich harvest of new species among the true ferns of the mountains."[13]

Elizabeth was no stranger to financial hardship, and she initially hesitated to purchase any ferns herself. "I am too poor to buy them," she confessed to William Darlington, "which for reasons more than one, I am extremely sorry." Money had become increasingly tight after her mother's death, and with only their inheritance to sustain them, she and Margaretta often worried about whether to make purchases like these, or even to continue some of their subscriptions to scientific and horticultural journals. Margaretta was primarily in charge of their finances, and perhaps she had instructed Elizabeth to limit unnecessary expenditures. Asa Gray was vaguely aware of Elizabeth's financial situation, and when she hesitated, he encouraged her to promote Fendler's collections with her friends and suggest they subscribe. In general, the pressed specimens went for ten cents a page, meaning a buyer could get roughly one hundred for ten dollars. Worried about her struggling friend in Venezuela, anxious to please Asa, and ever eager to expand her herbarium, Elizabeth eventually succumbed and purchased one of the sets.[14]

Elizabeth took it upon herself to keep Fendler connected to the American botanical community, soliciting William and others for "as much botanical gossip" as possible "to amuse the warm hearted, excellent exile." Asa Gray often delegated the work of shipping materials to Venezuela to Elizabeth, likely also saddling her with the cost despite her finite income. But she found ways to take advantage of this arrangement, tucking in her own notes and gifts into his packages, including the one she bestowed upon her closest botanical friends: forget-me-not seeds. "The seeds Fendler is in such raptures about were or you will guess," Elizabeth boasted to William, "the *Myosotis palustris* from the Pontine Marshes: he is grateful for small favors!" Fendler, in turn, sent her gifts of his own, such as "forty species," which was actually a jumble of forty miscellaneous seeds wrapped in paper. "So I shall sow them in the *same box* and a nice *mess* they will make if they should grow!" She was clearly amused by it all.[15]

Fendler represented, in a broad sense, the explorer's life that Elizabeth would never have. As a child, she had dreamed of combing the remote corners of the world in search of rare plants, but now well into her sixties with increasingly tenuous health, it was becoming clear that the ambition would remain unfulfilled. Her connections to younger men like Fendler allowed her to travel vicariously, with stories of his tribulations and adventures radiating from the pressed ferns and envelopes of exotic seeds from Venezuela's forests.[16]

Asa Gray also connected Elizabeth with one of his students, the twenty-five-year-old Daniel Cady Eaton. Though based in New Haven, Eaton was helping his former mentor process and distribute Fendler's ferns. Elizabeth was delighted by the prospect of a new correspondent, not least of all because she would almost certainly receive more native ferns for her garden from him. Daniel—both a grandson of the botanist Amos Eaton and a nephew of Elizabeth Cady Stanton—had as a child climbed through a window of a locked church in order to commence the Seneca Falls women's rights convention of 1848. By 1860, he was in his twenties, and firmly devoted to collecting and describing North American ferns.[17]

With the confidence of age, Elizabeth's tone with Daniel recalled Asa Gray's in the early stages of their relationship decades prior. In her letters, she repeatedly sent him her list of desired specimens and put him to work identifying some of the ferns she and her friends had purchased from Fendler or found in cultivation at a local nursery. "I have *almost* a passion for cultivating ferns, & much desire living specimens of rare or beautiful species," she wrote. Daniel was an enthusiastic correspondent and gladly responded to her many requests, including gifts from his botanical collection along with the information she requested. Sounding just like Asa, Elizabeth warned him: "You will I fear find me a troublesome, but not I trust an ungrateful correspondent."[18]

Despite its implicitly hierarchical nature, Elizabeth nurtured her relationship with Daniel by offering several rare ferns native to Pennsylvania, including the American climbing fern, or *Lygodium palmatum*. Her favorite, of course, was the one that had garnered her so much praise and gratitude from Asa Gray and William Hooker decades earlier: the *Asplenium pinnatifidum,* or lobed spleenwort. Urbanization, though, had begun to threaten this fern and not just because of the smoky skies. "The rocks on which it grew have been blown up to make place for roads, and I now do not know where to find it," Elizabeth reported to Daniel. Still, she had several pressed specimens in her collection, and she offered to send some to him. She also planned to try to cultivate the fern in her fernery, and proudly reported her success to both Daniel Eaton and William Darlington as the ferns sprouted new fronds in the years that followed.[19]

While her *Asplenium pinnatifidum* thrived anew, Elizabeth herself was growing progressively less mobile as age continued to take its toll. As she suffered with health issues from reoccurring attacks of liver complaints, she found "botanizing out of the question." Still, she could entertain herself by tending her indoor greenhouses and studying her sizable herbarium. She told Daniel Eaton how she amused herself through a long winter by cultivating a variety of different ferns and mosses in her indoor fernery, referring to her successes as her "pretty pets." She politely requested not only pressed specimens from Eaton but also live plants when possible. She was eager to add a shield fern or *Aspidium aculeatum* to her Wardian case, for instance, and asked him how she might get it in a "live state," clearly hoping he might read between the lines.[20]

ELIZABETH'S INCREASINGLY POOR HEALTH MAY HAVE LIM-ited the scope of new fieldwork she could undertake, but her previous research began to take off. Edward Tatnall, a botanist and businessman from Delaware who corresponded with both Asa Gray and William Darlington, published a catalogue of plants available in New Castle County, Delaware. This unillustrated list of scientific names for plants was a survey of the botanical landscape of Delaware in the mid-nineteenth century. Aimed at beginning enthusiasts, Tatnall's book would have been a cursory overview of the regional flora for nonspecialists, but those looking even for basic details would need to cross-reference more descriptive and illustrated volumes like Gray's textbook. One reviewer of the book generously touted the importance of local lists like Tatnall's for beginners: "A young botanist should first confine himself or herself—for botany is particularly a science for the lady—to a township or even smaller district." After developing a literacy of the local flora, they could then increase the scope of their study. Tatnall's list would not have been a particularly welcoming or instructive handbook for beginning botanists, but it would have at least been a useful reference.[21]

The most informative part of the book was its footnotes where Tatnall provided snippets of contextual information about some plants, noting that a particular flower was toxic or that a certain weed was the bane of farmers. One such footnote mentioned Elizabeth Morris's observations about dandelions, likely relayed by Asa or William: "Miss Morris, of Germantown, has noticed that after flowering, the head lies flat on the ground until the fruit has matured, when it again becomes erect." All these years later, she continued to admire and observe these flowers.[22]

When a reviewer for *Gardener's Monthly* picked up Tatnall's book, the footnotes took center stage: "There are numerous notes in the margin giving much novel and interesting information." The reviewer then quoted Elizabeth's discovery in full, adding, "We do not think this beautiful provision of nature for permitting the seed to blow readily away has been before noted by anyone." Elizabeth may have been uncomfortable with the magazine's prominent attribution and public declaration of its novelty, but she left no record of her reaction. Maybe she was instead delighted to share one of the flower's idiosyncrasies—at one time challenged by Gray but now very publicly promoted—with a broader audience, broader still than the limited readers of Tatnall's list.[23]

Elizabeth would have certainly seen the review in *Gardener's Monthly*, as she had begun publishing in that magazine regularly, dusting off her old pseudonyms and creating a few new ones. Thomas Meehan, the editor of the *Gardener's Monthly* who regularly celebrated Margaretta's discoveries and achievements in the pages of his journal, was a recent British transplant who had opened a thriving nursery in Germantown. Elizabeth was not only a regular customer but also a frequent contributor to the magazine. One of her first articles focused on the Rose of Jericho, a fascinating fernlike plant adapted to desert life that could lie dormant for years at a time and then immediately revive with an infusion of moisture. Now that more readers had access to the means and tools to do so, she shared advice about how to cultivate exotic plants like camellias indoors, and Meehan reinforced her wisdom with a final note vouching for her success. Some of her articles repeated material she had published in the *American Agriculturist* decades earlier, but mostly she took on a new tone while continuing to weave in her usual witty asides. Avoiding the domestic science and lifestyle advice she had relished in the 1840s, she instead stuck mostly to botanical information and gardening successes, speaking from a place of authority about pear varieties and shrub choices.[24]

Still, Elizabeth had not lost her sense of humor. Over the course of two issues, she celebrated the land tortoise and its benefit for gardens. She had a half-dozen tortoises herself that she kept as pets, setting them out to terrorize the slugs and snails that plagued her garden, cellar, and milk house. "The pretty foreigners, with striped shells," she wrote, "are multiplying fast in this neighborhood." She believed they had been accidentally imported from England in the dirt of plants—the Wardian cases that successfully transported plants around the globe also transported many unintended stowaways—but that they now served as a kind of pest control. She published a humorous poem about them one month and then an article about their many uses the next. "Epicureans say that, as a *bon bouche* they are quite as good as terrapins." She added at the end with a smirk, "But this is a treasonable whisper."[25]

In addition to the thrill of publishing, Elizabeth continued to delight in introducing her various correspondents to one other. The role she played connecting botanists with each other was integral to the vitality and cohesion of these scientific networks. She introduced Meehan to both William Darlington and Asa Gray, for instance, and Asa went on to occasionally publish in *Gardener's Monthly*. "I do not remember if I told you of my last exploit— that of introducing Thomas Meehan to our dear friend Dr. Gray," Elizabeth reported to William Darlington. "I had often acted as a medium in communicating between them—and I thought that it would not be an improper liberty for me to take with one who professes to be my *friend*, & whom I really love." Thanks to Elizabeth, the two men developed a connection that would last decades. Meehan, like many botanists and horticulturists, held Asa Gray up as a scientific celebrity, referring to him in the *Gardener's Monthly* as the American botanist "whose name and fame rise above all the rest." Elizabeth was thrilled that she was able to facilitate their friendship, finding that her carefully maintained botanical connections over the course of her life had real value. "Thomas is delighted and grateful beyond measure to me, for the hand I had in bringing it about."[26]

Elizabeth also connected novice botanists who were just starting out with her high-profile friends. When a "young friend" in Northampton, Massachusetts, reached out, anxious to continue his botanical education and hoping to find employment helping botanists with their herbariums, Elizabeth sent a note to Daniel Eaton, who was busy establishing himself at Yale. This young Massachusetts botanist, whom Elizabeth never named outright, was "unobtrusively modest, & I believe well calculated to fill the place he seeks." She hoped that Eaton might have some ideas, or

that perhaps the young man could go to St. Louis to help Engelmann and Fendler, who was newly back from Venezuela. Eaton,
in turn, suggested that the newly formed Botanical Society of
Canada might have a space for him. These networks that Elizabeth had cultivated and tended for decades became crucial for
those getting their start in botany. She worked quietly behind the
scenes to connect up-and-coming botanists with rare opportunities for employment.[27]

This network that made Elizabeth invaluable had been something she had created out of necessity to establish her credibility
in the profession just as Margaretta had cultivated allies to endorse her own research. After years of work and exchanges, she
had made herself indispensable to William Darlington, Asa Gray,
and others both as a friend and as a supplier of specimens, and in
turn she was able to access botanical conversations and resources
that would have otherwise been out of reach. Now the social infrastructure she had cultivated was being put to use by the next
generation of white male scientists so that they could establish
themselves in the profession. A different scientist might have re-
sented this, but Elizabeth was genuinely happy that her assistance
to others would be her ultimate legacy. She prioritized her relationships above all else.

IN ALL OF THESE LETTERS CHANGING HANDS BETWEEN
Elizabeth Morris, Thomas Meehan, Asa Gray, William Darlington, and Daniel Cady Eaton, there was increased discussion of
Asa's work with Japanese plants. He often circulated copies of
his articles and books with his botanical connections, taking advantage of the free copies he got from the publishers to distribute
them widely, sometimes asking friends to pass them along to another colleague after they were done. He had promised Elizabeth
a copy of his Japan paper, as everyone called it, titled "Diagnostic
Characters of New Species of Phaenogamous Plants," but with

demand for his copies so high, he admitted he had "naughtily" withheld hers in favor of giving it first to "working botanists," suggesting she might be able to borrow a copy from someone else. While Elizabeth no doubt took this in stride, and while Asa had expressed his regret, this was most certainly a statement about her place in the scientific hierarchy. Early in his career when he was so dependent on Elizabeth's gifts of specimens, he would have rushed to ply her with copies of his latest work. But with the paid support of professional collectors like Fendler, and the power afforded by scientific celebrity achieved partially thanks to the early help of those like Elizabeth, he was no longer reliant on her favors.[28]

Asa's paper on Japanese plants was in high demand for good reason. Naturalists had been puzzling over the similarities between Japanese and eastern North American plants since the late eighteenth century, and Asa himself had been captivated for decades, peppering many of his publications with mentions of the surprising botanical parallels. Louis Agassiz—Asa's colleague at Harvard—believed the global distribution of species could be explained theologically, concluding that the adaptations and distribution of species could only be "introduced, maintained, and regulated by the continuous intervention of the Supreme Intelligence, which from the beginning laid out the plan for the whole, and carried it out gradually in successive times." Agassiz's strongly expressed perspective helped Asa Gray, who himself was a devout Presbyterian, have more clarity about how his own thoughts differed.[29]

Asa's understanding of species distribution was evolving. While Charles Darwin nervously watched the public's response to Robert Chambers's anonymously published *Vestiges*, Asa wrote a scathing review of the author's theories and methods in 1846. He argued that *Vestiges* gave "birth to conclusions as incongruous with any common theistic scheme as they are revolting alike to our religious and common sense." Asa remained a staunch defender of religious natural history. Elizabeth, however, had been more willing to entertain the thought exercise proposed in *Vestiges* and considered it to be both well written and enjoyable. When she was about to mail a copy of the book to William Darlington after promising him as much, she was intercepted at the bookstore by her friend, the geologist Henry Darwin Rogers. He warned her against sending *Vestiges*, suggesting she send another book about agricultural chemistry instead, and either assuming her instincts about the controversial book were wrong or wary of offending her friend, Elizabeth took his advice.[30]

Still, despite Asa's deep dislike of *Vestiges*, he was increasingly interested in species distribution as he wrestled with how identical plants could be found in Japan and northeastern North America, but not the American West. Exactly how the plants got to these distant locations had the potential to be an incendiary topic. As botanists and zoologists probed into why species were in different locations around the world, they sometimes found themselves contradicting the biblical story of creation. This was something that Asa wrestled with. "I should like to write an essay on species, some day," Asa wrote to Charles Darwin in 1855, four years before Darwin published *On the Origin of Species*. As he struggled to substantiate Agassiz's theistic explanations, he found himself aligning more and more with his British correspondents, including Darwin and the botanist Joseph Hooker, who by this time had also published essays about global plant distribution. "I am most glad to be in conference with Hooker & yourself, on these matters," Gray wrote to Darwin, "and I think we may, or rather you may, in a few years settle the question as to whether Agassiz's—or Hooker's views are correct: they are certainly widely different."[31]

While Asa had been writing about the parallels he had identified between East Asian and northeastern North American plants for years, it was not until after he had received materials from the professional botanical collector Charles Wright that he was ready to make a definitive statement about *why* the plants in such distant locations were so similar. Wright had been in Japan from 1853 to 1855 as part of the United States Navy's North Pacific Exploring Expedition, and he had witnessed botanical connections with North American flora firsthand, making a point to collect both passingly familiar and unknown specimens and send them to Asa for a fee. In 1858, after years of exchanging letters with Darwin, Hooker, and the Yale geologist James Dana, Asa entered into the fray, arguing that glacial movements coupled with changing climates propelled species into temperate zones, like those in New England and Japan. "I cannot resist the conclusion, that the extant vegetable kingdom has a long and eventful history, and that the explanation of apparent anomalies in the geographic distribution of species may be found in the various and prolonged climatic or other physical vicissitudes to which they have been subject in earlier times," he wrote in his essay. He explicitly contradicted Agassiz and his "primordial explanation" that nature could not travel away from God's assigned place, arguing that biblical narratives were nearly impossible to test scientifically, let alone prove. Gray felt that his explanation was scientific and therefore testable. Agassiz watched from the audience while Gray delivered his findings as a speech at the American Academy of Arts and Sciences.[32]

When Elizabeth finally received her copy of Asa's essay, she was suitably impressed. She wrote to William Darlington, asserting, "Of course you have read, and enjoyed Dr. Gray's paper on the Plants of Japan." Unconcerned with the theological implications his theory might have, she was eager to adopt this new understanding of species distribution. "It seems to me he takes the philosophical and true view, in his theory of the causes of such marvelous distribution of plants—not continuous but so widely interrupted." She mused, "Strange that our eastern shore should correspond so much with the far off Japan and Eastern Asia."[33]

In the midst of all of this, Charles Darwin and Asa Gray found a natural affinity in their discussions of geographic distributions of species. Their friendship, which had begun carefully with professional correspondence noting similarities between North American and European species—as well as Asa's endorsement of Margaretta Morris's skills—blossomed as they exchanged data and philosophies. Darwin had shared his new theory only with a small set of confidants. However, when Asa probed for Darwin's thoughts on species distribution, Darwin went so far as to respond with a summary of his thesis on evolution. Worried about Asa's reception, he self-consciously added, "I know that this will make you despise me." He need not have worried. Even luckier for Darwin, he kept a personal copy of his September 1857 letter to Asa where he had laid out his core argument. When Alfred Russel Wallace independently came to the same conclusion and published his own essay on evolution the following year, Darwin and his friends were able to use that document to establish that Darwin had priority, having written about the evolution of species first.[34]

After Darwin published *On the Origin of Species* in 1859, Asa soon became his North American spokesperson. Asa not only negotiated with an American publisher to ensure profits returned to Darwin but also wrote a powerful and positive review of the book in Silliman's *American Journal of Science and Arts* in March 1860. "Darwin's aim and processes are strictly scientific," Asa wrote, distinguishing *On the Origin of Species* from the controversial *Vestiges*, which he found unscientific and weakened by novelistic techniques and numerous mistakes. "And his endeavor, whether successful or futile, must be regarded as a legitimate attempt to extend the domain of natural or physical science." The review was as much a celebration of Darwin's book and the magnitude of its impact on professional science as it was an attack on Louis Agassiz, who, Gray contended, assumed "the scientifically unexplained to be inexplicable." Still, Asa himself wrestled with the religious implications of Darwin's theory, finding some solace by reinforcing the coexistence of God and natural processes. He contended that the natural law Darwin described was itself an act of "continued and orderly Divine action." As he did with all his publications, Asa distributed copies of his review to friends, publicizing Darwin's controversial thesis. This review, along with his Japan paper, marked a significant personal and professional shift for Gray—one that clearly pitted him against his Harvard colleague, Agassiz.[35]

There was a power struggle brewing between Harvard's scientists, both of whom had supported the careers of the Morris sisters at various stages. The persona Agassiz sought to construct—with his adoring students, public acclaim, prolific publications, and growing influence at the university—had cast a shadow over Gray as Agassiz rose to become the face of American science. But far from simply being a battle of egos, it was more profoundly a battle of ideas. Gray bristled at Agassiz's stubborn insistence upon divine design, and Darwin's theory of evolution was the first significant alternative with a scientific methodology Gray could support. So Gray took aim in each review, essay, and speech, calling Agassiz out by name. He also readily embraced any opportunity to debate Agassiz publicly at the Cambridge Scientific Club and the American Academy of Arts and Sciences.[36]

FROM THE SCATTERED EVIDENCE THEY LEFT BEHIND, IT seems that Margaretta and Elizabeth would have accepted Darwin's theory of evolution, particularly as evidence mounted and it garnered widespread support in the scientific community. Elizabeth, after all, had recommended *Vestiges* to William Darlington, a sign that she was at least intrigued by its premise. Decades before, around 1830, Elizabeth had copied long passages about the struggle between species and the "chain of beings in the universe" out of William Smellie's *Philosophy of Natural History*—a book scholars have deemed a predecessor to Darwin's—judging it to be "a most delightful book, which I would read again if I had time." Still, Agassiz had been a staunch supporter of Margaretta's work, nominating her for membership in the AAAS and citing her work in his books. He was not a close friend or even a mentor, by any stretch, but he had played an essential role in her rise in stature among fellow scientists. Asa Gray, though, was a good friend, especially of Elizabeth, but also of Margaretta, hosting the two regularly at his home. Perhaps they heard his perspective on Darwin's theory firsthand as they sat with him in his parlor at Harvard's Botanic Garden during one of their many trips to Cambridge.[37]

Their grandmother had been a Quaker preacher and their mother had been a devoted member of St. Luke's Episcopal Church, but Margaretta's and Elizabeth's personal religious beliefs rarely emerge in their writings. In letters to friends, they would speak about religion predominantly in the context of a relative's death. Margaretta and Elizabeth had occasionally written about the Creator's influence on insects and plants in their anonymous articles—Margaretta casually referred to the parasites that attacked agricultural pests as being a divine gift, for instance. But even Asa Gray, Darwin's fervent American spokesperson, had advanced such views in his own textbooks. Like many nineteenth-century naturalists of his generation, Asa did not see science and faith as incompatible. He argued that the theory of evolution did not necessarily lead to atheism, but could be squared with a Creator establishing natural laws. He found solace there for his own religious beliefs. Perhaps the Morris sisters felt similarly.[38]

While Darwin's theory would not necessarily affect Elizabeth's and Margaretta's observational methodology, it would have shifted the context for their findings. They might have come to understand how varieties of plant species or insects were incipient species in formation. They might have thought more globally about their very local work. Whether the origin of species was natural or supernatural, Elizabeth and Margaretta and many scientists like them were still primarily concerned with characteristics of the creatures and plants that populated their everyday lives. The tectonic plates of natural history were beginning to shift as scientists digested the ramifications of Darwin's book. These philosophical and cultural shifts would continue as debates unfolded between naturalists, educators, politicians, and religious leaders in the decades and centuries to follow.

American science was beginning to grapple with the implications of Darwin's thesis when Margaretta's and Elizabeth's scientific work was mostly behind them. Still, as they settled into old age, their passions never waned. They continued to collect insects and tend ferns, discuss cutting-edge books with friends, visit the Wissahickon whenever they could, linger in the garden, and write articles. And Elizabeth, true to the promise she had made herself years prior, continued to prioritize her friends, supporting their work and celebrity, while connecting them with promising young botanists. Elizabeth's and Margaretta's failing health got in the way of certain projects, but it did not always limit what they could do. Surprisingly, illness also brought its own kind of opportunities.

Ten

She Is Everything Now, to Me

As Margaretta boarded the train back home after a brief but wonderful visit celebrating the start of 1857 with her niece's family, she did not feel quite right. She could not shake that feeling even with the gush of fresh air as she walked down the cobblestone streets of Germantown from the train station, or as she climbed the steps and fumbled with the cast-iron key to Morris Hall. The change of seasons and the brisk winds always seemed to bring on some sort of illness, and it had been exceptionally cold that month—so cold that seven inches of ice covered the local ponds. Still, arriving home and smelling its familiar aroma as if with a new nose, seeing her sister who greeted her first with glee and then concern, must have brought some comfort before she collapsed in her bed.[1]

Elizabeth was beside herself as Margaretta's illness persisted for days and then weeks. Half-finished specimen sheets sat forgotten on her desk, and her houseplants waited longer between waterings. Elizabeth even neglected to respond to letters from friends. When she finally found a moment to sit at her desk and write a note to William Darlington, she tried to articulate her fears. "She is everything now, to me," Elizabeth confessed, "and when she is confined to her room the parlor looks so desolate without her, and my solitary meals are so forlorn, that I too often yield to the feelings of sadness that steal over my head,—and—then of course, I am good for nothing." As Margaretta and Elizabeth aged, their health was increasingly fragile and each recovery became a shorter and shorter reprieve. Elizabeth had always taken for granted that she would have Margaretta by her side at the dinner table, in the parlor, and on their hikes through the woods. They had rarely been apart their entire lives.[2]

How the Morris sisters and their contemporaries understood their health and what would accelerate their recovery would seem in some ways foreign today. Nineteenth-century Americans saw their bodies as intimately connected to their surrounding environment. Changing weather patterns, rising or falling heat or humidity, a poorly drained field, a recently cleared forest, could all wreak havoc on one's health. Achieving a balance, both internally and externally, involved constant calibration. So when Margaretta and Elizabeth fell ill but were strong enough to manage it, they followed the standard medical advice from the period to travel to healthier environments, whether that involved cool mountain air, the salty breezes of the sea, warmer climates, or simply a change of scenery. While there were plenty of cultural restrictions that kept women like the Morris sisters from traveling widely, poor health opened doors for adventure.[3]

A FEW YEARS BEFORE MARGARETTA GOT SICK, ELIZABETH had been struggling to regain her own health after a prolonged illness associated with her liver. To do that, she visited friends up and down the East Coast, inhaling the sea air while still finding time to botanize in new environments. "My health has been so very precious since my fearful illness four years ago, that it seems as if my life may be spent, even to its end, in fruitless endeavors to regain it," she explained to William Darlington. She traveled partly with friends, partly on her own, and along the way loneliness gave way to confidence. She assured William that, while unpleasant, it was increasingly safe for women to travel unaccompanied. "At my age," Elizabeth wrote, "I shall mind my own business, keep quiet, and have no doubt of being unmolested." While it was potentially dangerous to travel, it was certainly also dangerous to live a life constrained by societal restrictions and anxieties. Old enough to enjoy some invisibility, Elizabeth struck out on her own.[4]

During that trip, Elizabeth stayed for long stretches at Asa Gray's home in Harvard's Botanic Garden. Asa set up a table for Elizabeth in his study, and the two examined specimens together, selecting plants for Elizabeth's herbarium. She collected materials from the botanical garden to dry and press, supplementing her collection with alpine plants she lacked. Leaving Cambridge, she continued on to the beaches of Massachusetts and Long Island with friends, searching for unique specimens of seaweed, while also swimming in the ocean and taking long walks. Traveling was both restorative and a rare opportunity for exposure to coastal flora. As the summer months waned, Elizabeth confessed to William Darlington that she was little better than when she left home: "I begin to be *very* home sick: and I have not yet recovered health or strength." Regardless of how much she may have enjoyed her summer getaway, she missed her sister.[5]

Years later in 1857 when Margaretta started to regain some of her strength a few months into her own illness, at least enough to travel, she planned to replicate the kind of trip Elizabeth had taken. Elizabeth may have even suggested as much. "The change will be of great service to her," Elizabeth hoped, "after having been shut up at home so much this winter." A trip away would at least lessen the burden on Elizabeth after months of caring for her sister.[6]

That summer, as good health seemed to evade her every effort, Margaretta made her way first to Cambridge to visit with Asa and Jane Gray, attend Harvard's commencement, and see a number of friends. She seems to have also gone to Newport, Rhode Island, where she got the chance to spend time with Dorothea Dix, and her cousin Harriet Hare, at the home of their mutual friend Sarah Gibbs. Not only was travel meant to be good for her health, it was also a chance to reconnect with far-flung friends.[7]

During the trip Margaretta suffered from intense seasickness, but it was not something to bemoan so much as celebrate. Nineteenth-century health advice often centered on circulating stagnant fluids, which generally meant purging or flushing the system, among other techniques. Elizabeth was pleased to learn that the seasickness "cleared [Margaretta's] system from much that had oppressed her for months—she now eats what she pleases, & has regained her strength." Perhaps Margaretta could actually enjoy herself.[8]

However, as soon as Margaretta had left Germantown, Elizabeth fell ill. She resisted the idea that it was because she was lonely, even after a friend called her "sick, solitary, and dumpy." The sisters were always together, and friends worried about how Elizabeth would fare without her. She might have been alone, she admitted, but tending to the garden kept her too busy to feel lonely. She had to harvest fruits and vegetables and set about preserving them for the winter. She had plenty to do indoors, too, not to mention the stacks of unanswered letters awaiting her reply. It was not the loneliness, she assured herself and her friends, but the weather that was to blame. "I hope you have borne the sudden change from damp cold weather to this extreme heat better than I have," she wrote to William Darlington. "I have not been well for an hour—not felt equal to the least exertion."[9]

Elizabeth recovered quickly, though, and Margaretta eventually returned, albeit not fully well. Hopeful that an entomological assignment might give her strength, Elizabeth encouraged Margaretta to return to a project she had long ago promised to create for William Darlington: a catalog of Philadelphia's insects. "I shall keep your request for 'that Catalogue' before Margaretta's very handsome black eyes!" Elizabeth promised William, "and have strong hope that before winter closes, she may be induced to extol her energies and address herself to the pleasant task: for pleasant I am sure she will find it when it is once begun." Over a year into her illness, Margaretta was still struggling. Severe back pain followed any kind of exertion, which hampered her ability to locate new specimens for her study. As she had supported countless other scientists, Elizabeth assisted Margaretta with her work, and as the insects began to buzz in their ears the very next summer, the two women joined forces. "As a matter of course, it is slow work, for the insects must first be caught—in that I help her vigorously," Elizabeth proudly reported to William.[10]

As Margaretta and Elizabeth made their way through Germantown together—sweeping their nets through the tall grasses, examining captives in spiderwebs, and prying up tree bark with their paring knives in search of specimens—they were not completely oblivious to the signs of human despair around them. After decades of prosperity fueled by the influx of gold from California, the expansion of railroads, thriving agricultural exports, and the Industrial Revolution, the Panic of 1857 shook the economic security of communities across the United States, rural and urban alike. Even by the summer of 1858, Philadelphians were still reeling from its aftershocks. In Independence Square, a rally of ten thousand unemployed workers inspired Philadelphia's mayor to begin a public works program. Elizabeth worried about the impact of the economic crisis on her neighbors and was grateful that her better health meant she was "enabled to do more for others." She joined efforts to help families in distress. "I am trying strenuously, to find places on farms or at least in the country for some of the children who are thrown out of employment in the factories," she wrote to William Darlington, hoping he might be able to connect them to farmers in need of farmhands. She also worked with friends at the Relief Association of Germantown, which she called the "Soup Society," helping those who could not afford to eat or buy necessities. Still, she was somewhat ambivalent about the long-term impact of her work. "I much doubt if we have not done more harm than good, by making them comfortable and contented in idleness," she wrote to Dorothea Dix, "in other words encouraging their pauperism. In many cases, I am sure it is so." Whatever Elizabeth's hesitations may have been, the organization managed to aid 286 struggling families in the first six months of its operation following the start of the economic depression.[11]

Many of the immigrant poor Elizabeth helped in Germantown were Irish Catholics, and she and Margaretta, like many of their contemporaries, passed a lot of judgment on these neighbors. When the *Germantown Telegraph* encouraged residents to hire farm help to ease economic despair among the poor, the editors mentioned how German, English, and Welsh immigrants were industrious and responsible employees, pointedly leaving out the Irish. When Elizabeth and Margaretta's unmarried niece Ann Johnson converted to Catholicism and entered a convent, the women were quietly appalled. "This has cast a gloom over the house, worse than her death would have done—and presses with a weight of lead on my heart," Elizabeth confessed. The women had similarly mourned the conversion of friends to the "thraldom of Rome," baffled by their decisions. Though Elizabeth and Margaretta sought to support their community during the financial crisis, they had a hard time disentangling their desire to do good from their prejudice.[12]

THROUGH ALL OF THIS—THE ECONOMIC CRISIS, THEIR worsening health, and their struggles to recuperate—life marched on. Elizabeth and Margaretta continued publishing on botany and entomology in agricultural and horticultural journals. They doted on their nieces and nephews—something that brought them pure joy. "I believe there are very few unmarried women at our ages who possess so much sincere love, and so many warm devoted friends among the young," Elizabeth boasted. "Nothing smoothes the path to old age, so much as to retain the affection and contribute to the wellbeing of the young & happy." No longer as quiet as it had been after their mother's death, Morris Hall bustled with the comings and goings of visiting family and friends, which delighted the sisters even if they sometimes complained about how all the activity kept them distracted from their scientific work.[13]

They continued to travel, too, for the sake of their health, family, and science. Elizabeth made trips to visit a distant cousin in New Jersey in 1858, taking advantage of the opportunity to slosh around in her waterproof gum boots to collect plants like the American climbing fern (*Lygodium palmatum*) so that she could foster its growth in her Wardian case. She also gathered algae specimens along the shoreline. The Irish botanist William Henry Harvey—whose *Sea-Side Book* sat on Elizabeth's shelf—trumpeted how time spent at the seashore studying natural history melded physical recovery with a search for new knowledge. Taking the perspective of an overworked businessman, Harvey described how beachcombing was a wonderful salve for modern life: "His pursuits lead him to take exercise of body, and, without fatiguing the mind, give it that pleasurable excitement which rapidly restores its tone when suffering from having been overwrought." Exhausted and unwell Americans followed his advice, scouring the coastline for shells and aquatic vegetation to restore their physical and mental health.[14]

A decade earlier, Asa Gray had mentioned to Elizabeth that Harvey, a professor at Trinity College in Dublin, was interested in North American algae. Elizabeth was glad to assist him by collecting specimens on her occasional trips to the Jersey Shore and Delaware Bay. In the years since, Harvey had cataloged and identified all of the materials he received from correspondents throughout the United States, Canada, and Mexico in preparation for publishing a series of volumes on North American algae for the Smithsonian, *Nereis Boreali-Americana*. One of the many specimens Elizabeth had sent him struck him as exceptional. While initially olive green in color when he unpacked it from the box, the dense, tangled tufts of algae turned a dark green once remoistened. He had never seen anything like it before in his many years studying algae specimens from around the world. He named it *Cladophora morrisiae* to honor Elizabeth.[15]

Harvey certainly was not obligated to name the new species for Elizabeth. He could have taken credit for himself, or bestowed it on Asa Gray as a sign of deference and respect. Instead, Harvey chose to express his gratitude for the collector herself, thus etching Elizabeth's name onto a piece of the natural world. It is not clear whether Elizabeth ever learned of this honor. While she certainly had a copy of Harvey's *Sea-Side Book*, she never mentioned reading his Smithsonian volumes. Even if she had known that Harvey named one of her specimens for her, Elizabeth would not have boasted about it in correspondence with friends. While other botanists might have congratulated themselves publicly, she would have brushed it all off in her signature humble, self-deprecating manner, and instead would have reveled in the thrill privately, perhaps sharing the news with her sister alone.

While Harvey paid tribute to Elizabeth with scientific taxonomy, Margaretta at long last was being recognized by her peers at the Academy of Natural Sciences. In 1859, after nearly two decades of contributing reports and donating specimens to

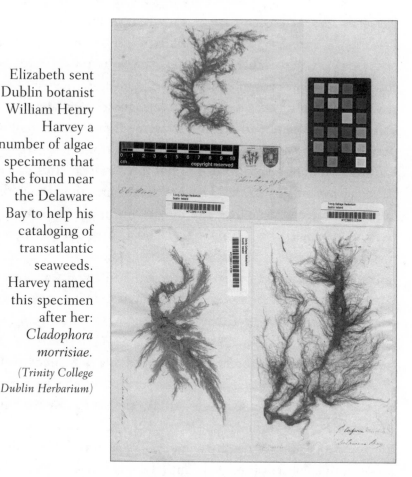

Elizabeth sent Dublin botanist William Henry Harvey a number of algae specimens that she found near the Delaware Bay to help his cataloging of transatlantic seaweeds. Harvey named this specimen after her: *Cladophora morrisiae.*

(*Trinity College Dublin Herbarium*)

the Philadelphia institution, a number of members—including James Coggswell Fisher and John Cassin, who had named the cicada species Margaretta discovered for themselves—sponsored her election to the Academy, nearly a decade after she had been elected to the American Association for the Advancement of Science. Given how many specimens and reports she had contributed to the Academy, this election would have been meaningful to Margaretta. Indeed, she would be only the second woman elected, after Lucy Say. Both Margaretta and Lucy Say, however, were never treated as equal members. Margaretta's membership listed her as "exempt," a designation typically given to honorary or corresponding members who lived overseas or out of state and therefore could not attend meetings.[16]

Perhaps the men who decided to categorize Margaretta as a corresponding member thought they were sparing her the cost of dues. However, she was summarily excluded from the meetings and from voting on the proceedings of the Academy. This designation also went against the bylaws of the organization, which required any member who lived in Philadelphia to be a dues-paying resident member. The entomologist John L. LeConte proposed that Margaretta be exempt from the requirements of the bylaws "as a mark of distinction," which the other members approved unanimously. Margaretta was elected, but not invited to participate. It was a distinction that further segregated her from her peers.[17]

The American Entomological Society, which shared many members with the Academy, began operation as the Entomological Society of Philadelphia in 1859—the same year that the Academy of Natural Sciences elected Morris. However, despite being both a leading agricultural entomologist and a Philadelphia resident, this collection of male entomologists never nominated Margaretta for a membership. Their records do not indicate that they even considered the possibility.[18]

As the antebellum American sciences began to centralize around universities and institutions—including the Smithsonian, Harvard's Museum of Comparative Zoology, the American Philosophical Society, and the Academy of Natural Sciences in Philadelphia—they were becoming increasingly competitive with their European counterparts. Still, remnants of old insecurities meant that the men who nominated others to join their organizations were not often eager to open their ranks to those they considered inferior because of their race, sex, or education. Optics outweighed the benefits of diverse membership or the contributions that might be made by the voices that they were silencing. The short-lived Entomological Society of Pennsylvania was so exclusive that they denied membership to all agricultural entomologists—let alone casual enthusiasts and collectors—instead only nominating their peers particularly interested in systematics and taxonomy. The American Entomological Society was ultimately more accessible, and included working-class collectors and enthusiasts, but they chose not to recognize Margaretta Morris as a peer.[19]

MARGARETTA WAS NOT SITTING STILL WHILE THESE MEN decided whether to bestow or withhold honors on her. Along with her sister Susan's twenty-four-year-old daughter, Harriet Hare Littell, Margaretta traveled south to Virginia in search of better health as the leaves began to turn in the fall of 1859. While Margaretta could only complain of a persistent cold, Harriet was suffering from a severe and mysterious back pain. Margaretta watched over her niece, administering hot plasters on her back in order to raise blisters that could be drained, monitoring her activities, and enforcing rest. That October they stayed for several weeks in Winchester, Virginia, on a plantation owned by John Page, a distant cousin and one of their great-aunt Mary Byrd's sons. The plantation was in the northernmost part of Virginia, close to Maryland, and about two hundred miles from Philadelphia. Harriet and Margaretta had a rejuvenating visit, joining their relatives in long conversations about family connections while playing backgammon in the parlor.[20]

While enslaved laborers tended to the table where Margaretta, Harriet, and their hosts dined, an uprising was stirring thirty miles northeast. The famous—even notorious—abolitionist John Brown had been quietly living in Maryland just over the border from Virginia for several months, growing increasingly frustrated with the passive tactics of abolitionist pacifism. He felt the anti-slavery movement needed more drastic action. After amassing a collection of weapons and like-minded men at the farm he rented, Brown planned to seize the federal armory at Harpers Ferry, and then travel through the South liberating the enslaved. From there he would head through Jefferson County in the direction of Winchester, where Margaretta was staying and where there were a significant number of enslaved people on plantations.[21]

Inspired by Black radicals and events like the Haitian Revolution and Nat Turner's rebellion, Brown hoped to ignite a slave revolt throughout the South. While he had the support of philanthropists, as well as fellow abolitionists like Harriet Tubman, he had a hard time recruiting others to join his cause. Frederick Douglass, whom he counted as a friend, declined to participate, deciding the mission was a "perfect steel trap" from which Brown and the others would not escape alive. Many potential allies hesitated to join him as the raid drew close and it became clear how impractical Brown's strategy was, given his failure to rally the local enslaved population ahead of time. Within thirty-six hours after Brown and his men seized the armory on October 16, 1859, militia companies battered down the doors and captured Brown and six other survivors. Nineteen men died in the struggle and its aftermath.[22]

Margaretta and Elizabeth rarely discussed politics in their letters and scrapbooks. Their stance on slavery and the growing fissures between North and South, let alone between abolitionists and slaveholders, is mostly lost. However, Margaretta's proximity to the foiled raid spurred her to recount the reality on the ground from her perspective as a visitor in the South. In a letter home to her younger sister Susan two weeks after the raid, Margaretta tried to ease her sisters' concerns. Having learned about the raid from newspapers as the story quickly spread like wildfire across the country, Susan was terrified about the experience her daughter and sister were having so close by. Margaretta's thoughts on the raid came several pages into the letter, after lengthy reports on Harriet's health and news of their distant cousins. "The excitement of the Harpers Ferry outrage is subsiding," Margaretta reported. She characterized John Brown as a fanatic. "I trust that the quiet conduct of the slaves will show the insane abolitionists in north and east, that they had better let them alone," she wrote. "The poor slaves have been more frightened than their Masters, and all who were feared to join made their escape home as soon as they were left unguarded." In her mind, Brown and his associates had badly misread the desires of the enslaved. Dorothea Dix similarly characterized the event as a "mad scheme" and evidence of the uncontrolled passions of Americans. Frederick Douglass, meanwhile, was busy penning articles countering the public opinions that Brown was a treasonous madman. He, along with Henry David Thoreau and others, argued that John Brown was a martyr for justice. "This age is too gross and sensual to appreciate his deeds, and so calls him mad," wrote Douglass, "but the future will write his epitaph upon the hearts of a people freed from slavery, because he struck the first effectual blow." Northerners' opinion would shift in Brown's favor in the weeks and months to come as his final interviews and speeches circulated.[23]

From Margaretta's privileged position as a guest at John Page's plantation, she tried to encapsulate what the enslaved truly wanted, though she was likely oblivious. When she told her nephew Gardiner Littell about the raid, referring to Brown and his comrades as "robbers," she reported, "Neither servant nor master believed the report of a rising among the Blacks, when it reached us, but went on as quietly hewing the wood and drawing the water as if nothing unusual was astir." Any discussion of the uprising among the enslaved would have been kept hushed. White southerners like Margaretta's cousin interpreted their inaction as contentment with the status quo, claiming that the lack of Black participation in the raid was testament to the affection between masters and the enslaved on plantations. It was further justification in their eyes for what they argued was a paternalistic system. Osborne Perry Anderson—a Black abolitionist who had visited slave quarters at other Virginia plantations during the raid to rally supporters and usher them to freedom—found accusations that "slaves were cowardly" and more in favor of staying with their masters "a gross imputation against them." He claimed instead to have been met with "general jubilee" when he arrived to free the enslaved. No one needed any coaxing to join the effort. There had been little time, however, for them to rally before the raid was stymied due to Brown's missteps.[24]

While John Brown may have failed, the legacy of his raid, trial, and subsequent hanging was incendiary, deepening the sectional crisis by both amplifying secessionist voices in the South and revealing stark divisions within the political parties throughout the country. Over the coming months, the sectional crisis and debates over the future of slavery in the United States would reach fever pitch. When Margaretta wrote home from Virginia that October, she certainly did not recognize how consequential John Brown's raid would prove to be when she dismissed the "madness of the whole prospect." However, as the abolitionist George William Curtis put it when Brown was hanged for treason, he was "not buried but planted." Foreshadowing what was to come, Curtis wrote: "He will spring up a hundred-fold."[25]

As the nation spiraled closer toward political crisis, Margaretta and Elizabeth continued to try to improve their health by traveling. In 1860 and 1861, Elizabeth spent months in the Allegheny Mountains near what is now State College, Pennsylvania, breathing in the pine-scented air and appreciating the "gorgeous autumnal traits & deep coloring" of the plants. She returned home, though, still sick and "as plump as a beanpole." She was suffering once again from "liver complaints" that caused debilitating nausea for months at a time. The effects were so severe that Margaretta at times feared that Elizabeth might die of it.[26]

The Morris sisters' travel would grind to a sudden halt, though, once the South Carolina militia and the United States Army exchanged gunfire at Fort Sumter on April 12, 1861, beginning the Civil War. Elizabeth, in a letter to William Darlington nearly three months later, discussed the "most unholy war." She made it clear that she did not identify with the "Black republicans and abolitionists" that she imagined southerners caricatured all northerners to be. Like many of the Philadelphia elite, the women had family all throughout the border states, and their wealth originated from the Willing and Shippen families who had trafficked and enslaved so many. As wealthy landowners, she and Margaretta were politically moderate and more likely to identify with landowning non-secessionist southerners, even if they were not specifically pro-slavery. Still, Elizabeth was horrified by the war and what its divisiveness meant for the country. "I do not believe the bitter animosity of the Southrons will ever permit them to return peaceably to their duties as members of our once glorious & beautiful Union—but—if they will not be good citizens of the Republic & stay quietly under the Flag that has protected them in every port in the civilized world, for so long—let them go!" Though the sisters did not take a radical stance against slavery in their letters, they passionately opposed secessionists breaking apart the Union. "They will find out their mistake one day or more properly, 'their sin will find them out,' & punishment will as surely follow," Elizabeth wrote.[27]

The Civil War shook the foundations of American society, socially, economically, and politically. It also temporarily muted the impact of the American publication of Charles Darwin's *On the Origin of Species* in late 1860. Some pirated copies had been circulating for months, but Asa Gray negotiated an arrangement for an official edition with D. Appleton and Company in New York on Darwin's behalf. The book would set off a revolution in the ways people understood the natural world. Murmurs of its impact were evident in the halls of Harvard and pages of science journals as Asa Gray and Louis Agassiz debated its theoretical underpinnings and theological implications. The *New York Times* reported that Darwin made "a series of arguments and inferences so revolutionary as, if established, to necessitate a radical reconstruction of the fundamental doctrines of natural history." However, with Americans' attention on the crisis facing the nation, the impact of Darwin's arguments would not be felt for years.[28]

The Civil War particularly rattled Philadelphia, given its proximity to the battlefields, and the political turmoil raging within the city itself. Philadelphia was very much a border city with segregated streetcars, anti-abolitionist rallies, and a strong Democratic Party presence. Still, the city rallied around Lincoln and the Union following the battle of Fort Sumter. Philadelphians decorated their homes, stores, and streetcars with red, white, and blue bunting. Elizabeth purchased patriotic stationery featuring a bald eagle perched on an American flag shield with "The Union and Constitution" written underneath, which she used liberally in her correspondence.[29]

More than expressing their patriotism with stationery, Margaretta and Elizabeth also aided Dorothea Dix, who had taken charge of the army's nurses once the war began. Though no nurse herself, Dorothea had risen to national prominence in recent decades thanks to her dogged devotion to reforming mental health asylums and prisons, which earned her some political standing with the Lincoln administration. She had been looking for a new cause to champion prior to the outbreak of war and saw an opportunity to support the Union army once the war broke out. Dorothea had long admired the work Florence Nightingale had done for England during the Crimean War, and perhaps hoped to become her American equivalent. She was, however, given very little funding to outfit the army's nurses and medical facilities, so she relied on the kind of personal fundraising she had been accustomed to tapping for her reform work. Margaretta, Elizabeth, their sister Susan, and their cousin Harriet Hare were all too eager to support Dorothea.[30]

Dorothea also found time to send notes to Margaretta, despite the demands of her position. Jokingly referring to herself as "General Dix," she sent two images of herself to Margaretta, with the instruction that Margaretta should choose the one she liked best and give the other to Elizabeth. Dorothea also asked if Margaretta might send her one pound of good coffee, roasted and ground. Signing the letter "yours affectionately," she turned the sheet over to scribble in her notoriously messy handwriting, "I enjoyed my visit with you greatly." While Dorothea's high standards and commanding personality may have vexed some of

Dorothea sent two images of herself to Margaretta, suggesting she choose her favorite and give the other to Elizabeth.

(*Houghton Library, Harvard University*)

the doctors and nurses she worked with, Margaretta and Elizabeth deeply admired her work and perseverance.[31]

Dorothea Dix was not the only person the Morris sisters knew who was directly involved in the war efforts, though she was perhaps their most intimate connection. Several of the young male naturalists from Philadelphia—including James Coggswell Fisher and John Lawrence LeConte—dusted off their medical degrees to become surgeons for the army. They were not alone. Many young scientists enlisted in both the Union and Confederate armies, while more established scientists occasionally commanded troops. Others deliberately avoided direct involvement. Regardless, many scientific projects were paused or lost funding, staffing, and other forms of support during the war. Louis Agassiz, for instance, complained that his assistants kept going off to war, leaving him short-staffed at Harvard's Museum of Comparative Zoology. Asa Gray, too, wrote to Charles Darwin, confessing, "I do not do so much scientific work as before the war,—but still I keep pottering away." Even the *Gardener's Monthly*—which had become the preeminent national magazine for horticulture enthusiasts over the last decade—struggled as it lost all its southern subscribers. Other scientific journals like Silliman's *American Journal of Science* similarly suffered from shrinking circulation. However, gardening and science could also serve as a welcome distraction for those safe enough to be able to momentarily put thoughts of the war aside. When the war started, Agassiz assured his anxious students at Harvard that "those who sought Nature always found relief if their minds were distressed."[32]

AS THE CIVIL WAR SHOWED NO SIGNS OF A QUICK CONCLUsion, the country wallowed in the violence, devastation, and loss of life plastered across the front pages of newspapers every day. It must have felt, then, like an extra blow to Elizabeth when William Darlington died in the spring of 1863, mere days before his eighty-first birthday. For years, William's morbid humor meant he constantly joked about his impending death and the indignities of aging, something Elizabeth always chided him for. "Never tell me you are getting too old for exertion," Elizabeth had written William a few years earlier. "I have known *many* aged persons, and have never seen one whose heart was fuller of kind and genial feelings, nor whose memory was more green." Still, both William and Elizabeth were aging, and jokes failed to halt the inevitable. Elizabeth's own health had been troubled, and their letters had slowed in the year leading up to his death. While Thomas Meehan, editor of *Gardener's Monthly*, hastily wrote an obituary for William in the magazine's May 1863 issue, he called on Elizabeth to help him write a more substantial and detailed account of her friend's life and work for later publication. Meehan would keep her anonymous, as she preferred, referring to her only as "one of his life-long friends."[33]

Given how popular William Darlington had been, it was not surprising that even in the midst of a war whose battlefront edged ever closer to Pennsylvania, his funeral was well attended. A long line of relatives and friends passed through his parlor to take a last look at his face and cover his chest in his favorite flowers. He had, after all, been a cheerful mentor and correspondent to countless botanists and collectors. As Elizabeth put it, "Dr. Darlington's constant effort was to render science practical and pleasing,—in every sense of the word, popular." Referred to in the obituary as "a lady friend from Philadelphia," Elizabeth gave an elaborate wreath of white flowers, directing his daughters to lay it on his chest, and then on his grave—the last botanical gift they would ever exchange. He had been one of her most cherished friends.[34]

Already in a fog of grief, Elizabeth and Margaretta read accounts in the *Philadelphia Inquirer* of the immense loss of life at Gettysburg at the start of July 1863. Even though the battlefield was more than one hundred miles away from Philadelphia, the

Confederate army had arrived at their doorstep, and the war was becoming less and less abstract. With more than fifty thousand dead after that battle, the sheer scale of lives lost at Gettysburg was inconceivable. The Morris sisters mourned the loss of their close friend alongside countless Americans mourning the loss of their loved ones.[35]

Determined to make a demonstrable impact on the war effort, Margaretta gladly attended Philadelphia's "Great Central Fair" in Logan Square in 1864 to support Dorothea Dix's work by raising funds for the Union's Sanitary Commission. The fair, run primarily by women, was inspired by others around the country, and featured local crafts, music, and food. The Black seamstress Emilie Davis visited the fair and was frustrated to find nothing by Black craftswomen for sale, but Margaretta likely did not even register that omission. She moved through the various displays, purchasing trinkets and souvenirs that included a slice of a branch from the Washington Elm in Cambridge, Massachusetts—a tree that had become part of Revolutionary War lore through stories about George Washington rallying troops beneath its canopy. Clinging to the patriotism of the Revolutionary era was a way to counter the divisiveness of a fractured nation. These fairs were wildly successful, raising millions of dollars for the Sanitary Commission, while also reinvigorating popular support for the war effort just as it was beginning to wane.[36]

Back in Germantown, however, Elizabeth's tenuous health prevented her from exploring the forests around the Wissahickon Creek, caught as they were in winter's icy grip. She longed to see the first tiny fronds of *Asplenium pinnatifidum* stretching out from rocky crevices as the earth began to thaw. "Every bright day makes me almost pant for the sweet breath of the woods," she wrote, "but the winds are yet too cold, and the ground too wet to make it quite safe for one who has so little to boast of in the way of health and strength as I—so I will try to be patient, and wait a little longer."

John Moran depicted the Wissahickon Creek around 1863 in this albumen print as an antidote to the Civil War ravaging the United States at the time. Margaretta and Elizabeth were not alone in their desire to escape to this space. *(National Gallery of Art)*

Her opportunities, however, grew fewer and fewer. Ever the consummate botanist, Elizabeth had seen nearly seven decades of plants bloom and decay, watching as her garden's vibrant green stems transformed into straw-yellow husks. In her last days, even as her health declined, she was at least able to admire the North American ferns she had cared for in the little Wardian case she kept in her bedroom. A heavy snowstorm began at four in the morning on February 12, 1865, and as it blanketed the Morris Hall garden, Elizabeth died from congestion of her lungs. She was sixty-nine years old.[37]

Elizabeth had fretted over what would be left of her life if Margaretta died. Now it was Margaretta who was alone. Elizabeth had always been at her side, collecting insects, combing the skies for meteors, and cultivating their backyard paradise together. Margaretta stood in the quiet parlor with the enormous potted camellias that Elizabeth wheeled into the house each winter, keeping them soaked in hopes of seeing their bright crimson flowers bloom in February. In the hours to follow, Margaretta, perhaps with her sister Susan, would have prepared Elizabeth's body in order to lay it out for the funeral. It was an intimate rite performed at home, and it would be Margaretta who washed her sister's body, scrubbing the dirt from her nails. Though Elizabeth prided herself on her meticulous hygiene, there would inevitably be some dirt—the dirt of a life spent making plants happy, digging in the garden, checking the moisture of pots on windowsills by wedging her finger deep into the cold soil. Until she could not physically manage it, Elizabeth would have been tending to her plants.[38]

Margaretta's care for Elizabeth's remains was far from unusual. The vast majority of Americans died at home, and their bodies were prepared by family rather than professional undertakers. Mourners held services with their rooms decorated in black crepe. The Civil War would change these rituals, with the introduction of embalming and funeral parlors, but Elizabeth's funeral would be held at Morris Hall. Margaretta purchased a casket for Elizabeth made by local cabinetmakers three days after her death and arranged it in the parlor beside her sister's beloved camellias. Once friends and family had paid their respects, she buried her sister beneath a young elm tree in the frozen earth blanketed in snow at the family plot at St. Luke's Episcopal Church, just down the street from their home.[39]

While Elizabeth had preferred the obscurity of pseudonyms to the limelight, her friend Thomas Meehan was determined to pay tribute to her. Not only had she written countless articles for the *Gardener's Monthly* up through the end of her life, but she had introduced him to Asa Gray and William Darlington. She was a reliable customer at his nursery and was generous with her expertise in helping plants thrive. In an article with the headline "Death of One of Our Contributors," he celebrated Elizabeth and all she had done for the botanical community. "Our readers have suffered a severe loss by the death of Miss Elizabeth C. Morris, of Germantown, who, under the signatures of E., E. C., M., and other initials, contributed some of the most valuable articles that have appeared in our pages." She would have hated being outed to readers. Still, Meehan went on, "Highly educated and accomplished,—a superior Botanist, and an excellent Horticulturist, we lose her from our circle with sincere regret; a feeling that will be shared in by numbers all over the Union, to whom she has been long known, and highly respected and esteemed." After a lifetime of hiding from both criticism and praise, Elizabeth was at last commemorated for her contributions to botanical science.[40]

Elizabeth left everything—her books, her herbarium, her clothing, stocks, properties, furniture, the house—to Margaretta. She stipulated in her brief will that Margaretta could then distribute their combined treasures after her own death or when she found it convenient. It would take Margaretta another year and a half to sit down and write her own will after she, too, turned sixty-nine, taking stock of the things that had accumulated in their house over the years, as it once again grew still and empty in Elizabeth's absence.[41]

Margaretta was mourning. The poems that she and Elizabeth had transcribed into their albums often tied inner emotional weather to the weather outside, not unlike their understandings of how human health was tied to their surrounding environments. As Margaretta made her way through the months to come—the end of the Civil War and Lincoln's assassination soon after, even the bittersweet blooming of Elizabeth's carefully tended garden that spring—it may have struck her as symbolic when a massive rainstorm swept through Philadelphia the following July. The *Philadelphia Inquirer* reported that it was the "hardest rain which has occurred for many years." The Wissahickon flooded dramatically, rising at least seven feet and taking with it several bridges, houses, trees, and lives. The flood caused immense damage to the many mills that lined the creek's shores, as the water rushed through machinery and swept away bundles of paper, wool, and lumber. The dye vats from a destroyed woolen mill careened down the creek like wayward boats, slamming into submerged trees in their path. While journalists and local politicians were predominantly concerned with the costs of replacing bridges and the financial ruin facing mill and lumberyard owners, Margaretta would have ached for the woods where she and her sister had spent so much time.[42]

That August, just six months after her sister had died and a few weeks after the flood, Margaretta took some time to copy part of a poem titled "Evening Hour" by the English poet Margaret Harries Wilson into her album.

> *This is the hour when memory wakes,*
> *Visions of joy that could not last.*
> *This is the hour when fancy takes,*
> *A survey of the past.*

It was a nostalgic poem meant for someone longing for a life already lived. "She brings before the pensive mind / The hallowed scenes of earlier years," the poem continued. While now doted upon by her many nieces and nephews, Margaretta still felt keenly all she had lost when she lost Elizabeth, and now even their favorite forest trails were weather-worn and irrevocably altered.

The few we liked, the one we loved,
A sacred band comes stealing on,
And many a friend for hence removed,
And many a pleasure gone.

Friendships in death that long are hushed,
And young affections broken chain,
And hopes that Fates too quickly crushed,
In memory bloom again.

After Elizabeth's death, Margaretta retreated inward. She never finished the catalog of Philadelphia's insects she had been working on. She even stopped publishing articles in agricultural journals. Perhaps her own precarious state of health wore away at her drive to publish. Perhaps it was her grief. It was harder to spend time at her worktable in the library that was now too quiet without her sister's chatter.[43]

On the first anniversary of Elizabeth's death, Margaretta sat down with her marbled album, a book she had been filling for decades with paintings and poems. At the bottom of one of the last pages of her book, Margaretta sketched and then painted a maple leaf at the height of its autumnal color. The leaf was golden with fiery red edges. A tiny hint of the leaf's youthful green lingered on a small part of the bottom edge. Both Elizabeth and Margaretta had come into their prime as scientists in midlife, reaching their peak colors just as they were about to fall from the tree. Perhaps Margaretta had chosen a leaf from a tree that was significant to Elizabeth, perhaps it was from a tree in their garden or along the Wissahickon. Or perhaps Margaretta was just thinking about the seasons of her sister's life and her own.[44]

After a long two years without her sister by her side, Margaretta died on May 29, 1867. Elizabeth could not be there to bear witness to her final words, or tend to her on her deathbed in the ways she had cared for her sister over the course of their lifetimes. Still, Margaretta would not have been alone, with her sister Susan and all of her nieces and nephews attending to her needs in her final days and weeks. Although her deteriorating health would have sapped her of the strength to walk along the Wissahickon, in death she could hopefully return to "haunt the leafy forest, where dwells my little friends the Insects, and there I have longed to be." As the family stood beneath the tree where they had recently buried Elizabeth, placing flowers on Margaretta's casket, perhaps there were some stirring cicada larvae in the pile of dirt the gravedigger had unearthed. Margaretta died one year shy of the seventeen-year cicada emergence in Philadelphia, an event she would have longed to witness just one more time.[45]

Margaretta carefully painted a leaf in honor of the first anniversary of Elizabeth's death.
(Collection of S. C. Doak)

Eleven

FORGETTING

IN THE CENTURY AND A HALF SINCE MARGARETTA AND ELIZ-abeth died, the Wissahickon Creek has continued to wind its way through the forest that they knew so well. Trees have fallen and saplings have grown, with newer introductions like Empress trees (*Paulownia tomentosa*) edging in amid the older oaks and beeches. The clanging and whirring mills that once drew their power from the creek and drowned out the sounds of songbirds are long gone, with the exception of those left in ruins or maintained as historic sites to host parties and teach children about the Industrial Revolution. Many of the mill owners decided to sell their properties once they realized the extensive damage from the 1865 flood made it impossible for them to recoup their losses. In the months after Margaretta's death, the Fairmount Park Commission would suggest expanding Philadelphia's biggest park to include the Wissahickon and the forest on either side, in hopes of securing pure water for the city's future needs. While the Commission struggled at first to get funding, by the next year they had successfully added a significant portion of the creek to the city's park system and ensured the creek's preservation. In many ways, the forest appears wilder now than it did when Margaretta and Elizabeth scaled its boulders and hiked the steep, root-tangled paths sparkling with glittery mica.[1]

While the Wissahickon is thriving, our collective memory of Margaretta and Elizabeth has faded. When I first stumbled upon the Morris sisters in the archives, I was genuinely surprised by Margaretta's obscurity. Elizabeth may have buried herself beneath her pseudonyms and preference for supporting the careers of others, but Margaretta was one of the first women elected to America's major scientific organizations. She published widely. She was in conversation with Charles Darwin, Louis Agassiz, and others. These were the types of accolades and connections that typically merit more than a passing reference in histories of science.

Margaretta and Elizabeth made very different decisions about how public they wanted to be with their scientific work, and yet both, nearly equally, disappeared from the historical record. This, in large part, has to do with the shifting scientific landscape over the nineteenth and twentieth centuries as the profession formalized. It also involved a number of idiosyncratic decisions made by entomologists, archivists, and historic preservationists. The Morris sisters' respective erasures were in some ways specific to them, thanks to the ways that botany and entomology developed in the centuries after their deaths, or even thanks to local Germantown politics. However, their absence also speaks to the reasons why so many marginalized scientists from this period remain unknown.

More curiously still, not every woman scientist disappeared from our stories. Maria Mitchell—the astronomer who was elected alongside Margaretta to the American Association for the Advancement of Science in 1850—is an example of someone who has been remembered. Though not exactly a household name, she has had multiple biographies, articles, children's books, and even a historic house dedicated to telling her story.

Margaretta and Elizabeth were not so lucky. Margaretta did not even have an obituary.[2]

WHEN NOTABLE SCIENTISTS DIED IN THE MIDDLE OF THE nineteenth century, it was customary for other scientists to honor their peers' lives and accomplishments in writing as a way of marking the significance of their intellectual contributions to science and culture. These memorials, or "memoirs" as they were typically called, were one of the many ways that scientists constructed the history of their profession and reflected upon the state of the field. They were also a way to enshrine their friends and colleagues— and, by association, themselves—as the "founding fathers" of American science. Scientists published these short biographies in journals like Silliman's *American Journal of Science and Arts*, as individual pamphlets or collected together with other memorials for that person in books.

Prominent scientists who published extensively and were members of professional organizations typically received this treatment. In the decade after the entomologist Thaddeus William Harris died in 1856, for instance, former students and colleagues from the Boston Society of Natural History not only wrote his biography but published it alongside a collection of the letters he exchanged with other entomologists, Margaretta included. Despite the spiritualism scandal that shadowed Robert Hare's final years, Margaretta and Elizabeth's cousin was honored with a number of obituaries and long-form memorials upon his death in 1858. William Darlington, who had a much less complicated legacy, also inspired several memorials. Elizabeth had likely prepared to contribute her letters from William Darlington to a collected volume since all of the letters he sent her—letters she cherished— are missing from the archives, perhaps lost to whoever had been organizing the ultimately unpublished volume. Scientists who left behind widows often fared best of all, as many of those women secured their husbands' legacies by writing books and organizing archival collections, as was the case with Jane Loring Gray, Elizabeth Cabot Agassiz, and many others.[3]

When Margaretta died in May 1867, she received no such tribute. The Academy of Natural Sciences announced her death at their next meeting but included no details about her life or accomplishments. To pass someone over in this way was both a statement about her place in the community and an assessment of her lasting significance. Entomologists and other naturalists did not consider her a peer, let alone a "founding father."[4]

These omissions have a rippling effect. For historians making sense of the state of early American science in the generations that followed, memorials were a jumping-off point, a first draft of the historical narrative, and a method of determining the contours of a life and the influence someone had on their field. Authors of these memorials inserted tidbits gathered from conversations and observations. They commented on personality quirks and physical appearance—the kinds of humanizing notes often absent in other types of sources. Without a memorial, historians would not know to include Margaretta in their narratives about the transformation of antebellum American science, the origins of agricultural entomology, or even notable American women of the period.

The erasure of the Morris sisters did not happen all at once. The absence of a memorial or obituary for Margaretta certainly contributed, but the omissions occurred unevenly, and some had even begun during their lifetimes. Many were not intentional, let alone malicious, but they gradually accumulated until the sisters' stories faded from view. While that forgetting might feel like a loss today, some men of science would not have considered the Morris sisters' contributions significant enough to lose. Although the process of professionalization depended heavily on women scientists' work, that work was generally viewed as supplemental or subordinate. In addition, the presence of scientists across lines of gender, race, and class presented a complication for men trying to advance their own careers. As scientific culture became more uniform and exclusive, it was in their best interest to homogenize the history of the profession in their own image by centering themselves. It is no wonder then that W. Conner Sorensen would title his 1995 history of American entomology *Brethren of the Net*, managing to write just six pages about the work of women in the field. It is not because women were not entomologists; they were just hard to locate in the sources he used. They were both left off the record and obscured in the archives.[5]

Part of Margaretta's erasure began with a lack of citation. References and citations are easy to overlook but they are how these entomologists passed along knowledge and validated one another. As we saw in 1841, Thaddeus William Harris had dismissed her observations about the wheat flies, claiming she had erroneously revived an "old discussion." This was further compounded in 1847, when Asa Fitch compiled his self-proclaimed "complete summary of all that is known" of the Hessian fly in the United States, only partially including Margaretta's publications. This had infuriated her because Fitch argued that she must have known her theory was wrong if she stopped publishing about it. Margaretta had contested these characterizations by writing to Harris and publishing an open letter in the *American Journal of Agriculture and Science* calling Fitch out, but these corrections ultimately had little effect.[6]

While Fitch would eventually apologize and Harris would come to admit that Margaretta had discovered a new species of *Cecidomyia*, their initial dismissals overshadowed their retractions. After she officially named the new species *Cecidomyia culmicola* (Morris) in 1849, entomologists in North America and Europe took notice, and many celebrated Margaretta's discovery as having been "triumphantly established." Entomologists included the *C. culmicola* in agricultural textbooks, state agricultural reports, and a Smithsonian report on *Diptera* (flies), among other publications. Still, when new infestations occurred in the decades that followed, they relied heavily on Fitch's definitive pamphlet when they referred to how ideas about wheat flies changed over time. With the specimens Margaretta deposited in the Academy of Natural Sciences long destroyed, there was nothing left to confirm her observations. When accounts differed, many men chose to trust Fitch's assessment.[7]

As increasing numbers of reports by the Department of the Interior and the Department of Agriculture cited Fitch's account, his legacy and authority grew. French entomologists—relying solely on publications rather than fieldwork in North America—judged Margaretta Morris as mistaken. That, in turn, influenced American entomologists in 1911 who once again suggested that Margaretta had conflated different insects. These reports, in turn, made the British entomologist H. F. Barnes confident in his decision to reiterate these claims in his heavily cited reference book for agricultural entomologists in 1954. He had more than a century's worth of skeptics to reference and no inclination to question their authority. By the time Raymond J. Gagne wrote *The Plant-Feeding Gall Midges of North America* in 1989, Morris and the *C. culmicola* had been omitted altogether. Fitch's and Harris's efforts to discredit her observations rippled outward across generations until the species and its researcher vanished from the literature and we lost the chance to better understand a wheat pest.[8]

Margaretta's wheat fly discoveries were not the only ones that were missed. Her 1850 theory that a tiny bark beetle (*Tomicus liminaris*) was the cause of the Peach Yellows proved incorrect, but she had been on the right track. Peach Yellows had been devastating American orchards since the end of the eighteenth century, sometimes killing an entire orchard of peach trees in just one or two seasons. It continued to be disastrous in the decades after

Margaretta's death. Several states, beginning with Michigan in 1875, enacted Peach Yellows Laws that required the destruction of any infected tree or fruit in hopes of containing the disease. The Department of Agriculture hired Erwin F. Smith to make a comprehensive report on the Peach Yellows in 1888 in hopes that he would aggregate enough information to identify a definitive cause. In nearly two hundred pages, Smith listed all documented theories, quoting sources ranging from the famous landscape architect Andrew Jackson Downing to little-known orchard owners who wrote in to magazines. Though he referenced other articles from the journals where Margaretta published her theory, and though her articles on the Peach Yellows were reprinted broadly, he chose not to include them. Smith's reasons for this are unclear, but it had the effect of invalidating and burying Margaretta's findings. He never did determine the source of the Yellows, but Margaretta's theory would have gotten him closer to the truth. Margaretta had argued that the bark beetles moved from tree to tree, explaining how they might make Peach Yellows seem like a contagion. In 1933, the plant pathologist L. O. Kunkel determined that a different insect, the plum leafhopper, spread the infection (which was still later determined to be a phytoplasma) as it flew between trees, serving as a vector.[9]

While Margaretta's peach bark beetles—now known as *Phloeotribus liminaris*—did not cause the Yellows, they remained incredibly destructive and had continued ravaging peach, cherry, and plum trees. In 1909, over half a century after she had initially published her articles about the powerful beetle's devastating tendency to burrow under bark and drain peach trees of their sap, the entomologist H. F. Wilson wrote a report on the creature for the USDA. He gave Margaretta full credit for being the first to write about it, citing her many articles in his bibliography. Her early work documenting the behavior of the beetle had given

Wilson and the entomologists at the Ohio Agricultural Experiment Station a head start as they sought ways to stave off disaster for American fruit crops.[10]

Unlike Margaretta, Elizabeth was instrumental in obscuring some of her own legacy. While Margaretta chose to engage in public debates with farmers and entomologists, Elizabeth preferred to stay anonymous. She felt freer to express herself, she argued, when protected by pseudonyms. In a culture where it was rare for women to step boldly into the public sphere, there would have been ambient if not explicit pressure to stay anonymous. Regardless, Elizabeth's decision to use pseudonyms and avoid credit for her articles and illustrations meant that her contributions are hard to trace back to her. She was less interested in attribution than in supporting the careers of men like William Darlington and Asa Gray through labor that, though essential, would ultimately be considered unremarkable.

Other erasures were outside of Elizabeth's control. When she collected specimens for Asa Gray or William Darlington, who in turn distributed those specimens to other botanists, sometimes the collector's information on herbarium sheets was lost. Botanists occasionally transferred sheets, relabeling them and losing details in the process. Other times, details about enclosed specimens were found in letters and the recipients had to label the pages themselves. Some specimen sheets list her as "E. C. Morris," others as "Miss Morris." As universities and institutions have digitized their herbaria, making their databases and images of specimens available globally, "Miss Morris" has sometimes been captured simply as "Morris," further obscuring the identity of the collector, particularly when other identifying details like dates or places are missing. Elizabeth, for instance, seems to have collected a number of lichens in 1846 along the Wissahickon that she gifted to Asa Gray who, in turn, gifted them to his mentor John Torrey. Torrey later donated his herbarium to Columbia College in 1860, which then moved their collection to the New York Botanical Garden in 1898. Through all these transfers, the lichen's origin with Elizabeth grew more tenuous. If a researcher were to start with the specimen sheet alone, the name "Morris" would mean very little.[11]

Still, even if it is hard to trace all of them back to her, Elizabeth's specimens—collected from her garden, grown from seeds sent to her from around the world, and picked along roadsides on trips to visit friends—persist in the archives. Thanks to the gift economy among botanists and the movement of collections from institution to institution, Elizabeth's herbarium sheets can still be found at Harvard University, Kew Gardens, the Field Museum, Trinity College Dublin, Delaware State University, Miami University, the New York Botanical Garden, West Chester University, the Missouri Botanical Gardens, and likely in many other herbaria. Biologists are now revisiting herbaria for new insight into species diversity changes and the impact of climate change, among other pressing issues, so there is a chance Elizabeth Morris's carefully collected specimens will continue to contribute to the advancement of science and our collective understanding of the environment.[12]

IN ADDITION TO THE SPECIFIC WAYS THAT THE SISTERS' CONtributions were obscured by uneven citation practices and sparse recordkeeping, larger cultural shifts within the field of biology in the century after their deaths further buried their work. By the mid-twentieth century, histories of biological sciences made natural history methods and the collection of specimens seem outdated and amateurish. These narratives privileged the shift toward experimentation taking place in twentieth-century laboratories, highlighting advancements in molecular biology most of all. In addition to this, male biologists felt pressure to resist the emasculating stereotypes associated with botany and biology more generally. The naturalists of the twentieth century, fighting for professional respectability and competing for limited resources in universities, therefore had little incentive to celebrate the lives of two women from the nineteenth century, let alone the many women who collected and donated specimens, illustrated articles and books, and participated in the nineteenth-century correspondence networks. Their stories would not serve the interests of those trying to be taken seriously by their colleagues.[13]

Beyond this, agricultural entomology—alternatively called applied or economic entomology—formally separated from biology departments in colleges and universities, and Margaretta's foundational work became even less visible in the larger story about American entomology. In 1862, in the middle of the Civil War, Republican legislators were able to pass the Morrill Act thanks to the secession of southern Democrats who could no longer block the vote. This provided support for the opening of land-grant universities that focused primarily on agricultural sciences. With the addition of the 1887 Hatch Act, which created state agricultural experiment stations, the United States government significantly elevated and helped to professionalize agricultural entomology, but it also had the effect of siloing different types of entomologists from each other, altering community dynamics.[14]

When agricultural entomologists wrote about the origin of their specific field, Margaretta's name was conspicuously absent. Francis Marion Webster actually mentioned Margaretta in his narrative, but he was an exception. More typical were histories like the one written by L. O. Howard, the chief of the USDA's Bureau of Entomology, that credited Thaddeus William Harris and Asa Fitch as founding fathers of the field. In 1947, after a century of omissions, the entomologist H. B. Weiss was shocked to encounter articles written by women about insects. In his brief essay "Early Feminine Entomologists," Weiss wrote about the work of Margaretta, Dorothea Dix, and Charlotte de Bernier

Taylor, a science writer from Georgia, and celebrated them as exceptional, but this did little to affect the dominant narratives about the brotherhood of entomologists.[15]

These narratives in the history of biology to some degree also help to explain why the astronomer Maria Mitchell is better remembered than Margaretta. As an astronomer, Mitchell did not face the same kinds of narrative erasures that would happen to naturalists who were women. Astronomy, as a higher-profile science, was not fighting for its legitimacy, and so perhaps there was room for inclusion of more diverse figures. Mitchell had also become an international celebrity in 1847 after she famously spotted her comet, winning not only a gold medal from the king of Denmark but also legitimacy for Harvard's observatory, where she and her father reported her findings. Other factors also played a role. Mitchell was a generation younger than Margaretta and outlived her by over twenty years. In the decades after the Civil War, she became a professor of astronomy at Vassar College, a credential that gave her a public platform. In addition, her advocacy for and mentoring of women in science, her involvement in the antislavery and suffrage movements, and her role in founding the American Association for the Advancement of Women all contributed to her public profile and lasting legitimacy. After her death, her former students joined with family members to establish the Maria Mitchell Association in 1902 to preserve her house and archives on Nantucket. They safeguarded her legacy.[16]

The Morris sisters, on the other hand, faced several archival erasures. While most of their letters are tucked in the papers of male scientists, sometimes their letters were not preserved at all. For instance, while Elizabeth saved fifty-five letters that she received from Asa Gray, there are only two from Elizabeth at Harvard. It is hard to determine precisely how Elizabeth's letters were lost. Maybe Asa destroyed them in order to conceal their more intimate character from his new wife. In one of his letters to Elizabeth shortly after his wedding, he mentioned how Jane "claims the privilege of looking over your late letters," which may have served as a warning. Jane Gray herself may have deliberately removed them, as she organized his archives after his death in 1888. Perhaps it was Mary Day, the librarian of the Gray Herbarium who curated the collection at the turn of the twentieth century and assessed which correspondents were notable enough to keep in the collection. It is also possible that the letters were not deliberately destroyed but instead accidentally misplaced, although Asa consistently saved his letters, and his collection at the Gray Herbarium contains many complete exchanges with scientists from around the world. The absence of Elizabeth's letters, in other words, is notable.[17]

Archives have a tendency to reflect cultural power structures at the moment when the collections are acquired. How archivists decide what to collect, how to name a collection, and who to highlight all have ramifications for the types of history that get told. Power is also easier to market. In 1999, John Johnson, a rare books dealer in Vermont who specialized in natural history, prepared a collection of letters Margaretta and Elizabeth had received from a number of scientists to sell to archives. Rather than call them the Margaretta and Elizabeth Morris Papers, he decided to name the collection "Letters by Dr. Asa Gray," since a sizable portion were letters from Asa to Elizabeth. The other letters, however, have absolutely nothing to do with Asa Gray. The collection's sole common denominator is the Morris sisters. Still, the best way to position the collection was to connect it to the most notable name in the pile, and Asa Gray, as a professor from Harvard and friend of Charles Darwin, was easily the most marketable choice. When the Library of Congress bought the papers, the science specialist who purchased the collection and the archivist who processed it decided to follow John Johnson's lead and call it the "Asa Gray Papers." Given that there is a very limited listing of the collection's contents, this particular archival naming decision further obscured the legacy of the sisters, making them even less discoverable.[18]

Historical sources include more than library and museum collections, letters and scientific specimens. Places and environments hold their own historical records in the form of landscapes and structures. When a community preserves these spaces and interprets the site's significance, it has an immediate impact on the afterlife of the history that took place there. The fate of Margaretta and Elizabeth's home and garden therefore also played a role in the erasure of their legacy. Margaretta left Morris Hall to her younger sister Susan Littell, at which point it became the "Morris-Littell House." While it stayed in the family for two decades more, a neighbor, Edgar H. Butler, bought it in 1885, happy to expand his estate by using the home to house his in-laws. The Butlers kept an impressive greenhouse on their property and seem to have maintained the Morris family's garden.[19]

By the turn of the twentieth century, however, Germantown had grown so large that local families began to lobby the city for a public high school so that nearby students would not have to travel all the way into central Philadelphia for school. Fears of what might happen particularly to Germantown's middle-class white adolescent girls traveling alone on streetcars fueled the public pressure for the school. In 1913, the school board identified the spacious lot that included the Morris-Littell House as an ideal site for Germantown High School, forcing a reluctant Butler to sell his estate for $150,000 by threatening him with condemnation proceedings.[20]

As soon as the school district purchased the property, the Germantown Site and Relic Society, the Germantown Horticultural Society, and the Germantown and Chestnut Hill Improvement Association intervened to ensure that the Board of Education's plans for the new school would preserve the Morris-Littell House and a number of healthy, thriving trees on the property. Journalists and preservationists celebrated the trees and the house for having witnessed Germantown's glorious past: German colonists settling in, Revolutionary War soldiers marching past, even the Morris sisters making scientific discoveries. Under the watchful eyes of so many local residents and organizations, the Board of Education promised that they would preserve the house and as many of the trees as possible.[21]

Not content to rely on promises alone, the Site and Relic Society pursued a strategic national publicity campaign, circulating a press release to a variety of newspapers in the summer of 1913. Published in dozens of papers, their article celebrated the Morris-Littell House's storied past, linking it as best they could to celebrities from Christopher Witt—the German pietist who originally planted the botanical garden in the eighteenth century—to the Morris family, who were the "descendants of Charles Willing, mayor of Philadelphia." Some of the Society's claims were exaggerated in order to underscore the house's significance. For instance, the article emphasized that Ann Willing Morris was friends with Dolley Madison (a stretch at best) and stated that Margaretta was "the first woman elected a member of the Academy of Natural Sciences" (Lucy Say was). The local preservationists were doing their best to drum up national enthusiasm to save the old house.[22]

At the turn of the twentieth century, historic preservation in the United States was fueled by Progressive Era reformers and women's groups, focused primarily on protecting sites tied to American patriotism. As is true today, historic preservation reflects the politics and sensibilities of preservationists at that moment, and in the early twentieth century they privileged homes connected to colonial or Revolutionary War history with a clear moral or patriotic message. If a property could claim that George

Washington had slept there, all the better. The Morris-Littell House could not claim a Washington or even Revolutionary War connection, so the Site and Relic Society was grasping at anything they could—colonial connections, mayoral connections, even Dolley Madison connections—in hopes of making their case.[23]

The thousands of community members in the audience at the Germantown High School cornerstone ceremony in 1914 might have believed the preservationists had won their campaign when they heard Martin Brumbaugh's speech. Brumbaugh, former superintendent of Philadelphia schools, gestured to the house and declared, "On a part of this ground, if I am correctly advised, still stands an interesting old building which I trust no hand will ever remove from its sacred setting." His speech emphasized Germantown's history of educational innovation and the various scholars who had lived there. However, when he spoke about the history and significance of the Morris-Littell House, he made no mention of the Morris sisters.[24]

Brumbaugh may have omitted them from his speech for any number of reasons. He was running for governor of Pennsylvania at the time and women's suffrage was a divisive topic. In Germantown alone, there were frequent meetings, lectures, and parades both for and against women getting the vote. While Brumbaugh himself supported suffrage, it may have been too politically polarizing for a candidate trying to draw broad support from voters. Spotlighting two pioneering women scientists might have edged too close to the contentious debates happening in 1914. It was safer and more uniformly popular to draw on nostalgia for Germantown's storied past.[25]

Another reason for omitting the Morris sisters involves the dentist and local antiquarian Naaman H. Keyser. One of the founders of the Site and Relic Society, Keyser was a preservationist and had participated in the campaign to save the house. He was something of a local leader and he had coauthored the

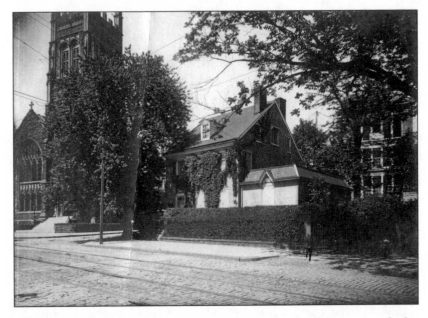

In early 1915 it seemed as though the Morris-Littell House might be preserved as a classroom space for Germantown High School, which is seen behind it. *(Germantown Historical Society)*

History of Old Germantown several years earlier. He therefore played a central role in the ceremony, choosing which documents would go inside the cornerstone. He also seems to have advised Brumbaugh on the history of the site and Germantown more broadly for the sake of his speech. Keyser was never particularly impressed by the stories people told about the women scientists, and he may have advised against foregrounding Margaretta, Elizabeth, and their scientific accomplishments in the speech.[26]

Omissions aside, the Site and Relic Society celebrated what they believed was a preservation win. Not only had they gotten the Board of Education to promise to preserve the house, they had secured the gubernatorial candidate's endorsement. The Society therefore commissioned and installed a copper plaque on the Morris-Littell House to testify to the site's history. The plaque not only spoke of Christopher Witt, the German mystic, but also informed passersby about Margaretta and Elizabeth Morris:

On this site a Botanical Garden
one of the first in America, was Planted by
DR. CHRISTOPHER WITT
Botanist, Mystic and Physician
Born, Wiltshire, England. Nov. 10, 1675
Died, Germantown. Jan. 30, 1765.

———

Later on the same site lived
ELIZABETH CARRINGTON MORRIS
Botanist and her sister
MARGARETTA HARE MORRIS
who here investigated and discovered the
life habits of the "Seventeen Year Locust"
and who was an active woman member of the
Academy of Natural Sciences of Philadelphia

Site and Relic Society of Germantown[27]

Unbeknownst to his fellow members of the Site and Relic Society, Naaman Keyser had different plans for the Morris-Littell House once Germantown High School had been fully built. He lived across the street from the new school and regretted his organization's fight to save the old house. He felt the juxtaposition of old and new took away from the grandeur of the high school's architecture. After learning that the school district's superintendent of buildings had complained about the expense of maintaining the historic structure, Keyser saw an opportunity to reverse course. In the spring of 1915, he circulated a petition calling for the Board of Education to demolish the home, shocking his fellow members of the Site and Relic Society. Ironically, he had been preparing the second volume of *History of Old Germantown* and had highlighted the significance of the Morris sisters to American science in his research notes. However, after further research into the house's deeds, he did not think that Christopher Witt had lived in the main part of the house, and without the Witt connection, he felt the house had "little historical significance." Still, as Keyser knocked on the doors of his neighbors, he was only able to collect sixty signatures. The leaders of the Methodist church right across the street from the Morris-Littell House were happy to add their support, in hopes of having better sightlines on Germantown Avenue.[28]

The Site and Relic Society quickly organized a pro-preservation postcard campaign with their eight hundred members. Both the Germantown Horticultural Society and the Germantown and Chestnut Hill Improvement Association wrote statements. This time, however, their efforts failed. The school district, with the support of Keyser and his petition, began to demolish the house a few months later in July 1915, saving an ornate mantel, the front door, and a set of corner cupboards from the dining room to donate to the Site and Relic Society in hopes of placating the preservationists. The Board of Education similarly failed to preserve most of the trees on the site, prioritizing ease of construction over plans that might have saved them. Without more substantive preservation laws, it was difficult for the local residents and civic groups to do more than secure handshake promises that were easily broken.[29]

So much of what remained of the Morris sisters' garden had already been trampled and torn up by steam shovels and workers. To make space for the school, workers had uprooted the apple and pear trees that had fed thousands of cicadas and smashed ferns and forget-me-nots under construction debris. However, the loss of the building and any geographic memory of the sisters' presence was an even larger erasure.

The construction of Germantown High School in 1914, seen here, had already involved the destruction of the Morris sisters' garden. The demolition of their house in 1915 was an even more significant erasure.

(Courtesy of PhillyHistory.org, a project of the Philadelphia Department of Records)

The house alone would not have necessarily preserved their legacy, but its physical existence with the plaque would have conveyed to students and visitors that there had been early American women scientists. It might have inspired historians and biographers to investigate their lives more fully. It likely never would have become a historic house museum since it was owned by the school district, but the initial plan had been for it to serve as a space for home economics classes. Given how much Elizabeth Morris had anonymously published on the topic in the 1840s, this would have been particularly fitting, even if only coincidentally so. A local journalist lamented the irony of the loss: "The quaint little house on the corner of the property will be swept aside, and girls of tomorrow will study biology where sixty years ago the first woman to be admitted to membership of the Academy of Natural Science lived and made her investigations."[30]

The plaque that the Site and Relic Society had put on the house just months earlier was taken down when demolition began and eventually attached to a stone post near where the house had been. The school removed the plaque in the final decades of the twentieth century and returned it to the Site and Relic Society, now known as the Germantown Historical Society, where it still remains tucked behind an HVAC unit.[31]

ALL OF THESE ERASURES—THE LACK OF OBITUARY, THE footnotes, the archival collections, the house—added up until the Morris sisters were more or less forgotten. Blame cannot be placed with any one individual, whether an insecure entomologist in 1847, a rogue preservationist in 1915, or a rare books dealer in 1999. If we make any specific person the villain, it is harder to view these erasures as part of larger patterns. The inequality was systemic, woven into the fabric of American culture. Power, whether in the form of sexism, racism, classism, colonialism, or nativism, impacted the creation of archival collections, the naming of species, decisions about which buildings to save and which to demolish, even snap judgments about which researchers seemed trustworthy. This is how so much knowledge and so many stories have been lost. This is why we keep uncovering hidden figures and discovering how we have undervalued the environmental knowledge of so many people in the past.[32]

Sometimes obscuring knowledge suits those in power. Historians Robert Proctor and Londa Schiebinger coined the term "agnotology" to mean the cultural production of ignorance, and it certainly applies in this case. We continue to be surprised by newly rediscovered women scientists because so few of the sciences' origin stories included them. No story can possibly contain everything. Storytellers, historians among them, necessarily need to edit the narratives they tell lest things get too unwieldy and hard to follow. Unfortunately, stories about Margaretta and Elizabeth and other women like them were some of those repeatedly discarded. Those omissions continued until few writers were aware of them at all. While some people may have found benefits to erasing the Morris sisters in the short term, whether to have a simpler story about a fly behaving a certain way, to earn better profits on an archival collection, to write a heroic narrative about male scientists, or to gain better sightlines down Germantown Avenue, mostly there was just a loss. There was a loss of environmental knowledge, a loss of observations about species diversity, even a loss of stories about how common it was for women to be passionate about the sciences in the nineteenth century.[33]

PHYSICAL EVIDENCE OF THE MORRIS SISTERS IS HARD TO find in Germantown. Their gravestones beneath the elm tree at St. Luke's Episcopal Church have worn down so completely that even they are unreadable. Today, the site of the Morris-Littell House is covered with a combination of overgrown grass, tree stumps, and litter. With the School District of Philadelphia facing a budget crisis, Germantown High School officially closed its doors in 2013. The community rallied to protect the school from demolition and with unanimous support, the Philadelphia Historical Commission designated it a historic landmark on March 13, 2020. There is hope of it being adaptively reused, but for now the shuttered school building as well as the surrounding property appears disheveled. There are several Trees of Heaven—the much-maligned weed tree that I had been researching when I first stumbled upon the Morris sisters—edging up along the chain-link fence that encloses much of the school grounds. While their garden was long ago uprooted to make way for grass and a few new trees, dandelions are one of the few plants that have managed to persist.[34]

Dandelions were the cheerful flowers that Elizabeth studied so closely, sharing her findings with Asa, William, and others, thereby influencing their descriptions of the *Taraxacum officinale* in their books. These were the humble flowers she delighted in, describing how "its beautiful feathery ball dig[s] into the sunlight before the seeds are loosened and fly off." These were also the tenacious weeds she paid local boys to uproot. While I trekked through the high school property looking for any lingering evidence of the Morris sisters and their lives there, I was amused—as I imagine Elizabeth would have been—to find evidence of the persistent flower that once captured her attention, outlasting even her beloved forget-me-nots.[35]

The dandelions are thankfully not the only remnant of the Morris sisters and although their legacies have been obscured in so many ways, there were still people who saved or uncovered parts of their stories. There had to be, after all, if I were to eventually find them.

In the 1970s and 1980s, scholars like Sally Gregory Kohlstedt and Margaret Rossiter revealed just how many women were part of American scientific practice, conversations, and discoveries. Though they wrote before most of Margaretta and Elizabeth's papers became available, they managed to find the sisters' names in the rosters and publications of the AAAS, the American Philosophical Society, and the Academy of Natural Sciences, and their words tucked into William Darlington's carefully bound letters at the New-York Historical Society. Kohlstedt and Rossiter provided the larger context about the uncountable numbers of American women exploring science beyond the few exceptional women we knew to celebrate. Still other historians have revealed how male scientists in the later nineteenth century categorized women as inherently unscientific and intellectually inferior, making their exclusion seem objective, natural, and indisputable. More recent scholars have probed into the ways gender, particularly masculinity, has impacted the creation of knowledge and the culture of science, further expanding the ways we understand power relations in labs, out in the field, and on the pages of journals.[36]

At the turn of the twentieth-first century, more of the Morris sisters' materials became available thanks to their family members. Jeanie Morse Littell Winslow, a descendant of their sister Susan Morris Littell, donated the family's papers to the University of Delaware's Special Collections. While the women's letters with scientists are primarily in the Library of Congress collection, the Littell Family Papers include their poetry albums, artwork, drafts, treasures, and photographs. These collections made it possible to tell a richer story of their personal lives in a way that incorporated more than just their connections to the network of scientists and their published writings.[37]

While writers looking to unearth the histories of twentieth-century scientists might have the chance to interview their subjects before it is too late, that is not an option for nineteenth-century subjects. Fortunately, archival digitization projects in the last two decades make it much more possible to uncover the lives and work of lesser-known people like the Morris sisters. Digital finding aids, as well as scanned archival materials, herbarium specimens, newspapers, and rare books, made it possible for me to search some of the deepest recesses of collections. The needles in the haystack and the random letters tucked in someone's collection of papers that would have been nearly impossible to locate even a generation ago become so much easier to find thanks to increased efforts of archivists to make their institutions' collections accessible.[38]

There are even changes happening with species names. When the descendants of Margaretta's cicadas, what we now refer to as Brood X, began to emerge in the spring of 2021, I published an article in *Scientific American* about Margaretta's discovery of a new cicada species in the 1840s and how John Cassin and James Coggswell Fisher named the creature for themselves. Eponymous naming of species reflects power structures, whether that involves a famous botanist like Joseph Hooker at Kew Gardens selecting the names of plants to the dismay of the original collectors who found them, or more modern examples of scientists honoring politicians and celebrities when they name new species. It is not particularly surprising that a large percentage of species bear the names of white men with European heritage. However, some biologists today are interested in righting the wrongs of the past. After the article was posted online, the entomologist John Cooley, a specialist in cicadas, began an effort to change the common name of the *Magicicada cassini* from "Cassin's cicada" to "Morris's cicada." Perhaps Margaretta will finally get credit for her discovery.[39]

MARGARETTA AND ELIZABETH MORRIS WERE BOTH EXCEPtional and unexceptional. They were closely connected and indispensable to powerful scientists, they published widely, and they were recognized for their accomplishments on a level that few were able to achieve. They were also part of a vast culture that embraced science in the early nineteenth century, and they rank among many marginalized scientists who were erased and forgotten over time. If we only talk about the exceptional women, the pathbreakers, we miss the bigger story. In this chaotic moment in the nineteenth century, when the boundaries and rules of the scientific profession were still being drawn, diverse voices contributed to the creation of scientific knowledge. It was far from a halcyon moment of equality—Margaretta and Elizabeth established their credibility against great odds—but it was still a critical moment on the path to modern science before many people were marginalized, silenced, or erased. Uncovering Margaretta's and Elizabeth's lives and work has simultaneously revealed how the process of professionalization edged out women like them, while also erasing them from historical narratives. Perhaps now, with their stories told, the Morris sisters might finally have the afterlives that were taken from them.

Margaretta and Elizabeth were never alone, even if they stood out in their single-minded devotion to entomology and botany. They were surrounded by women and men who were enchanted with the sciences and eager to become literate in their environments—to look at a plant or animal and know what species it was, to name the geological formations, to predict a meteor's path. Telling Margaretta and Elizabeth's story is not just about restoring them to the larger history of antebellum science; it is also about restoring the countless women who were a part of this scientific community for whom so little remains: the anonymous science writers, the uncredited illustrators, the regional environmental experts, the close observers, and those who filled their waking hours with explorations and experiments, fitting them in around their other responsibilities.

Through the highs and lows of their lives—the losses of family and friends, the adjustments of dreams and desires, the professional slights and triumphs, even profound national political crises—Margaretta and Elizabeth had each other. And they had the Wissahickon, their "book of nature," spread open before them. Just as they had as girls, as older women they held tight to the wonder that would come with a hike through the forest and a slow afternoon in search of beetles or orchids. There was always something more to find. Something new to astonish them. Something they had not noticed before that they could paint or write about. Something they could share with each other and the world. Margaretta and Elizabeth were working within the limits of their moment in history. They had finite time, and other demands on it, but they never lost the thrill of paying close attention to the world around them—bright, verdant, and alive.

Acknowledgments

THIS BOOK CARRIED ME THROUGH THE PANDEMIC IN WAYS that I never expected. When COVID-19 paused everything in the spring of 2020, I had to cancel a crucial research trip as well as conference presentations. Whatever time I might have devoted to writing seemed to vanish as my family adjusted to remote school and working together at home. While at first it felt like an obstacle course, I came to realize what a gift it was to have a project to focus on as I carved out time to keep writing. For all of these reasons, I want to start by thanking my husband, Dan, and my children, Carter and Julia, who have all lived with this book as much as I have. I'm so grateful for Carter's botanical interests and thoughtful questions about my progress, Julia's deadpan humor and commitment to making sure I don't become too boring, and Dan's steady hand on the wheel as we navigated the pandemic and remapped our lives. I'm also thankful for our sweet pup Mabel and all those walks and nudges, even if she may have interrupted several half-written sentences. How could I have ever done this without you all?

In some of the most harried weeks of the pandemic, Alicia Cahill Ha and I began scheduling weekly phone calls where we would walk and talk through whatever was going on at the moment. These walks became a way for me to keep my head above water. Though no hike on the Wissahickon, I'm grateful we had the technology to log all those miles together despite being on opposite sides of the continent. Thank goodness for the friends who lift us up.

Pandemic or not, it can be difficult to research a story that takes place nearly three thousand miles away from where you live. During my visits to Philadelphia, it was a treat to stay with my dear friend Francesca Russello Ammon and her family, Peter, Isabella, Oliver, and Phoebe. They welcomed me into their home and lives as I searched for the Morris sisters in the archives and retraced their steps as best I could. I hope that all of my future projects have some Philadelphia component so I have a good excuse to stay with them again.

I am so very thankful for the archivists and librarians who made it possible for me to explore materials both on-site and remotely. My discovery of the Morris sisters was due to the New-York Historical Society's then curator of manuscripts, Ted O'Reilly, suggesting that I take a look at the Darlington Papers as I sought out the Tree of Heaven. Little did I know that suggestion would change everything. When my research took me to the University of Delaware, L. Rebecca Johnson Melvin's enthusiasm for my project spilled forth, and her support ever since has been incredible. Laura Keim, the curator at Stenton, Deborah Norris Logan's house, not only gave me a personalized tour of the museum and access to the documents on-site, but also traded information about all the Germantown women we had both come to know. What a delight that was. My travel to Philadelphia was paid for by a Portland State University Faculty Enhancement Grant. I'm very thankful for the support I received not only from my university but also from the History Department chairs, Tim Garrison and John Ott, who supported my applications over the years. Ann Fabian, Karl Jacoby, Matt Klingle, and Bill Deverell have also gone above and beyond supporting this project as I applied for grant after grant.

Given how so much of the Morris sisters' papers were scattered in dozens of archives around the country and overseas, I had to rely on archivists and reference librarians to scan materials for me. I am so thankful for that labor, as it made it possible to tell the story of these women in ways that I otherwise couldn't. Thanks are due in particular to Maurita Baldock, Penny Baker, Alex Bartlett, Dana Fisher, David Gary, Mary Haegert, Ron McColl, and Jayne Ptolemy, among so many others. I'm also grateful for Portland State's amazing librarians, like Joan Petit, who sought out ways to get me the full run of microfilm for the *Germantown Telegraph*; Min Cedillo, who made my Interlibrary Loan dreams come true; and Jill Emery, who gladly purchased all the books I needed. I owe thanks to my students Madelyn Miller and Taylor Rose, who served briefly as research assistants as I transitioned from a project about urban trees to a project about Philadelphia scientists.

One of my favorite research finds felt particularly magical. I honestly don't know how else to describe it. After having made small talk on the school playground about our work, Andrew Wood, the father of my daughter's friend, returned several months later to tell me he thought he was related to Margaretta and Elizabeth Morris. Once we determined it was true, he connected me with his uncle Sam Doak, who lives just an hour away from me in Oregon. Sam and his wife Ann kindly opened their home to me in the midst of the pandemic and shared Margaretta's album of paintings and poetry. I continue to be baffled by how a brief conversation could lead to such a discovery, how one of the Morris sisters' descendants is my child's close friend, and even how something so relevant for my Philadelphia story was in driving distance of Portland. I owe a great deal of thanks to the Wood and Doak families for realizing the connection and sharing that album with me.

In my mind, there is no greater generosity than reading and commenting on drafts. From the very start of this project, Ann Fabian has served as a mentor, enthusiastically brainstorming with me in a hotel lobby at a conference, recommending countless fiction and history books, and reading every single chapter with amazing feedback to boot. It will be impossible for me to repay this generous gift of time and attention, though I will keep trying. My father, John McNeur, also read the entire book and cheered me on with every draft, lifting me up along the way. My beloved writing groups kept me buoyed first through our sabbaticals and then the pandemic. Fran Ammon, Julia Guarneri, Sara Hudson, Kathryn Gin Lum, and Robin Morris are the real thing and have been since our grad school days. I feel like the Morris sisters are as alive to them as they are to me. I hope we continue sharing our writing for decades to come. Andrew Robichaud and Bart Elmore formed an environmental history writing group with me as we worked through new book projects. Their suggestions were a bouquet I returned to again and again, as Elizabeth Morris would have put it. Many thanks, also, to Richard Beyler, Jared Farmer, Charlotte Leib, Brian McCammack, Dael Norwood, Patricia Schechter, Ashanti Shih, and Karen Thompson Walker, who all gave me invaluable suggestions and advice on drafts at crucial moments.

I had the fortune to workshop chapters over the last several years, and I'm so thankful for the feedback these groups have given me. I owe a special thanks to Marsha Weisiger, who organizes not only the Cascadian Environmental History Collective but also the University of Oregon and Mellon Foundation's Workshop on Writing Environmental Humanities for Public Audiences, where I shared parts of my project. The Society for Historians of the Early American Republic hosted working groups for Second Book Writers and Biography Writers, and I'm grateful for the suggestions I got from both. At the very start of this project, a junior faculty writing group in the History Department at Portland State vetted an early conference paper. What a particularly wonderful bookend to have my colleagues read and respond to a chapter draft as I completed the manuscript, through the Department's colloquium sponsored by the Portland State Friends of History. I presented portions of chapters at conferences at the American Society for Environmental History meetings, York University's Traces of the Animal Past symposium, as well as the Smithsonian Institution's symposium on American Women of Science. The feedback I received from co-panelists and audience members was phenomenal and helped shape the way I thought about the project. Many thanks, in particular, to Jennifer Bonnell, Tina Gianquitto, Sean Kheraj, Don Opitz, Adrienne Petty, Harriet Ritvo, Andy Robichaud, and Ashanti Shih.

From the birth of this project when it was merely a drafty chapter and a sketch of a book proposal, my agent Wendy Strothman saw the potential, gave me fantastic advice, and advocated for both me and the Morris sisters. I'm thankful for Claire Potter, who brought the project to Basic Books, enthusiastic as she was for the story that could be told through the lives of two sisters, and for Brian Distelberg, who has seen it to completion with amazing suggestions and editorial advice. Marissa Koors's line edits were masterful, and this book is so much better for them. I am incredibly thankful for the care and attention that all of these editors have given to *Mischievous Creatures*.

As I finished writing this book, my mom's health began to decline. I sat beside my dad in the ICU room as she rested following surgery, editing the seventh chapter where Margaretta and Elizabeth faced their own mother's illnesses and death. It was a difficult chapter to work through, but it also felt like I had company in my grief. Along with my father, my mother was my greatest cheerleader. My parents shared with me their love of stories and encouraged me to tell my own. I am so very lucky to have had that support, and it's for that that I dedicate this book to them, as well as to my children. What are we, if not for our family and friends.

NOTES

CITATION ABBREVIATIONS

AA   *American Agriculturist*

AGCF   Asa Gray Correspondence Files of the Gray Herbarium, 1838–1892, Archives of the Gray Herbarium, Harvard University

AGP   Asa Gray Papers, Library of Congress

AJSA   *American Journal of Science and Arts*

ANSP   Academy of Natural Sciences of Philadelphia, Drexel University, Philadelphia, PA

APS   American Philosophical Society

DCEP   Daniel Cady Eaton Papers, Yale Manuscripts and Archives, MS 581

DCP   Darwin Correspondence Project, Cambridge University

DDP   Dorothea Dix Papers, Houghton Library, Harvard University

DL   Darlington Letters, West Chester University Library

DNLD   Deborah Norris Logan Diaries, Historical Society of Pennsylvania

ECHP   Edward Claudius Herrick Papers, Yale University Manuscripts and Archives

GHS   Germantown Historical Society

GM   *Gardener's Monthly*

GT   *Germantown Telegraph*

HSP   Historical Society of Pennsylvania

JANSP   *Journal of the Academy of Natural Sciences of Philadelphia*

JCS   Jane Campbell Scrapbooks, Germantown Historical Society

JFP   James Family Papers, Schlesinger Library, Harvard University

LFP   Littell Family Papers, University of Delaware Special Collections

NYBG   C. V. Starr Herbarium, New York Botanical Gardens

PAAAS   *Proceedings of the American Association for the Advancement of Science*

Citation Abbreviations

PANSP   *Proceedings of the Academy of Natural Sciences of Philadelphia*
PAPS   *Proceedings of the American Philosophical Society*
PFP   Powel Family Papers, Historical Society of Pennsylvania
TAPS   *Transactions of the American Philosophical Society*
TWHP   Thaddeus William Harris Papers, Museum of Comparative Zoology, Harvard University
WAP   Wyck Association Papers, American Philosophical Society
WDP   William Darlington Papers, New-York Historical Society
WLCL   William L. Clements Library, The University of Michigan

INTRODUCTION: SISTER SCIENTISTS

1. E. C. Morris to William Darlington, 1–17 April 1851, WDP; E. S., "The Garden," *AA* 5.4 (April 1846): 127; William Darlington, *Agricultural Botany* (J. W. Moore, 1847), 142–143; "Meteorological Observations, for April 1851," *GT*, 7 May 1851.

2. Margaretta Hare Morris, manuscript on cicada, [n.d.], Box 1, Folder 9, LFP; "Stated Meeting, December 15, 1846," *PANSP* 3.6 (November–December 1846): 131–134.

3. "The Seventeen Year Locust," *The Friend: A Religious and Literary Journal* 24.25 (8 March 1851): 193–194; Pennsylvania Horticultural Society Minute Book, vol. 5, August 1848–December 1850 (2010.03.16): 231, 263, Pennsylvania Horticultural Society McLean Library; *PAAAS, Fourth Meeting, Held at New Haven, Conn., August 1850* (SF Baird, 1851): v, ix, xiii, xv, 354.

4. E. C. Morris to William Darlington, 1–17 April 1851, WDP; Margaretta Hare Morris to Thaddeus William Harris, 6 November 1850, TWHP; "Stated Meeting, December 15, 1846," *PANSP* 3.6 (November–December 1846): 131–134; "Stated Meeting, March 16th, 1847," *PANSP* 3.8 (March–April 1847): 189–192.

5. "Donations to the Museum, in July and August, 1853," *PANSP* 6 (1852–1853): lxviii; "Donations to the Museum, in March and April, 1849," *PANSP* 4 (1848–1849): 175.

6. Charles Darwin to J. D. Hooker, 7 April 1855, "Letter no. 1661," DCP.

7. Sara Stidsone Gronim, "What Jane Knew: A Woman Botanist in the Eighteenth Century," *Journal of Women's History* 19.3 (Fall 2007): 33–59; Joan Hoff Wilson, "Dancing Dogs of the Colonial Period: Women Scientists," *Early American Literature* 7.3 (1973): 225–235; Margaret W. Rossiter, *Women Scientists in America: Struggles and Strategies to 1940* (Johns Hopkins University Press, 1982), 1:1–8; Tina Gianquitto, *"Good Observers of Nature": American Women and the Scientific Study of the Natural World* (University of Georgia Press, 2007), 15–56, 136–176; Britt Rusert, *Fugitive Science: Empiricism and Freedom in Early African American Culture* (New York University Press, 2017), 181–218; Renee Bergland, *Maria Mitchell and the Sexing of Science: An Astronomer Among the American Romantics* (Beacon Press,

2008). For women working in medicine, see, for instance, Myra C. Glenn, *Dr. Harriot Kezia Hunt* (University of Massachusetts Press, 2018); Kabria Baumgartner, *In Pursuit of Knowledge: Black Women and Educational Reform in Antebellum America* (New York University Press, 2019), 181–189; Janice Nimura, *The Doctors Blackwell: How Two Pioneering Sisters Brought Medicine to Women and Women to Medicine* (W. W. Norton, 2021).

8. Both Isabella Batchelder and Sarah Coates corresponded with William Darlington, and their letters are preserved in the William Darlington Papers at the New-York Historical Society. Sally Gregory Kohlstedt, "Physiological Lectures for Women: Sarah Coates in Ohio, 1850," *Journal of the History of Medicine and Allied Sciences* 33.1 (January 1978): 75–90; Frances W. Kaye, "The Ladies' Department of the 'Ohio Cultivator,' 1845–1865: A Feminist Forum," *Agricultural History* 50.2 (April 1976): 414–423; Asa Gray to Miss Batchelder, [n.d.], AGCF; Asa Gray to Samuel Batchelder, 10 December 1866, AGCF; Isabella Batchelder to William Darlington, 17 September 1849, WDP; Isabella Batchelder to William Darlington, 26 September 1849, WDP.

9. [William Whewell], "Article III: *On the Connexion of the Physical Sciences* by Mrs. Somerville," *The Quarterly Review* 51 (March and June 1834): 54–68; Kathryn A. Neeley, *Mary Somerville: Science, Illumination and the Female Mind* (Cambridge University Press, 2001), 1–10, 33–34; Bergland, *Maria Mitchell and the Sexing of Science*, 146–149; Renee Bergland, "Urania's Inversion: Emily Dickinson, Herman Melville, and the Strange History of Women Scientists in Nineteenth-Century America," *Signs* 34.1 (Autumn 2008): 78; Londa Schiebinger, *The Mind Has No Sex? Women in the Origins of Modern Science* (Harvard University Press, 1989); Cynthia Eagle Russert, *Sexual Science: The Victorian Construction of Womanhood* (Harvard University Press, 1989); Evelleen Richards, "Redrawing the Boundaries: Darwinian Science and Victorian Women Intellectuals," in *Victorian Science in Context*, ed. Bernard Lightman (University of Chicago Press, 1997), 119–142.

10. Paul White, *Thomas Huxley: Making the "Man of Science"* (Cambridge University Press, 2002), 33–45; Ruth Barton, "'Men of Science': Language, Identity, and Professionalization in the Mid-Victorian Scientific Community," *History of Science* 41 (2003): 73–119; Jim Endersby, *Imperial Nature: Joseph Hooker and the Practices of Victorian Science* (University of Chicago Press, 2008), 20–28; Paul Lucier, "The Professional and the Scientist in Nineteenth-Century America," *Isis* 100.4 (2009): 699–732; Melinda Baldwin, *Making Nature: The History of a Scientific Journal* (University of Chicago Press, 2015), 4–5.

11. Asa Gray to Elizabeth Carrington Morris, 19 July 1859, AGP.

12. See, for instance, George H. Daniels, *American Science in the Age of Jackson* (Columbia University Press, 1968), 34–62; Chandos Michael Brown, *Benjamin Silliman: A Life in the Young Republic* (Princeton University Press, 1990), 70, 100–154; W. Conner Sorensen, *Brethren of the Net: American Entomology, 1840–1880* (University of Alabama Press, 1995), 82; Kara W. Swanson, "Rubbing Elbows and

Blowing Smoke: Gender, Class, and Science in the Nineteenth-Century Patent Office," *Isis* 108.1 (March 2017); Jeffrey K. Barnes, *Asa Fitch and the Emergence of American Entomology* (University of the State of New York, 1988).

13. Nathan Reingold, "Definitions and Speculation: The Professionalization of Science in America in the Nineteenth Century," in Alexandra Oleson and Sanborn C. Brown, eds., *The Pursuit of Knowledge in the Early American Republic* (Johns Hopkins University Press, 1976), 33–69; Simon Baatz, "'Squinting at Silliman': Scientific Periodicals in the Early American Republic," *Isis* 82.2 (1991): 223–244; Elizabeth B. Keeney, *The Botanizers: American Scientists in Nineteenth-Century America* (University of North Carolina Press, 1992); Barton, "'Men of Science,'" 73–119; Paul Lucier, "The Professional and the Scientist in Nineteenth-Century America," *Isis* 100.4 (2009): 699–732.

14. Rossiter, *Women Scientists in America*, 1:51–72; Dava Sobel, *The Glass Universe: How the Ladies of the Harvard Observatory Took the Measure of the Stars* (Penguin Books, 2016); Helen Veit, *Modern Food, Moral Food: Self-Control, Science, and the Rise of Modern American Eating in the Early Twentieth Century* (University of North Carolina Press, 2013); Ann Fabian, "The Travails of a 'Lady Scientist,'" *Raritan* 41.2 (Fall 2021): 35–59.

15. Baldwin, *Making* Nature; Melinda Baldwin, "Scientific Autonomy, Public Accountability, and the Rise of 'Peer Review' in the Cold War United States," *Isis* 109.3 (September 2018): 538–558; Noah Moxham and Aileen Fyfe, "The Royal Society and the Prehistory of Peer Review, 1665–1965," *History Journal* 61.4 (December 2018): 863–889; Emily Pawley, *The Nature of the Future: Agriculture, Science, and Capitalism in the Antebellum North* (University of Chicago Press, 2020); Ariel Ron, *Grassroots Leviathan: Agricultural Reform and the Rural North in the Slaveholding Republic* (Johns Hopkins University Press, 2020).

16. Schiebinger, *The Mind Has No Sex?*, 20, 245–277; Sally Gregory Kohlstedt, *The Formation of the American Scientific Community: The American Association for the Advancement of Science, 1848–1860* (University of Illinois Press, 1976), 21–22, 102–104; Robert V. Bruce, *The Launching of Modern American Science, 1846–1876* (Alfred A. Knopf, 1976), 135–139; Ann B. Shteir, *Cultivating Women, Cultivating Science* (Johns Hopkins University Press, 1996), 157–158; Rossiter, *Women Scientists in America*, 1:73–99.

17. Keeney, *Botanizers*, 1–8; Ann B. Shteir, "Botany in the Breakfast Room: Women and Early Nineteenth-Century British Plant Study," in *Uneasy Careers and Intimate Lives: Women in Science, 1787–1979*, ed. Pnina G. Abir-Am and Dorinda Outram (Rutgers University Press, 1987), 31–44; Adrian Desmond, "Redefining the X Axis: 'Professionals,' 'Amateurs,' and the Making of Mid-Victorian Biology; A Progress Report," *Journal of the History of Biology* 34.1 (2001): 3–50; "Amateur," *Oxford English Dictionary* (Oxford University Press, 2022).

18. J. F. A. Adams, "Is Botany Suitable for Young Men?" *Science* 9.209 (4 February 1887): 116–117; Schiebinger, *The Mind Has No Sex?*, 240–244; Rossiter, *Women*

Scientists in America, 1:73–99; Erika Lorraine Milam and Robert A. Nye, eds., "Scientific Masculinities," *Osiris* 30 (2015).

19. Bruce, *Launching of Modern American Science*, 25–26; Kohlstedt, *Formation of the American Scientific Community*, 22–25, 132–153; Susan Scott Parrish, *American Curiosity: Cultures of Natural History in the Colonial British Atlantic World* (University of North Carolina Press, 2006), 103–135; Robert A. Nye, "Medicine and Science as Masculine 'Fields of Honor,'" *Osiris* 12 (1997): 60–79; Shteir, *Cultivating Women, Cultivating Science*, 175.

20. Margaretta Hare Morris to Thaddeus William Harris, 6 November 1850, TWHP.

21. Margaretta Hare Morris to Thaddeus William Harris, 12 September 1847, TWHP; Nicholas B. Wainwright, ed., *A Philadelphia Perspective: The Diary of Sidney George Fisher* (Historical Society of Pennsylvania, 1967), 107–108, 328.

22. For more on erasure as part of the creation of ignorance and as an act of power, see Robert N. Proctor and Londa Schiebinger, eds., *Agnotology: The Making and Unmaking of Ignorance* (Stanford University Press, 2008); Michel-Rolph Trouillot, *Silencing the Past: Power and the Production of History* (Beacon Press, 2015).

23. See, for instance: Sally Gregory Kohlstedt, "Parlors, Primers, and Public Schooling: Education for Science in Nineteenth-Century America," *Isis* 81.3 (1990): 424–445; Dorinda Outram, "New Spaces in Natural History," in *Cultures of Natural History*, ed. N. Jardine, J. A. Secord, and E. C. Spary (Cambridge University Press, 1996), 249–265; Debra Lindsay, "Intimate Inmates: Wives, Households, and Science in Nineteenth-Century America," *Isis* 89.4 (December 1998): 631–652; Barbara T. Gates, *Kindred Nature: Victorian and Edwardian Women Embrace the Living World* (University of Chicago, 1998); Donald L. Opitz, "Domestic Space," in *A Companion to the History of Science*, ed. Bernard V. Lightman (Wiley & Sons, 2016), 306–323; Elaine Leong, *Recipes and Everyday Knowledge: Medicine, Science and the Household in Early Modern England* (University of Chicago Press, 2018).

24. Elizabeth Gilbert, *The Signature of All Things* (Riverhead Books, 2013); Esi Edugyan, *Washington Black* (Alfred A. Knopf, 2018); Barbara Kingsolver, *Unsheltered* (Harper, 2018); Tracy Chevalier, *Remarkable Creatures* (Dutton Adult, 2010).

CHAPTER ONE: WORLD OF WONDERS

1. Deborah Norris Logan, DNLD, 8 March 1827, XI: 24; E. S., "The Garden," *AA* 5.4 (April 1846): 127; Margaretta Hare Morris to Richard Chandler Alexander, 17 June 1855, "Letter no. 1701," DCP.

2. Elizabeth C. Morris to William Darlington, 14 October 1842, WDP.

3. Margaretta Hare Morris to Thaddeus William Harris, 12 September, 1850, TWHP; Margaretta Hare Morris to Thaddeus William Harris, 6 November, 1850, TWHP; Mary Donaldson to Ann Haines, 13 October 1841, Series 2, Box

267, Folder 418, WAP; Catharine E. Beecher, *A Treatise on Domestic Economy* (Thomas H. Webb & Co., 1843), 323–324.

4. Thomas Jefferson, *Notes on the State of Virginia* (Printed for John Stockdale, 1787), 33–120; Philip Pauly, *Biologists and the Promise of American Life: From Meriwether Lewis to Alfred Kinsey* (Princeton University Press, 2000), 15–22; Andrea Wulf, *Founding Gardeners: The Revolutionary Generation, Nature, and the Shaping of the American Nation* (Vintage, 2011); W. Conner Sorensen, *Brethren of the Net: American Entomology, 1840–1880* (University of Alabama Press, 1995), 1–14; Chandos Michael Brown, *Benjamin Silliman: A Life in the Young Republic* (Princeton University Press, 1989), 102, 107–108; Edgar Fahs Smith, *Robert Hare: An American Chemist* (Lippincott Company, 1917), 6–8; Timothy W. Kneeland, "Robert Hare: Politics, Science, and Spiritualism in the Early Republic," *Pennsylvania Magazine of History and Biography* 132.3 (July 2008): 245–260; Robert V. Bruce, *The Launching of Modern American Science, 1846–1876* (Alfred A. Knopf, 1976), 14–15.

5. Thomas Wilson, *Picture of Philadelphia for 1824, containing the "Picture of Philadelphia for 1811, by James Mease, M.D." with all its improvements since that era* (T. Town, 1823), 348; Richard G. Miller, "The Federal City, 1783–1800," in *Philadelphia: A 300-Year History*, ed. Russell F. Weigley (W. W. Norton, 1982): 173–174; John C. Greene, "Science, Learning, and Utility: Patterns of Organization in the Early American Republic," in *The Pursuit of Knowledge in the Early Republic: American Scientific and Learned Societies from the Colonial Times to the Civil War*, ed. Alexandra Oleson and Sanborn C. Brown (Johns Hopkins University Press, 1976), 2–9; Brown, *Benjamin Silliman*, 102–103; Aaron Sachs, *The Humboldt Current: Nineteenth-Century Exploration and the Roots of American Environmentalism* (Viking, 2006), 1–7; Andrea Wulf, *The Invention of Nature: Alexander von Humboldt's New World* (Vintage, 2015); Patrick Gass, *A Journal of the Voyages and Travels of a Corps of Discovery* (Printed by Zadok Cramer for David McKeehan, 1807); Pauly, *Biologists and the Promise of American Life*, 15–17; Charles Coleman Sellers, *Mr. Peale's Museum: Charles Willson Peale and the First Popular Museum of Natural Science and Art* (W. W. Norton, 1980); Edward Potts Cheyney, *History of the University of Pennsylvania, 1740–1940* (University of Pennsylvania Press, 1940), 165, 203–207; Bruce, *Launching of Modern American Science*, 47–48; Simon Baatz, "'Squinting at Silliman': Scientific Periodicals in the Early American Republic," *Isis* 82.2 (1991): 224–225; Richard Judd, *The Untilled Garden: Natural History and the Spirit of Conservation in America, 1740–1840* (Cambridge University Press, 2009), 100–103.

6. Some of the relatives who served as mayors include Edward Shippen, Anthony Morris I, William Hudson, Anthony Morris II, Edward Shippen II, Charles Willing, Thomas Willing, and Samuel Powel.

7. "Luke Morris," Philadelphia (Pennsylvania) Register of Wills, Pennsylvania, Wills and Probate Records, 1683–1993, 1802, Case Number 68.

8. Elizabeth Hudson Morris, Diary 1743–1778, Haverford College Quaker and Special Collections; Robert Charles Moon, *Morris Family of Philadelphia, Descendants of Anthony Morris, born 1654–1712 Died* (R. C. Moon, 1898), 1:228–229; Margaret Hope Bacon, ed., *Wilt thou Go on My Errand? Journals of Three 18th Century Quaker Women Ministers* (Pendle Hill Publications, 1994), 121–282; "To be Let," *Pennsylvania Gazette*, 6 April 1758; Mary Harrison, *Annals of the Ancestry of Charles Custis Harrison and Ellen Waln Harrison* (J. B. Lippincott Company, 1932), 254–255.

9. Nicholas Scull and George Heap, *A Map of Philadelphia, and Parts Adjacent* (1752); Moon, *Morris Family of Philadelphia*, 1:223, 224; The Original Papers of Manumission by the Monthly Meetings Composing the Quarterly Meeting of Philadelphia, 1772–1790, Haverford College Quaker Collection, Manuscript Collection 1250; Jean Soderlund, *Quakers and Slaves: A Divided Spirit* (Princeton University Press, 1985); Gary Nash, *Forging Freedom: The Formation of Philadelphia's Black Community, 1720-1840* (Harvard University Press, 1988).

10. "Abstract of Will of Anthony Morris," in Moon, *Morris Family of Philadelphia*, 1:233; Elizabeth C. Morris to William Darlington, 22 January 1860, WDP.

11. Moon, *Morris Family of Philadelphia*, 1:233–236, 2:399–404; Philadelphia Monthly Meeting, Southern District, *Philadelphia Yearly Meeting Minutes*, 1773–1789, 26 December 1781, Haverford College Quaker Collection, Haverford, Pennsylvania.

12. *MacPherson's Directory for the City and Suburbs of Philadelphia* (Francis Baily, 1785), 148; John Fanning Watson, *Annals of Philadelphia and Pennsylvania in the Olden Time* 3 (Edwin S. Stuart, 1899), 448; Nancy Slocum Hornick, "Anthony Benezet: Eighteenth Century Social Critic, Educator and Abolitionist," (PhD diss., University of Maryland, 1974), 123–124; Zara Anishanslin, *Portrait of a Woman in Silk: Hidden Histories of the British Atlantic World* (Yale University Press, 2016), 174–175; Darold D. Wax, "Negro Imports into Pennsylvania, 1720–1766," *Pennsylvania History: A Journal of Mid-Atlantic Studies* 32.3 (July 1965): 254–287; Donald Brooks Kelley, "'A Tender Regard to the Whole Creation': Anthony Benezet and the Emergence of an Eighteenth-Century Quaker Ecology," *Pennsylvania Magazine of History and Biography* 106.1 (January 1982): 69–88. For more about the kind of elite education available to Ann Willing as a wealthy girl in Philadelphia, see Sarah E. Fatherly, "'The Sweet Recourse of Reason': Elite Women's Education in Colonial Philadelphia," *Pennsylvania Magazine of History and Biography* 128.3 (July 2004): 229–256.

13. S. F. Hotchkin, *Ancient and Modern Germantown, Mount Airy, and Chestnut Hill* (P. W. Ziegler & Co., 1889), 143; John Fanning Watson, *Annals of Philadelphia and Pennsylvania: In the Olden Time* (Edwin S. Stewart, 1891), 448–450; Moon, *Morris Family of Philadelphia*, 2:399–404, 581–583; R. Winder Johnson, *The Ancestry of Rosalie Morris Johnson* (Printed for Private Circulation by Ferris & Leach, 1905–8), 2:163.

14. John Fanning Watson, *Annals of Philadelphia, and Pennsylvania, in the Olden Time* (Edwin S. Stuart, 1887), 3:448–450; Moon, *Morris Family of Philadelphia,* 2:400–403; Susan Branson, *These Fiery Frenchified Dames: Women and Political Culture in Early National Philadelphia* (University of Pennsylvania Press, 2001), 135.

15. Watson, *Annals of Philadelphia,* 3:448; David S. Shields and Fredrika J. Teute, "The Meschianza: Sum of All Fetes," *Journal of the Early Republic* 35.2 (Summer 2015): 185–214.

16. *Marriage Record of Christ Church, Philadelphia, 1709–1806* (Lane S. Hart, State Printer, 1878), 183; 1800 United States Federal Census, Census Place: Southwark, Philadelphia, Pennsylvania, Series M32, Roll 42, Page 45; "Luke Morris," Tax and Exoneration Lists, 1762–1794. Series No. 4.61, Microfilm Roll 337, Records of the Office of the Comptroller General, Pennsylvania Historical and Museum Commission, Harrisburg; "Luke Morris," United States Direct Tax of 1798, Tax Lists for the State of Pennsylvania, Microfilm Roll M372, Records of the Internal Revenue Service, National Archives and Records Administration, Washington, DC.

17. "Luke Morris," Philadelphia (Pennsylvania) Register of Wills, Pennsylvania, Wills and Probate Records, 1683–1993, 1802, Case Number 68; Edward Hogan, *The Prospect of Philadelphia and Check on the Next Directory* (Francis & Robert Bailey, 1795), 124, 154; William Stevens Perry, *Journals of General Conventions of the Protestant Episcopal Church in the United States, 1785–1835* (Claremont Manufacturing Company, 1874), 2:337.

18. Benjamin Rush, *Thoughts Upon Female Education* (Samuel Hall, 1787).

19. *The Rise and Progress of the Young-Ladies' Academy* (Stewart & Cochran, 1794); J. A. Neal, *An Essay on the Education and Genius of the Female Sex* (Jacob Johnson & Co., 1795); Sally Gregory Kohlstedt, "Parlors, Primers, and Public Schooling: Education for Science in Nineteenth-Century America," *Isis* 81.3 (1990): 424–445; Mary Kelley, *Learning to Stand and Speak: Women, Education, and Public Life in America's Republic* (University of North Carolina Press, 2006), 34–65; Lucia McMahon, "'Of the Utmost Importance to Our Country': Women, Education, and Society, 1780–1820," *Journal of the Early Republic* 29.3 (Fall 2009): 475–506; Margaret Nash, "Rethinking Republican Motherhood: Benjamin Rush and the Young Ladies' Academy of Philadelphia," *Journal of the Early Republic* 17.2 (Summer 1997): 171–191.

20. Neal, *An Essay on the Education and Genius of the Female Sex;* Mary Wollstonecraft, *A Vindication of the Rights of Woman, with Strictures on Political and Moral Subjects* (Mathew Carey, 1794); Branson, *These Fiery Frenchified Dames,* 35–39, 162n64.

21. "Luke Morris," Philadelphia (Pennsylvania) Register of Wills, Pennsylvania, Wills and Probate Records, 1683–1993, 1802, Case Number 68; Elizabeth C. Morris to William Darlington, 22 January 1860, WDP; *Minutes of the Philadelphia Society for the Promotion of Agriculture, from its institution in February, 1785, to March, 1810* (J. C. Clark & Son, 1854), 8.

22. Wilson, *Picture of Philadelphia for 1824, containing the "Picture of Philadelphia for 1811, by James Mease, M.D.,"* 25–26, 218, 241–243; Edgar P. Richardson, "The Athens of America, 1800–1825," in *Philadelphia: A 300-Year History*, ed. Russell F. Weigley (W. W. Norton, 1982), 208–257; Clare A. Lyons, *Sex Among the Rabble: An Intimate History of Gender and Power in the Age of Revolution, Philadelphia, 1730–1830* (University of North Carolina Press, 2006), 323–395.

23. Wilson, *Picture of Philadelphia for 1824, containing the "Picture of Philadelphia for 1811, by James Mease, M.D.,"* 311–314; Charles Coleman Sellers, *Mr. Peale's Museum: Charles Willson Peale and the First Popular Museum of Natural Science and Art* (W. W. Norton, 1979), 18–19, 48, 101; David R. Brigham, *Public Culture in the Early Republic: Peale's Museum and Its Audience* (Smithsonian Institution Press, 1995); Patricia Tyson Stroud, *Thomas Say: New World Naturalist* (University of Pennsylvania Press, 1992), 29; Christoph Irmscher, *The Poetics of Natural History* (Rutgers University Press, 2019), 64–110.

24. "Lectures on Natural History," *Claypoole's American Daily Advertiser*, 26 September 1799, in *The Selected Papers of Charles Willson Peale and His Family*, vol. 2, pt. 1 (Yale University Press, 1988), 256–258; Charles Willson Peale, *Discourse Introductory to a Course of Lectures on the Science of Nature* (Zachariah Poulson, 1800), 13, 50; Jessica C. Linker, "The Fruits of Their Labor: Women's Scientific Practice in Early America, 1750–1860" (PhD diss., University of Connecticut, 2017), 92–93; Susan Branson, *Scientific Americans: Invention, Technology, and National Identity* (Cornell University Press, 2022), 8–38.

25. Record of Births and Burials, Kept by the Direction of the Monthly Meeting of Friends in Philadelphia for the Southern District, I: 38, Quaker Meeting Records, Friends Historical Library, Swarthmore College, Swarthmore, PA.

26. Lisa Wilson, *Life After Death: Widows in Pennsylvania, 1750–1850* (Temple University Press, 1992), 25–58; "Luke Morris," Philadelphia (Pennsylvania) Register of Wills, Pennsylvania, Wills and Probate Records, 1683–1993, 1802, Case Number 68.

27. Mary Byrd to T. M. Willing, 27 July 1811, Willing-Hare Collection (MS Coll. 104), Box 1, APS; Sara B. Bearss, "Byrd, Mary Willing," in *The Dictionary of Virginia Biography*, ed. Sara B. Bearss et al. (Library of Virginia, 2001), 2:457–459; "To Thomas Jefferson from Mary Willing Byrd, 28 February 1781 [redated]," in *The Papers of Thomas Jefferson*, vol. 4, *1 October 1780–24 February 1781*, ed. Julian P. Boyd (Princeton University Press, 1951), 690–692; "The Will of Mrs. Mary Willing Byrd, of Westover, 1813, with a List of the Westover Portraits," *Virginia Magazine of History and Biography* 6.4 (April 1899): 346–358; Amy Pflugrad-Jackish, "'What Am I but an American?': Mary Willing Byrd and Westover Plantation During the American Revolution," in *Women in the American Revolution: Gender, Politics, and the Domestic World*, ed. Barbara B. Oberg (University of Virginia Press, 2019), 171–191; Wilson, *Life After Death*, 3–5, 116–119; David W. Maxey, "A Portrait of Elizabeth Willing Powel (1743–1830)," *TAPS*, New Series 96.4 (2006): 1–91.

28. "Luke Morris," Philadelphia (Pennsylvania) Register of Wills, Pennsylvania, Wills and Probate Records, 1683–1993, 1802, Case Number 68; Christina D. Wood, "'A Most Dangerous Tree': The Lombardy Poplar in Landscape Gardening," *Arnoldia* 54.1 (Winter 1994): 24–30.

29. Wilson, *Picture of Philadelphia for 1824, containing the "Picture of Philadelphia for 1811, by James Mease, M.D." with all its improvements since that era* (T. Town, 1823), 350–351.

30. Seventeen-year-old Ann's friends were Maria Abercrombie and Harriet Barclay, daughters of religious and political leaders in the city. Ann Willing Morris to "My Dear Friend," 12 January 1809, Hare-Willing Papers, Series 1, APS; Townsend Ward, "The Germantown Road and Its Associations: Part Fifth," *The Pennsylvania Magazine of History and Biography* 6.1 (1882): 20; Hotchkin, *Ancient and Modern Germantown, Mount Airy, and Chestnut Hill*, 61; Naaman H. Keyser et al., *History of Old Germantown* (Horace F. McCann, 1907), 220–223; "Howell House," National Register of Historic Places Nomination Form, Pennsylvania Historical and Museum Commission, 1971.

31. Elizabeth Powel to Ann Willing Morris, 12 May 1808, Series 3, Box 4, Folder 4, PFP; Elizabeth Powel to Ann Willing Morris, 16 June 1808, Series 3, Box 4, Folder 4, PFP.

32. Deed: William J. Miller, of the city of Philadelphia, merchant, and Frances Bartholomew, his wife, to Elizabeth Powel, of the city of Philadelphia, widow, 26 April 1808, Philadelphia Deed Book E.F., No. 28, p. 718, City Archives of Philadelphia; Deed: Peter Kline, of Marlborough Township, Montgomery County, yeoman, and Ann, his wife, to Elizabeth Powel, of the city of Philadelphia, widow, for $2,700, 18 February 1814, Philadelphia Deed Book I.C., No. 29, p. 404, City Archives of Philadelphia; 1810 United States Federal Census, Census Place: Germantown, Philadelphia, Pennsylvania, Roll 56, Page 9; Elizabeth Powel to Nancy Morris, 15 September 1809, PFP, Series 3, Box 4, Folder 5, HSP; Elizabeth Powel to Ann W. Morris, 30 September 1811, Series 3, Box 4, Folder 7, PFP; Elizabeth Powel to Thomas W. Morris, 16 January 1814, Series 3, Box 4, Folder 11, PFP.

33. M. I. Wilbert, "Dr. Christopher Witt: An Early American Botanist and a Man of Many and Varied Attainments," *American Journal of Pharmacy* 77 (July 1905): 311–323; William Darlington, *Memorials of John Bartram and Humphry Marshall* (Lindsay & Blakiston, 1849), 86–87; Nancy Everill Hoffmann and John C. Van Home, *America's Curious Botanist: A Tercentennial Reappraisal of John Bartram* (American Philosophical Society, 2004), 28, 167; William Shainline Middleton, "John Bartram, Botanist," *The Scientific Monthly* 21.2 (August 1925), 191; Edwin Jellett, *Germantown Gardens and Gardeners* (Horace F. McCann, 1914), 13–14, 78–79; Hotchkin, *Ancient and Modern Germantown*, 144; P. Collinson to John Bartram, 3 February 1735/6, in *The Correspondence of John Bartram*, ed. Edmund Berkeley and Dorothy Smith Berkeley (University Press of Florida, 1992), 18.

34. Luke Morris would eventually be reinterred with his family at St. Luke's Church in Germantown. Rev. B. Wistar Morris, *Sermon at the Semi-Centennial*

Celebration of the Consecration of Saint Luke's Church, Germantown, August 27, 1868, With a Sketch of the History of the Parish from Its Origins (Lamar, Printer, 1868), 67; "William White's Private Baptisms, begun 1801," Historic Pennsylvania Church and Town Records, Reel 242, HSP.

35. Elizabeth Powel to A. W. Morris, 9 August 1817, Series 3, Box 5, Folder 3, PFP.

36. [Elizabeth Gibbs Willing Alleyne?] to Ann Willing Morris, 12 May 1818, Ann W. Morris Letters, WLCL.

CHAPTER TWO:
IN A TANGLED WILDERNESS WITHOUT A GUIDE

1. Elizabeth Carrington Morris, "To the Wissihiccon," in Offerings of Friendship, LFP, Box 1, Folder 12.

2. Frances Anne Butler, *Journal of Frances Anne Butler* (John Murray, 1835), 2:92–93; Edgar A. Poe, "Morning at the Wissahiccon," in N. P. Willis, *The Opal: A Pure Gift for the Holy Days* (J. C. Riker, 1844), 249–256.

3. Ralph Waldo Emerson celebrated the connection between poetry and science in his essay "The Poet," *Essays: First and Second Series* (Houghton Mifflin, 1865), 2:25. For more on the division of the sciences and humanities in the later nineteenth century, see Renee Bergland, "Urania's Inversion: Emily Dickinson, Herman Melville, and the Strange History of Women Scientists in Nineteenth-Century America," *Signs* 34.1 (Autumn 2008): 75–99; Laura Dassow Walls, *Seeing New Worlds: Henry David Thoreau and Nineteenth-Century Natural Science* (University of Wisconsin Press, 1995), 147–157.

4. Elizabeth Carrington Morris, Offerings of Friendship, LFP, Box 1, Folder 12; Margaretta Hare Morris, album, collection of S. C. Doak, Hood River, Oregon; Amy Matilda Cassey Album, The Library Company of Philadelphia; Mary Kelley, "'Talents Committed to Your Care': Reading and Writing Radical Abolitionism in Antebellum America," *New England Quarterly* 88.1 (2015): 37–72; Britt Rusert, *Fugitive Science: Empiricism and Freedom in Early African American Culture* (New York University Press, 2017).

5. Elizabeth Carrington Morris, "To the Wissihiccon."

6. Elizabeth Carrington Morris, "To the Wissihiccon"; T. A. Daly, *The Wissahickon* (Garden Club of Philadelphia, 1922), 22–23, 66–80; David Costa and Carol Franklin, *Metropolitan Paradise: The Struggle for Nature in the City* (Saint Joseph University Press, 2010).

7. Elizabeth Carrington Morris, "To the Wissihiccon"; Deborah Norris Logan, 1 December 1821, DNLD 5:7.

8. Mrs. Ann Willing Morris, probate 16 February 1853, Wills no. 1–60, 1853, Pennsylvania Probate Record, Philadelphia, Pennsylvania; Deborah Norris Logan, 8 July 1815, DNLD, 1:116; John Fanning Watson, *Annals of Philadelphia and Pennsylvania in the Olden Time* (E. S. Stuart, 1844), 448–450.

9. Thomas W. Morris to Ann W. Morris, 24 September 1814, War of 1812 Papers, WLCL; S. F. Hotchkin, *Ancient and Modern Germantown, Mt. Airy, and Chestnut Hill* (P. W. Ziegler, 1889), 142–144; Robert Winder Johnson, *The Ancestry of Rosalie Morris Johnson* (Printed for private circulation by Ferris & Leach, 1905–1908), 165–166; Mary to Ann W. Morris, 29 September [1817], Ann W. Morris Letters, WLCL.

10. Mary to Ann W. Morris, 4 March 1818, Ann W. Morris Letters, WLCL; Mary to Ann W. Morris, 29 September [1817], Ann W. Morris Letters, WLCL; "Airy Nothing" in LFP, Box 1, Folder 17.

11. Lucy Leigh Bowie, "Madame Grelaud's French School," *Maryland Historical Magazine* 39 (1944): 141–148.

12. Kim Tolley, *The Science Education of American Girls: A Historical Perspective* (Routledge, 2003), 13–54; Margaret A. Nash, *Women's Education in the United States, 1780–1840* (Palgrave Macmillan, 2005), 38, 54, 77–98.

13. Tolley, *Science Education,* 39–42; Frederika Bremer, *America of the Fifties: Letters of Frederika Bremer,* ed. Adolph B. Benson (American-Scandinavian Foundation, 1924), 285; Frances Trollope, *Domestic Manners of the Americans* (Whittaker, Treacher, 1832), 112; Alexis de Tocqueville, *Democracy in America,* 4th ed. (J. & H. G. Langley, 1840), 2:209–211.

14. Robert Hare Jr., *Memoir on the Supply and Application of the Blow-Pipe* (Printed for the Chemical Society, 1802); "Doctor Robert Hare," *Scientific American* 13.38 (29 May 1858): 301; John C. Greene, "The Development of Mineralogy in Philadelphia, 1780–1820," *PAPS* 113.4 (15 August 1969): 283–295; Edgar Fahs Smith, *The Life of Robert Hare, An American Chemist* (1781–1858), 1–64; Robert V. Bruce, *The Launching of Modern American Science, 1846–1876* (Alfred A. Knopf, 1976), 47–48; Sally Gregory Kohlstedt, *The Formation of the American Scientific Community: The American Association for the Advancement of Science, 1848–1860* (University of Illinois Press, 1976), 28–29.

15. WAP, Box 20, Folders 262–265, APS, Philadelphia, PA; Reuben Haines to Hannah Haines, 26 August 1817, Series 2, Box 15, Folder 146, WAP; John M. Groff, "'All That Makes a Man's Mind More Active': Jane and Reuben Haines at Wyck, 1812–1831," in *Quaker Aesthetics: Reflections on a Quaker Ethic in American Design and Consumption,* ed. Emma Jones Lapsansky and Anne A. Verplanck (University of Pennsylvania Press, 2003), 90–121; W. Edmunds Claussen, *Wyck: The Story of an Historic Home, 1690-1970* (Mary T. Haines, 1970), 53, 69.

16. Regarding Thomas's lifelong interest in natural history despite taking another path, see, for example: Elizabeth C. Morris to William Darlington, 18 August 1842, WDP; "Stated Meeting, December 15, 1846," *PANSP* 3.6 (November–December 1846): 131–134; [Elizabeth Gibbs Willing Alleyne] to Ann Willing Morris, 12 May 1818, Ann W. Morris Letters, William L. Clements Library, The University of Michigan.

17. Patricia Tyson Stroud, *Thomas Say: New World Naturalist* (University of Pennsylvania Press, 1992), 20, 28–43; Robert McCracken Peck and Patricia Tyson

Stroud, *A Glorious Enterprise: The Academy of Natural Sciences of Philadelphia and the Making of American Science* (University of Pennsylvania Press, 2012), 2–23; *List of the Members of the American Philosophical Society* (American Philosophical Society, 1890); *Members and Correspondents of the Academy of Natural Sciences of Philadelphia* (Printed for the Academy, 1877).

18. Stroud, *Thomas Say*, 40–41; Jeanette E. Graustein, *Thomas Nuttall, Naturalist: Explorations in America, 1808–1841* (Harvard University Press, 1967), 120; Lesueur quoted in E.-T. Hamy, *The Travels of the Naturalist Charles A. Lesueur in North America, 1815–1837* (Kent State University Press, 1968), 29.

19. Thomas Nuttall, *The Genera of North American Plants and A Catalogue of the Species, to the Year 1817* (Printed for the author by D. Heartt, 1818); Graustein, *Thomas Nuttall*, 114–131, 454n56; Peck and Stroud, *A Glorious Enterprise*, 2–23.

20. Margaretta Hare Morris to Thaddeus William Harris, 6 November 1850, TWHP.

21. Lesueur to Anselme Desmarest, [c. 1818], quoted in R. W. G. Vail, "The American Sketchbooks of a French Naturalist, 1816–1837," *Proceedings of the American Antiquarian Society* (April 1938): 54.

22. Say quoted in Peck and Stroud, *A Glorious Enterprise*, 13; W. Conner Sorensen, *Brethren of the Net: American Entomology, 1840–1880* (University of Alabama Press, 1995), 34; Stroud, *Thomas Say*, 34–36; Graustein, *Thomas Nuttall*, 114–131; Margaretta Hare Morris to Thaddeus William Harris, September 12, 1847, TWHP.

23. Stroud, *Thomas Say*, 129–130; Margaretta Hare Morris to Thaddeus William Harris, 6 November 1850, TWHP.

24. John D. Godman, *Rambles of a Naturalist* (Thomas T. Ash, Key & Biddle, 1833), 7; John D. Godman, *American Natural History* (H. C. Carey & L. Lea, 1826), 1:84–96.

25. Edgar Fahs Smith, *The Life of Robert Hare, An American Chemist (1781–1858)* (J. B. Lippincott, 1917); *Robert Hare, M.D.* ([n.p.], 1858).

26. Lee Chambers-Schuller, *Liberty, a Better Husband: Single Women in America: The Generations of 1780–1840* (Yale University Press, 1984), 10–28, 43–45.

27. [Elizabeth Gibbs Willing Alleyne] to Ann Willing Morris, 12 May 1818, Ann W. Morris Letters, WLCL; Mary Roberdeau to Elizabeth C. Morris, 26 May 1827, LFP, Box 1, Folder 11; Mary Roberdeau to Elizabeth C. Morris, 5 July 1827, LFP, Box 1, Folder 11; Mary Roberdeau to Elizabeth C. Morris, 3 January 1828, LFP, Box 1, Folder 11; Page of hearts and initials in Elizabeth C. Morris's copybook 1829–1831, LFP, Box 1, Folder 13; Deborah Norris Logan, DNLD, 27 February 1827, XI: 15; Elizabeth Morris, "Questions to be answered by a gentleman," Box 1, Folder 15, LFP.

28. Deborah Norris Logan, 28 December 1820, DNLD, IV: 44; 6 July 1821, IV: 82; 5–16 May 1825, VIII: 25–28; 3 June 1821, IV: 74; 6–9 July 1821, IV: 82; 11 December 1821, V: 13; 23–24 December 1821, V: 17; 16 January 1822, V: 33; 2 March 1822, V: 49; Nina Baym, "Between Enlightenment and Victorian:

Toward a Narrative of American Women Writers Writing History," *Critical Inquiry* 18.1 (Autumn 1991): 22–41; Marleen Barr, "Deborah Norris Logan, Feminist Criticism, and Identity Theory: Interpreting a Woman's Diary Without the Danger of Separatism," *Biography* 8.1 (Winter 1984): 12–24; Terri L. Primo, *Winter Friends: Women Growing Old in the New Republic, 1785–1835* (University of Illinois Press, 1990).

29. Deborah Norris Logan, 1 January 1817, DNLD, II: 82; 5 February 1821, IV: 52–53; 30 October 1823, VI: 139; 7–10 January 1824, VI: 136-b, 137-b; 17–24 January 1824, VI: 140-b–141-b; 15 December 1828, XII: 123; 27 February 1830, XII: end pages; Claussen, *Wyck*, 134.

30. Deborah Norris Logan, [15] September 1822, DNLD, V: 147; [10] April 1823, VI: 40; 31 May 1824, VI: 199; 5–16 May 1825, VIII: 25–28; 6 July 1825, VIII: 59; 11 September 1825, VIII: 128; 1 March 1827, XI: 16; 3 January 1828, XI: 234; 15 December 1828, XII: 123; 8 February 1829, XII: 161–162; 27 February 1830, XII: end pages; 24–27 January 1831, XIII: 156–158; 6 and 10 August 1831, XIII: 255–257; 17 January 1832, XIII: 335, 1 July 1838, XVI: 301; 29 January 1838, XVI: 213.

31. Deborah Norris Logan, 16 January 1822, DNLD, V: 33; [November] 1823, VI: 114-b–115-b; 10 October 1824, VII: 123–124; 6 June 1825, VIII: 41; 20 July 1825, VIII: 71; Claussen, *Wyck*, 98–100.

32. Deborah Norris Logan, 17–20 June 1826, DNLD IX: 172–175; Elizabeth Carrington Morris, "To the Wissahiccon," in Offerings of Friendship, Box 1, Folder 12, LFP.

33. Deborah Norris Logan, 1 December 1821, DNLD, V: 7.

34. Nash, *Women's Education in the United States*; Margaret W. Rossiter, *Women Scientists in America: Struggles and Strategies to 1940* (Johns Hopkins University Press, 1982), 1:2–4; Kohlstedt, *Formation of the American Scientific Community*, 7–17; Bruce, *Launching of Modern American Science*, 75–93.

35. Deborah Norris Logan, DNLD, 1 April 1825, VIII: 9.

36. "Herschell's Great Telescope," in Elizabeth C. Morris's copybook 1829–1831, LFP, Box 1, Folder 13; Londa Schiebinger, *The Mind Has No Sex? Women in the Origins of Modern Science* (Harvard University Press, 1989), 260–264; Renee Bergland, *Maria Mitchell and the Sexing of Science: An Astronomer Among the American Romantics* (Beacon Press, 2008), 42–44, 110–114; Richard Holmes, *The Age of Wonder* (Vintage Books, 2008), 60–124.

37. Elizabeth Carrington Morris, "Meteors of the 13 November 1833," in Elizabeth Morris's copybook, 1832–1856, Box 1, Folder 14, LFP.

38. James C. Brandow, compiler, *Genealogies of Barbados Families* (Clearfield Company, 2001), 61–67; "John Forster Alleyne: Profile and Legacy Summary," *Centre for the Study of British Slavery*, University College of London, www.ucl.ac.uk/lbs/person/view/2146641201 [accessed 7 September 2021]; Mary Snodgrass, *The Underground Railroad: An Encyclopedia of People, Places, and Operations* (Taylor & Francis, 2015), 295. The Johnson House Historic Site is located at 6306 Germantown Road, Philadelphia.

39. E. C. Morris to William Darlington, 16 June 1851, WDP; Elizabeth Powel to Susan Sophia Morris, 1 January 1822, Box 1, Folder 19, LFP; Philadelphia, Pennsylvania Wills and Probate Records, 1683–1993 for Ann Willing Morris, Case Number 154 (1820); Mrs. Ann Willing Morris, probate 16 February 1853, Wills no. 1–60, 1853, Pennsylvania Probate Records, Philadelphia, Pennsylvania; E. C. Morris, "Stanzas Copied from 'Etonians' and Addressed to the Memory of my sister Anne W. Morris," in Susan S. Morris's Album and Scrapbook, 1826–1860, Box 2, Folder 36, LFP; "Stanzas," *The Etonian*, 2nd ed. (London: Henry Colburn & Co., 1822), 2:200; Ann Willing Morris, "Stanzas Copied from the Etonian and adopted for Ann Willing Morris," 3 November 1826, in Margaretta Hare Morris, album, collection of S. C. Doak, Hood River, Oregon.

40. John W. Jordan, ed., *Colonial and Revolutionary Families of Pennsylvania* (Clearfield, 1978), 1:72; Margaret Law Calcott, ed., *Mistress of Riversdale: The Plantation Letters of Rosalie Stier Calvert, 1795–1821* (Johns Hopkins University Press, 1991), preface, 317, 321, 326, 336, 343; George Calvert to Charles J. Stier, 2 November 1822 and 25 January 1823, in *The Ancestry of Rosalie Morris Johnson*, comp. R. Winder Johnson (Printed for private circulation by Ferris & Leach, 1905–1908), 1:169, 2:54; Elizabeth C. Morris to Ann Haines, 8 July 1822, Series 2, Box 26, Folder 396, WAP; Elizabeth Carrington Morris to Ann Willing Morris, 9 July 1823, "Spared & Shared 2," https://sparedandshared2.wordpress.com /letters/1823-elizabeth-carrington-morris-to-ann-willing-morris/ [accessed 8 March 2018].

41. John Stockton Littell, "The war of the Anemonies," in Elizabeth Carrington Morris, "Offerings of Friendship," LFP, Box 1, Folder 12; John Stockton Littell, "Twilight," in Elizabeth Carrington Morris, "Offerings of Friendship," LFP, Box 1, Folder 12; John Stockton Littell, "Spring breathes her first kisses . . . ," in Elizabeth Carrington Morris, "Offerings of Friendship," LFP, Box 1, Folder 12; Deborah Norris Logan, DNLD, 11 May 1826, IX: 147.

42. The poem, penned in 1826, was later copied into Elizabeth's album in 1849 by her nephew, John's son, Gardiner Littell. J. S. Littell, "The Wissahiccon," LFP, Box 1, Folder 17; Mary Roberdeau to Elizabeth C. Morris, 26 May 1827, LFP, Box 1, Folder 11.

43. Margaretta Hare Morris, "The Breath of Spring," in Margaretta Hare Morris's copybook, LFP, Box 1, Folder 7; "The Breath of Spring," *Analectic Magazine* 5 (April 1815): 345.

44. Margaretta Hare Morris, "The Arbor Saturni," in Margaretta Hare Morris's copybook, LFP, Box 1, Folder 7.

45. For more on same-sex love among women in the nineteenth century and the silences and erasures in archives, see Carroll Smith-Rosenberg, "The Female World of Love and Ritual: Relations Between Women in Nineteenth-Century America," *Signs* 1.1 (1975): 1–29; John D'Emilio and Estelle B. Freedman, *Intimate Matters: A History of Sexuality in America*, 3rd ed. (University of Chicago Press, 2012), 125–127; Rachel Hope Cleves, *Charity and Sylvia: A Same-Sex Marriage in Early*

America (Oxford University Press, 2014); Jen Manion, *Female Husbands: A Trans History* (Cambridge University Press, 2020); Estelle Freedman, "'The Burning of Letters Continues': Elusive Identities and the Historical Construction of Sexuality," *Journal of Women's History* 9.4 (Winter 1998): 181–200; Lillian Federman, *Surpassing the Love of Men: Romantic Friendship and Love Between Women from the Renaissance to the Present* (William Morrow, 1981), 147–230; Leila J. Rupp, *A Desired Past: A Short History of Same-Sex Love in America* (University of Chicago Press, 1999), 37–72.

46. Leonard Warren, *Maclure of New Harmony* (Indiana University Press, 2009), 71, 124–130; *Members and Correspondents of the Academy of Natural Sciences of Philadelphia*, 10; Stroud, *Thomas Say*, 23–24, 152–154.

47. Donald E. Pitzer, "The New Moral World of Robert Owen and New Harmony," in *America's Communal Utopias*, ed. Donald E. Pitzer (University of North Carolina Press, 1997), 88–134; Kohlstedt, *Formation of the American Scientific Community*, 13; Stroud, *Thomas Say*, 167–209; Warren, *Maclure of New Harmony*, 153–178; Josephine Mirabella Elliott, *Partnership for Posterity: The Correspondence of William Maclure and Marie Duclos Fretageot, 1820–1833* (Indiana Historical Society, 1994); Carol A. Kolmerten, *Women in Utopia: The Ideology of Gender in the American Owenite Communities* (Indiana University Press, 1990).

48. "From the Philadelphia Gazette, Nov. 29, the Owen Meeting on Friday," *Salem Gazette*, 6 December 1825; "The Philadelphia Gazette states . . . ," *Boston Commercial Gazette*, 8 December 1825.

49. Deborah Norris Logan, DNLD, 13 November 1825, IX: 6; 25 November 1825, IX: 13; 16 September 1826, X: 72; 7 January 1827, X: 148–149; Groff, "'All That Makes a Man's Mind More Active,'" 90–121;

50. Deborah Norris Logan, DNLD, 13 November 1825, IX: 6; Kolmerten, *Women in Utopia*.

51. Margaretta Hare Morris to Thaddeus William Harris, 6 November 1850.

52. Elizabeth C. Morris, "A Sister's Love," in Margaretta Hare Morris, album, collection of S. C. Doak, Hood River, Oregon; "A Sister's Love," *Etonian* 3 (1824): 278; Sara Coleridge, *Collected Poems*, ed. Peter Swaab (Carcanet Press, 2007), 112.

53. Peck and Stroud, *A Glorious Enterprise*, 38–44; Graustein, *Thomas Nuttall, Naturalist*, 216; Donald E. Pitzer, "William Maclure's Boatload of Knowledge: Science and Education into the Midwest," *Indiana Magazine of History* 94.2 (June 1998): 110–137; Warren, *Maclure of New Harmony*, 179–202.

54. DNLD, 1815–1839, Barbara Jones, transcriber, 7 January 1827, X: 148–149.

55. Stroud, *Thomas Say*, 216–217, 257; Hamy, *Travels of the Naturalist*, 51–52; Warren, *Maclure of New Harmony*, 269; Harry B. Weiss and Grace M. Ziegler, *Thomas Say: Early American Naturalist* (Charles C. Thomas, 1931).

56. Margaretta Hare Morris to Thaddeus William Harris, 6 November 1850, TWHP.

57. Deborah Norris Logan, DNLD, 30 October 1829, XII: 210-b; 6 March 1830, XIII: 11; Jane B. Haines to Ann Haines, 19 September 1830, Series 2, Box

20, Folder 275, WAP; M. A. Donaldson to Ann Haines, 14 December 1836, Series 2, Box 27, Folder 409, WAP; Deborah Norris Logan, 5 June 1831, DNLD, XIII: 222; 4 March 1833, XIV: 236; Mary Donaldson to Ann Haines, 5 August 1832, Series 2, Box 267, Folder 406, WAP; Elizabeth Hartshorne Haines to Ann Haines, 8 December 1833, Series 2, Box 29, Folder 437, WAP; Judith Callard, "The Alcotts in Germantown," *Germantown Crier* 70.1 (Spring 2020): 1–33, "Earliest Kindergarten," *Germantown Crier* 26.1 (Winter 1974): 19–20; Claussen, *Wyck*, 110–116, 133.

58. Deborah Norris Logan, 8 March 1828, DNLD, XI: 263–264; 25 May 1828, XI: appendix; 30 July 1830, XIII: 63–65; 2 August 1830, XIII: 79; 21 and 27 February 1831, XIII: 172, 175–177; 20 March 1832, XIV: 16.

59. "Littell" clipping, 15 July 1868, Box 2, Folder 39b, LFP. For contemporary expectations for wives and mothers, see Charles Butler, *The American Lady* (Hogan & Thompson, 1836); Lydia Maria Child, *The Mother's Book* (Carter, Hendee & Babcock, 1831). On marriage and the personal lives of women scientists, see Pnina G. Abir-Am and Dorinda Outram, ed., *Uneasy Careers and Intimate Lives: Women in Science, 1789–1979* (Rutgers University Press, 1987).

60. Deborah Norris Logan, DNLD, 14 December 1829, XII: 239; 17 January 1830, XII: 261–262; 24 January 1830, XII: 265; 23 February 1830, XII: end pages; 1 May 1830, XIII: 18–19; Mrs. Ann Willing Morris, probate 16 February 1853, Wills no. 1–60, 1853, Pennsylvania Probate Record, Philadelphia, Pennsylvania; PFP, Series 3, Box 7.

61. Godman's poem "To such as gaze with heedless eye" that he wrote in Elizabeth's album was published anonymously just a few months later in *The Friend: A Religious and Literary Journal*, making reference to it being in an album (clearly Elizabeth's based on the details provided). Perhaps Elizabeth published it with his permission, perhaps not, but she thought it ought to be shared widely. "Stanzas," in "Offerings of Friendship," LFP, Box 1, Folder 12; J. D. Godman, "To such as gaze with heedless eye . . . ," in "Offerings of Friendship," LFP, Box 1, Folder 12; G. "Lines," *The Friend: A Religious and Literary Journal* (John Richardson, 1829), II: 311; "Stanzas" in "Offerings of Friendship," LFP, Box 1 Folder 12; "The Baptism," in "Offerings of Friendship," LFP, Box 1, Folder 12; John D. Godman, "Mon Bienvenue au premier Bouquet d'un Printemps Tardif," in LFP, Box 1, Folder 12; Deborah Norris Logan, DNLD, 14 January 1829, XII: 146–147; 1 May 1830, XIII: 18–19; Thomas Sewall, *Memoir of Dr. Godman* (Tract Society of the Methodist Episcopal Church, 1832); Susan A. C. Rosen, "John D. Godman, M.D.," in *Early American Nature Writers*, ed. Daniel Patterson (Greenwood Press, 2008), 173–177; Lawrence Buell, *The Environmental Imagination: Thoreau, Nature Writing, and the Formation of American Culture* (Harvard University Press, 1995), 399–400.

62. Deborah Norris Logan, 18–23, 29 October 1831, DNLD, XIII: 292–294, 298.

63. Harrison Ellery, *The Memoirs of Gen. Joseph Gardner Swift, LL.D., U.S.A.* (Printed privately, 1890), 216; Master Hankes, "Family of John Quincy Adams,"

1829, Silhouette Collection, Massachusetts Historical Society; Deborah Norris Logan, [10] December 1833, DNLD, XIV: 386; M. A. Donaldson to Ann Haines, 8 December 1833, Series 2, Box 27, Folder 407, WAP; Elizabeth Hartshorne Haines to Ann Haines, 8 December 1833, Series 2, Box 29, Folder 437, WAP; Elizabeth C. Morris to Ann Haines, 3 June 1840, Series 2, Box 267, Folder 417, WAP; Lady Eleanor Butler, "The old Maid's Prayer to Diana," *The Port Folio* XIV (1822): 444; Mary Tighe, *The Collected Poems of Mary Tighe*, edited by Paula R. Feldman and Brian C. Cooney (Johns Hopkins University Press, 2016), 507–509; "Song," in Elizabeth Morris's copybook, 1829–1831, LFP, Box 1, Folder 13.

64. Mary Donaldson and Jane Haines to Ann Haines, 16 August [1830], Series 2, Box 26, Folder 404, WAP; Elizabeth C. Morris to Ann Haines, 8 July 1822, Series 2, Box 26, Folder 396, WAP.

65. Elizabeth C. Morris to Ann Haines, 3 June 1840, Series 2, Box 267, Folder 417, WAP; Deborah Norris Logan, DNLD, [28] December 1836, XVI: 22; 30 and 31 December 1838, XVII: 27, 29; Elizabeth Hartshorne Haines et al. to Jane B. Haines, 1 May 1834, Series 2, Box 29, Folder 437, WAP; Mary Donaldson to Ann Haines, 16 August [1830], Series 2, Box 26, Folder 404, WAP.

CHAPTER THREE: AN OBJECT OF PECULIAR INTEREST

1. Miss M. H. Morris, "On the Cecidomyia Destructor, or Hessian Fly," read 2 October 1840, *TAPS* Series 2 VIII (1843): 49–51; Margaretta Hare Morris to Thaddeus William Harris, 12 September 1847, TWHP; M., "The Army Worm," *AA* 6.2 (February 1847): 50; Anne Larsen, "Equipment for the Field," in *Cultures of Natural History*, ed. N. Jardine, J. A. Secord, and E. C. Spary (Cambridge University Press, 1996), 373–377; William Kirby and William Spence, "On Entomological Instruments and the Best Methods of Collecting, Breeding, and Preserving Insects," in *An Introduction to Entomology* (Logman, Rees, Brown, and Green, 1826), 4:515–546.

2. Morris to Harris, 12 September 1847, TWHP.

3. Steven Stoll, *Larding the Lean Earth: Soil and Society in Nineteenth-Century America* (Hill and Wang, 2002); Emily Pawley, *The Nature of the Future: Agriculture, Science, and Capitalism in the Antebellum North* (University of Chicago Press, 2020); Ariel Ron, *Grassroots Leviathan: Agricultural Reform and the Rural North in the Slaveholding Republic* (Johns Hopkins University Press, 2020).

4. A small selection of the long-running debate can be seen in these articles and others: R., "Hessian Fly," *American Farmer* 2.24 (September 1820): 192; Hezekiah M'Clelland, "Hessian Fly," *American Farmer* 2.30 (October 1820): 234; S., "Hessian Fly," *American Farmer* 5.18 (July 1823): 139; John H. Cocke, "Hessian Fly," *American Farmer* 5.31 (October 1823): 241; "Simple Method of Destroying the Hessian Fly," *New England Farmer and Horticultural Register* 12.6 (August 1833): 44; "Chich Bug—Scab—Hessian Fly—Smut," *Farmer and Gardener* 2.41 (February

1836): 323; Jefferson Shields, *Farmer and Gardener* 3.21 (September 1836): 163; A. B., "Hessian Fly and Wheat Worm," *Genesee Farmer and Gardener's Journal* 6.45 (November 1836). For American grains in 1840, see United States Department of the Interior, *Report of the Secretary of the Interior* (1849), 519–522.

5. Paul D. N. Hebert, Sujeevan Ratnasingham et al., "Counting Animal Species with DNA Barcodes: Canadian Insects," *Philosophical Transactions of the Royal Society B* 371.1702 (5 September 2016); Raymond Gagné also talks about the vast number of *Cecidomyia* that remain unnamed and unstudied in Raymond J. Gagné, *The Gall Midges of the Neotropical Region* (Comstock Publishing Associates, Cornell University Press, 1994), 2.

6. Thomas Say, "Some account of the Insect known by the name of Hessian Fly, and of a parasitic Insect that feeds on it," *JANSP* 1.3 (July 1817): 45–48; Charles Alexandre Lesueur, "Plate III," *JANSP* 1.4 (August 1817): 64–65; Philip Pauly, *Fruits and Plains: The Horticultural Transformation of America* (Harvard University Press, 2007), 33–50.

7. Morris to Harris, 12 September 1847; Margaretta Hare Morris to Thaddeus William Harris, 6 November 1850, TWHP; Mary Donaldson to Ann Haines, 13 October 1841, Series 2, Box 267, Folder 418, WAP; Deborah Norris Logan, DNLD, 18, 21, 23 October 1831, XIII: 292–294.

8. Thomas Say, "Some account of the Insect known by the name of Hessian Fly," 45–48.

9. Say, "Some account of the Insect known by the name of Hessian Fly," 47; Miss M. H. Morris, "On the Cecidomyia Destructor, or Hessian Fly," read 2 October 1840, *TAPS* Series 2, VIII (1843): 49–51.

10. Morris, "On the Cecidomyia Destructor, or Hessian Fly"; Patricia Tyson Stroud, *Thomas Say: New World Naturalist* (University of Pennsylvania Press, 1992), 256–257. For more on how farmers adapted their planting cycles to best the wheat fly, see Alan L. Olmsted and Paul W. Rhode, "Biological Innovation in American Wheat Production: Sciences, Policy, and Environmental Adaptation," in *Industrializing Organisms: Introducing Evolutionary History*, ed. Susan R. Schrepfer and Philip Scranton (Routledge, 2004), 43–85.

11. "A Preventive of the Wheat Fly," *Farmers' Cabinet* 1.5 (September 1836): 73; Jefferson Shields, *Farmer and Gardener* 3.21 (September 1836): 163; "The Hessian Fly," *Farmer and Gardener* 3.22 (September 1836): 169; "Lime as a Manure for Wheat," *Farmers' Cabinet* 1.6 (October 1836): 85–85; "Hessian Fly from the Pennsylvania Republican," *New England Farmer and Horticultural Register* 15.22 (December 1836): 172; "Hessian Fly," *Farmers' Cabinet* 1.12 (January 1837): 185–186; "The Observer—No. 5," *Farmers' Cabinet* 1.19 (April 1837): 290; "The Observer—No. 6," *Farmers' Cabinet* 1.20 (May 1837): 306–307; "Wheat—Important Discovery," *Farmers' Cabinet* 1.23 (June 1837): 359–361.

12. Henry Green, "The Wheat-Worm," *Cultivator* (June 1836): 53; "The Harvest Prospect," *Cultivator* (July 1836): 59; "A Preventive of the Wheat Fly," *Farmers'*

Cabinet 1.5 (September 1836): 73; William Penn Kinzer, "The Cut Worm and Hessian Fly," *Cultivator* (October 1836): 109; "The Crops," *Cultivator* (September 1836): 87; J. Hathaway, "Italian Spring Wheat," *Cultivator* (September 1836): 94; "Enemies of the Wheat Crop," *Cultivator* (September 1836): 93; "Lime as a Manure for Wheat," *Farmers' Cabinet* 1.6 (October 1836): 85–85; "Hessian Fly," *Farmers' Cabinet* 1.12 (January 1837): 185–186; "The Observer—No. 5," *Farmers' Cabinet* 1.19 (April 1837): 290; "The Observer—No. 6," *Farmers' Cabinet* 1.20 (May 1837): 306–307; "Wheat—Important Discovery," *Farmers' Cabinet* 1.23 (June 1837): 359–361; "Price of Flour—Comparative Table," *Farmers' Cabinet* 1.19 (April 1837): 303; Jessica M. Lepler, *The Many Panics of 1837: People, Politics, and the Creation of a Transatlantic Financial Crisis* (New York: Cambridge University Press, 2013), 67–70; Sean Wilentz, *The Rise of American Democracy: Jefferson to Lincoln* (W. W. Norton, 2005), 456–457.

13. Morris to Harris, 12 September 1847. On women authors in the public sphere and the perils of "appearing in public," see Mary Kelley, *Private Woman, Public Stage: Literary Domesticity in Nineteenth-Century America* (Oxford University Press, 1984); Robert Gunn, "'How I Look': Fanny Fern and the Strategy of Pseudonymity," *Legacy* 27.1 (2010): 23–42; Nicole Tonkovich, *Domesticity with a Difference: The Nonfiction of Catharine Beecher, Sarah J. Hale, Fanny Fern, and Margaret Fuller* (University Press of Mississippi, 1997): 48–51.

14. Mary Donaldson to Ann Haines, 13 October 1841, Series 2, Box 267, Folder 418, WAP; "Stated Meeting, October 2," *PAPS* 1.13 (1838–1840): 282; "Stated Meeting, December 18," *PAPS* 1.14 (November–December 1840): 318–319; Miss M. H. Morris, "On the Cecidomyia Destructor, or Hessian Fly," *TAPS* 8 (1843): 49–51. On women having men present their findings, see Sally Gregory Kohlstedt, "In from the Periphery: American Women in Science, 1830–1880," *Signs* 4.1 (Autumn 1978): 90.

15. "Stated Meeting, October 2," *PAPS* 1.13 (1838–1840): 282; Morris, "On the Cecidomyia Destructor, or Hessian Fly."

16. "On the Cecidomyia Destructor, or Hessian Fly."

17. Morris, "On the Cecidomyia Destructor, or Hessian Fly"; "Stated Meeting, December 18," *PAPS* 1.14 (November–December 1840): 318–319.

18. Benjamin H. Coates, "Hessian Fly," *Farmers' Cabinet* 5.6 (January 1841): 201–205. On the role of agricultural journals as spaces for public science, see James E. McWilliams, *American Pests: The Losing War on Insects from Colonial Times to DDT* (Columbia University Press, 2008), 16–24.

19. Coates, "Hessian Fly."

20. "Observer—No. 24: 'Who Shall Decide, When Doctors Disagree?'" *Farmers' Cabinet* 5.7 (February 1841): 237–239; "Hessian Fly," *Southern Planter* 1.1 (January 1841): 14; Barbara Welter, "The Cult of True Womanhood: 1820–1860," *American Quarterly* 18.2 (1966): 151–174; Leila Rupp, ed., "Women's History in the New Millennium: A Retrospective Analysis of Barbara Welter's 'The Cult of

True Womanhood, 1820-1860,'" *Journal of Women's History* 14.1 (2002): 149–173; Robert Nye, "Medicine and Science as Masculine 'Fields of Honor,'" *Osiris* 12 (1997): 60–79.

21. M. H. Morris, "Controversy Respecting the Hessian Fly," *American Journal of Agriculture and Science* 5.12 (April 1847): 208; "Hessian Fly—A Lady Observer," *Cultivator* 8.3 (March 1841): 46; W. L. H., "Observations on the Hessian Fly," *Farmers' Cabinet* 5.7 (February 1841): 229–230.

22. William Darlington, *A Discourse on the Character, Properties, and Importance to Man of the Natural Family of Plants Called Gramineae, or True Grasses* (Printed for the Chester County Cabinet of Natural Science, 1841), 5–6.

23. James J. Levick, "Benjamin Hornor Coates, M.D., One of the Founders of the Historical Society of Pennsylvania and for Many Years Its Senior Vice-President," *Pennsylvania Magazine of History and Biography* 6.1 (1882): 33. Benjamin Hornor Coates, M.D. (1797–1882), can easily be confused with his cousin, also living at the time in Philadelphia, Benjamin Coates (1808–1887), a dry goods merchant who was an outspoken proponent of African colonization efforts. Emma J. Labsansky-Werner and Margaret Hope Bacon, eds., *Back to Africa: Benjamin Coates and the Colonization Movement in America, 1848–1880* (Pennsylvania State University Press, 2005), ix; B. H. Coates, "Cecidomyia," *Farmers' Cabinet* 5.8 (March 1841): 263–264; B. H. Coates to William Darlington, 18 June 1841, WDP; "Observer—No. 24," *Farmers' Cabinet* 5.7 (February 1841): 237–239.

24. B. H. Coates, "Cecidomyia."

25. On the early professionalization of American entomology and the limitations of the field, see W. Conner Sorensen, *Brethren of the Net: American Entomology, 1840–1880* (University of Alabama Press, 1995), 60–91; McWilliams, *American Pests*, 26–55.

26. Thomas Anthony Thacher, *Sketch of the Life of Edward C. Herrick* (Printed by Thomas J. Stafford, 1862); "Death of Edward C. Herrick, Treasurer of Yale College," *New York Times*, 22 June 1862; ECHP.

27. TWHP; Clark A. Elliott, *Thaddeus William Harris (1795–1856): Nature, Science, and Society in the Life of an American Naturalist* (Lehigh University Press, 2008), 33, 48–50, 62–65.

28. A selection of these letters is excerpted in Samuel H. Scudder, ed., *The Entomological Correspondence of Thaddeus William Harris* (Boston Society of Natural History, 1869), 181–207. Many of the originals as well as additional letters are in the TWHP and the ECHP; Erica McAlister, *The Secret Life of Flies* (Firefly Books, 2017), 139–141; Janet Browne, "Corresponding Naturalists," in *The Age of Scientific Naturalism*, ed. Bernard Lightman and Michael S. Reidy (Pickering & Chatto, 2014), 157–170.

29. "Stated Meeting, December 18," *PAPS* 1.14 (November–December 1840): 318–319; Thaddeus William Harris to Edward Claudius Herrick, 6 March 1841, TWHP.

30. "Stated Meeting, December 18," *PAPS* 1.14 (November–December 1840): 318–319; Harris to Herrick, 6 March 1841, TWHP.

31. Edward C. Herrick, "Art. XV.—A Brief, Preliminary Account of the Hessian Fly and Its Parasites," *AJSA* 41 (1841): 153–158.

32. Herrick, "Art. XV.—A Brief, Preliminary Account of the Hessian Fly and Its Parasites," 155.

33. Benjamin Silliman to Margaretta Hare Morris, 7 July 1841, Box 1, Folder 10, AGP; B. H. Coates to William Darlington, 18 June 1841, WDP; [Benjamin Silliman], *Letters of Shahcoolen, A Hindu Philosopher* (Russell and Cutler, 1802); Chandos Michael Brown, *Benjamin Silliman: A Life in the Young Republic* (Princeton University Press, 1989), 71–81; Margaret W. Rossiter, *Women Scientists in America: Struggles and Strategies to 1940* (Johns Hopkins University Press, 1982), 1:2. The handful of women who published in the *American Journal of Science and Arts*, perhaps excluding those writing pseudonymously, include the following: Miss D. L. Dix, "ART IX—Notice of the Aranea Aculeata, the Phaloena antiqua and some species of the Papilio," *AJSA* 19.1 (January 1831): 61; Mrs. Graham, "ART IV—On the Reality of the Rise of the Coast of Chile, in 1822," *AJSA* 28.2 (January 1835): 236; Mary Griffith, "ART III—On the Halo or Fringe which surrounds all Bodies," *AJSA* 38.1 (April 1840): 22. For more on female scientists during this period, see Kohlstedt, "In from the Periphery."

34. Thaddeus William Harris, *A Report on the Insects of Massachusetts, Injurious to Vegetation* (Folsom, Wells, and Thurston, 1841), 423–424, 426, 427, 429–432, 440–441.

35. Thaddeus William Harris to E. C. Herrick, Esq., copy, 24 November 1841, Series 1, Box 2, Folder 19, ECHP.

36. Margaretta Hare Morris to Benjamin Hornor Coates, 9 June 1841, AGP, Box 1, Folder 11; Morris to Harris, 12 September 1847. Coates also looked to reinforce Morris's findings by heading into the fields around Philadelphia in search of compromised wheat. "Verbal Communications," *PANSP* 1 (22 June 1841): 44–45.

37. "Verbal Communications," *PANSP* 1 (13 July 1841): 54–56.

38. Margaretta Hare Morris to Dr. B. Coates, 14 July 1841, Special Collections, Pennsylvania State University Library, 104 Paterno, Vault, 1986-0104R VF Lit; "Written Communications," *PANSP* 1 (10 August 1841): 66–68.

39. Margaretta Hare Morris to Dr. B. Coates, 14 July 1841; Margaretta Hare Morris to Thaddeus William Harris, 6 November 1850, TWHP.

40. Jeanette E. Graustein, *Thomas Nuttall, Naturalist: Explorations in America, 1808–1841* (Harvard University Press, 1967), 356–361, 374.

41. "Written Communications," *PANSP* 1 (22 June 1841): 57; "Written Communications," *PANSP* 1 (10 August 1841): 66–68; Morris to Harris, 12 September 1847.

42. "Written Communications," *PANSP* 1 (22 June 1841); "Written Communications," *PANSP* 1 (10 August 1841); Morris to Harris, 12 September 1847.

43. Mary Donaldson to Ann Haines, 13 October 1841, Series 2, Box 267, Folder 418, WAP.

44. Margaretta Hare Morris to Thaddeus William Harris, 31 August 1843, TWHP.

45. Elliott, *Thaddeus William Harris*, 64–65; Morris to Harris, 12 September 1847.

46. Jeffrey K. Barnes, *Asa Fitch and the Emergence of American Entomology* (University of the State of New York/New York State Museum, 1985), 21, 32–33, 45–49. For more on Fitch, see McWilliams, *American Pests*, 31–33.

47. Asa Fitch, *The Hessian Fly: Its History, Character, Transformations, and Habits* (C. Van Benthuysen, 1847): 23, 42–43; "Stated Meeting, August 10, 1841," *PANSP* 1.5 (August 1841): 65–68.

48. Morris to Harris, 12 September 1847.

49. Morris, "Controversy Respecting the Hessian Fly," *American Journal of Agriculture and Science* 5.12 (April 1847): 206–208.

50. Morris, "Controversy Respecting the Hessian Fly."

51. Morris, "Controversy Respecting the Hessian Fly."

52. Morris to Harris, 12 September 1847.

Chapter Four: Webs of Correspondence

1. E. S., "Ladies' Department: Reply to Reviewer on Knitting, &c.," *AA* 6.2 (February 1847): 65; "For Farmers' Daughters," *AA* 4.4 (April 1845): 47; Asa Gray to Elizabeth Carrington Morris, 24 September 1844, Box 1, Folder 3, AGP.

2. E. S., "Lard Lamps," *AA* 6.4 (April 1847): 111–112; Elizabeth C. Morris to Ann Haines, 3 June 1840, Series 2, Box 267, Folder 417, WAP; Asa Gray to Elizabeth Carrington Morris, 25 July 1847, Box 1, Folder 4, AGP.

3. On delights, see Ross Gay, *The Book of Delights: Essays* (Algonquin Books, 2019), which itself is a delight.

4. Janet Browne, "Asa Gray and Charles Darwin: Corresponding Naturalists," *Harvard Papers in Botany* 15.2 (2010): 173–186; Richard R. John, *Spreading the News: The American Postal System from Franklin to Morse* (Harvard University Press, 2009).

5. "The Late Dr. William Darlington," *GM* 5.6 (June 1863): 182–183.

6. William Darlington to Ann Haines, 3 September 1842, Series 2, Box 267, Folder 419, WAP; T. P. Thomas, "An Obituary Notice of Dr. William Darlington," *PAPS* 9.70 (June 1863): 330–342; William Darlington, *Flora Cestrica* (S. Siegfried, 1837).

7. Elizabeth C. Morris to William Darlington, 5 July 1842, WDP.

8. Elizabeth C. Morris to William Darlington, 9 August 1842, WDP; Elizabeth C. Morris to William Darlington, 18 August 1842, WDP; M. H. Morris to Dr. Darlington, 18 August 1842, WDP; Elizabeth C. Morris to William Darlington, 7 September 1842, WDP; Elizabeth C. Morris to William Darlington, 14 October 1842, WDP.

9. A few examples of their book exchanges: Elizabeth C. Morris to William Darlington, 14 October 1842, WDP; Elizabeth C. Morris to William Darlington, 15 November 1842, WDP; Elizabeth C. Morris to William Darlington, 24 April 1843, WDP; Elizabeth C. Morris to William Darlington, 28 April 1843, WDP; Elizabeth C. Morris to William Darlington, 19 July 1843, WDP; Elizabeth C. Morris to William Darlington, 19 February 1844, WDP; Elizabeth C. Morris to William Darlington, 2 March 1846, WDP. For more on the railroad development around Philadelphia, see Nicholas B. Wainwright, "The Age of Nicholas Biddle, 1825–1841," in *Philadelphia: A 300-Year History*, ed. Russell F. Weigley (W. W. Norton, 1982), 272–274.

10. Elizabeth C. Morris to William Darlington, 18 August 1842, WDP.

11. Elizabeth C. Morris to William Darlington, 27 January 1843, WDP; Elizabeth C. Morris to William Darlington, 24 April 1843, WDP; Augustin Pyramus de Candolle and Kurt Polycarp Joachim Sprengel, *Elements of the Philosophy of Plants* (W. Blackwood, 1821); Elizabeth B. Keeney, *The Botanizers: Amateur Scientists in Nineteenth-Century America* (University of North Carolina Press, 1992), 62–69; Ann B. Shteir, *Cultivating Women, Cultivating Science* (Johns Hopkins University Press, 1996), 155–158; Ann B. Shteir, "Gender and 'Modern' Botany in Victorian England," *Osiris* 12 (1997): 29–38; Nina Baym, *American Women of Letters and the Nineteenth-Century Sciences* (Rutgers University Press, 2002), 30–35. For more on power and classification, see Harriet Ritvo, "Zoological Nomenclature and the Empire of Victorian Science," in *Victorian Science in Context*, ed. Bernard Lightman (University of Chicago Press, 1997), 334–353.

12. Lawrence Farber, *Finding Order in Nature: The Naturalist Tradition from Linnaeus to E. O. Wilson* (Johns Hopkins University Press, 2000); Bruno J. Strasser, "Collecting Nature: Practices, Styles, and Narratives," *Osiris* 27.1 (2012): 303–340; Robert E. Kohler, "Finders, Keepers: Collecting Sciences and Collecting Practices," *History of Science* 45 (2007): 428–454; Robert E. Kohler, *All Creatures: Naturalists, Collectors, and Biodiversity* (Princeton University Press, 2006), preface and chapter 1.

13. Elizabeth C. Morris to William Darlington, 18 March 1843, WDP.

14. Elizabeth C. Morris to William Darlington, 7 September 1842, WDP; Elizabeth C. Morris to William Darlington, 15 November 1842, WDP; Elizabeth C. Morris to William Darlington, 27 January 1843, WDP.

15. WDP; William Darlington Letter Books, West Chester University.

16. Elizabeth C. Morris to William Darlington, 3 November 1845, WDP.

17. Asa Gray, *The Botanical Text-book* (Wiley & Putnam, 1842); Elizabeth C. Morris to William Darlington, 27 August 1842, WDP; A. Hunter Dupree, *Asa Gray: American Botanist, Friend of Darwin* (Johns Hopkins University Press, 1959, 1988), 30, 106–114, 127.

18. Dupree, *Asa Gray*, 115–118; Elizabeth C. Morris to William Darlington, 15 November 1842, WDP; Asa Gray to William Darlington, 13 July 1842, WDP;

Asa Gray to William Darlington, 3 September 1842, WDP; William Darlington to Asa Gray, 23 September 1842, AGCF.

19. William Darlington to Asa Gray, 23 September 1842, AGCF; Asa Gray to William Darlington, 10 October 1842, WDP.

20. William Darlington to Asa Gray, 18 October 1842, AGCF.

21. Elizabeth C. Morris to William Darlington, 15 November 1842, WDP.

22. William Darlington to Asa Gray, 29 November 1842, AGCF; Elizabeth C. Morris to William Darlington, 15 November 1842, WDP; Asa Gray to Elizabeth Carrington Morris, 5 December 1842, Box 1, Folder 3, AGP; Elizabeth C. Morris to William Darlington, 12 December 1842, WDP; Ritvo, "Zoological Nomenclature and the Empire of Victorian Science," 334–353.

23. Elizabeth C. Morris to William Darlington, 27 January 1843, WDP.

24. "The Late William F. Harnden," *Living Age* 9 (1846): 17; "Origin of Wells, Fargo & Company, 1841–1852," *Bulletin of the Business Historical Society* 22.3 (June 1948): 70–71; William Darlington to Asa Gray, 29 November 1842, AGCF; Asa Gray to Elizabeth Carrington Morris, 9 April 1843, Box 1, Folder 3, AGP; Asa Gray to Elizabeth Carrington Morris, 22 November 1843, Box 1, Folder 3, AGP; Asa Gray to Elizabeth Carrington Morris, 25 March 1844, Box 1, Folder 3, AGP.

25. Thomas Nuttall, *The Genera of North American Plants* (D. Heartt, 1818), 2:251; Elizabeth C. Morris to William Darlington, 24 April 1843, WDP; Elizabeth C. Morris to William Darlington, 28 April 1843, WDP.

26. Anne Larsen, "Equipment for the Field," in *Cultures of Natural History*, ed. N. Jardine, J. A. Secord, and E. C. Spary (Cambridge University Press, 1996), 366–367; E. S., "The Garden," *AA* 5.4 (April 1846): 127.

27. Asa Gray, "Of Collecting and Preserving Plants for An Herbarium," in *Elements of Botany*, ed. Asa Gray (G. & C. Carvill, 1836), 396–405; Charles W. Short, "Instructions for the Gathering and Preservation of Plants for Herbaria," *Journal of the Philadelphia College of Pharmacy* 5 (1836): 218–227. Examples of Elizabeth's herbarium specimens can be found, though altered, at the West Chester University William Darlington Herbarium, the New York Botanical Garden, and the Asa Gray Herbarium at Harvard University, among other locations.

28. Gray, *Elements of Botany*, 403. For more on the complex relationship between the deep environmental knowledge of local collectors and more powerful cosmopolitan collectors, see Jim Endersby, "'From Having No Herbarium': Local Knowledge vs. Metropolitan Expertise: Joseph Hooker's Australasian Correspondence with William Colenso and Ronald Gunn," *Pacific Science* 55.4 (2001): 343–358; Jeremy Vetter, "Lay Participation in the History of Scientific Observation," *Science in Context* 24.2 (2011): 127–141; Anne Secord, "Corresponding Interests: Artisans and Gentlemen in Nineteenth-Century Natural History," *British Journal for the History of Science* 27.4 (December 1994): 383–408; Strasser, "Collecting Nature."

29. Prof. Rafinesque, *New Flora of North America* (Printed for the Author and Publisher, 1836).

30. Elizabeth C. Morris to William Darlington, 14 October 1842, WDP; William Cronon, "The Trouble with Wilderness: Or, Getting Back to the Wrong Nature," *Environmental History* 1.1 (1996): 7–28; Amy Kohout, *Taking the Field: Soldiers, Nature, and Empire on the American Frontier* (University of Nebraska Press, 2023). Women did occasionally make it onto expeditions, though not without courage and defiance, as was the case for Maria Sibylla Merian and Jeanne Baret. Kim Todd, *Chrysalis: Maria Sibylla Merian and the Secrets of Metamorphosis* (Harvest Books, 2007); Glynnis Ridley, *The Discovery of Jeanne Baret: A Story of Science, the High Seas, and the First Woman to Circumnavigate the Globe* (Broadway Paperbacks, 2010).

31. Asa Gray to Elizabeth Carrington Morris, 24 April 1844, Box 1, Folder 3, AGP; Asa Gray to William Darlington, 27 May 1844, WDP; Asa Gray to Elizabeth Carrington Morris, 24 September 1844, Box 1, Folder 3, AGP.

32. Asa Gray to William Hooker, 16 May 1844, Director's Correspondence, Royal Botanic Gardens, Kew; William J. Hooker to Asa Gray, 29 May 1844, AGCF; William Jackson Hooker, *A Century of Ferns* (W. Pamplin, 1854), Tab. 27; William Jackson Hooker, *Species Filicum, being descriptions of the known ferns* (W. Pamplin, 1846–1864), 3:91–92.

33. Asa Gray, *A Manual of the Botany of the Northern United States* (J. Munroe, 1848), 627.

34. Asa Gray to Elizabeth C. Morris, 31 August 1845, Box 1, Folder 3, AGP; Elizabeth C. Morris to William Darlington, 29 September 1845, WDP; Asa Gray to Elizabeth Carrington Morris, 21 October 1845, Box 1, Folder 3, AGP; Edith Denner, "Dress Reform: Science and Long Skirts," *Water-Cure Journal* 20.1 (July 1855): 7; E. S., "The Garden," *AA* 5.4 (April 1846): 127.

35. Asa Gray to Elizabeth Carrington Morris, 24 September 1844, Box 1, Folder 3, AGP; Asa Gray to Elizabeth C. Morris, 31 August 1845, Box 1, Folder 3, AGP; Gray, *A Manual of the Botany*, 337–338. For more on the changing context of the language of flowers, see Shteir, *Cultivating Women, Cultivating Science*, 158–160.

36. Elizabeth C. Morris to William Darlington, 23 December 1844, WDP; Keeney, *The Botanizers*, 64–68.

37. Asa Gray to Elizabeth Carrington Morris, 9 June 1845, Box 1, Folder 3, AGP.

38. Asa Gray to Elizabeth Carrington Morris, 9 April 1843, Box 1, Folder 3, AGP; Asa Gray to Elizabeth Carrington Morris, 16 June 1843, Box 1, Folder 3, AGP; Asa Gray to Elizabeth Carrington Morris, 9 June 1845, Box 1, Folder 3, AGP.

39. Asa Gray to Elizabeth Carrington Morris, 9 April 1843, Box 1, Folder 3, AGP; Elizabeth C. Morris to William Darlington, 16 April 1844, WDP; Asa Gray to Elizabeth Carrington Morris, 15 January 1846, Box 1, Folder 3, AGP; Asa Gray to Elizabeth Carrington Morris, 9 June 1845, Box 1, Folder 3, AGP; Asa Gray to Elizabeth C. Morris, 31 August 1845, Box 1, Folder 3, AGP.

40. Asa Gray to Elizabeth Carrington Morris, 9 April 1843, Box 1, Folder 3, AGP.

41. Asa Gray to Elizabeth Carrington Morris, 9 April 1843, Box 1, Folder 3, AGP; Elizabeth C. Morris to William Darlington, 15 May 1859, WDP; W. Edmunds Claussen, *Wyck: The Story of an Historic Home, 1690-1970* (Mary T. Haines, 1970), 139.

42. Asa Gray to Elizabeth Carrington Morris, 9 April 1843, Box 1, Folder 3, AGP; Asa Gray to Elizabeth Carrington Morris, 16 June 1843, Box 1, Folder 3, AGP; Gray, *A Manual of the Botany*, 250–251.

43. Asa Gray to Elizabeth Carrington Morris, 15 June 1846, Box 1, Folder 3, AGP; Asa Gray to Elizabeth Carrington Morris, 25 July 1847, Box 1, Folder 4, AGP.

44. Asa Gray to Elizabeth Carrington Morris, 26 November 1845, Box 1, Folder 3, AGP; Dupree, *Asa Gray*, 127–131; Asa Gray to Elizabeth Carrington Morris, 25 March 1844, Box 1, Folder 3, AGP; Asa Gray to Elizabeth Carrington Morris, 24 September 1844, Box 1, Folder 3, AGP; Asa Gray to Elizabeth Carrington Morris, 9 November 1844, Box 1, Folder 3, AGP; Asa Gray to Elizabeth Carrington Morris, 14 April 1845, Box 1, Folder 3, AGP; Asa Gray to Elizabeth Carrington Morris, 21 October 1845, Box 1, Folder 3, AGP; Asa Gray to Elizabeth Carrington Morris, 15 January 1846, Box 1, Folder 3, AGP.

45. Asa Gray to Elizabeth Carrington Morris, 8 June 1846, Box 1, Folder 3, AGP; Asa Gray to Elizabeth Carrington Morris, 29 June 1846, Box 1, Folder 3, AGP; Asa Gray to Elizabeth Carrington Morris, 25 July 1847, Box 1, Folder 4, AGP; Asa Gray to Elizabeth Carrington Morris, 11 August 1849, Box 1, Folder 4, AGP; Asa Gray to Elizabeth Carrington Morris, 18 January 1850, Box 1, Folder 4, AGP.

46. Asa Gray, *The Botanical Text-book* (Wiley & Putnam, 1842); Asa Gray to Elizabeth Carrington Morris, 8 June 1846, Box 1, Folder 3, AGP; Mrs. Folsom, "For the Birth-Day of Linnaeus," in Offerings of Friendship, Box 1, Folder 12, LFP.

47. Elizabeth C. Morris to William Darlington, 16 June 1846, WDP; Asa Gray to Elizabeth Carrington Morris, 25 August 1848, Box 1, Folder 4, AGP. Elizabeth also gladly paid and gathered subscriptions for William Darlington's books: Elizabeth C. Morris to William Darlington, 19 July 1843, WDP.

48. The Italian specimens that Elizabeth shared with William are in his herbarium at West Chester University. Asa Gray to Elizabeth Carrington Morris, 24 April 1844, Box 1, Folder 3, AGP; Elizabeth C. Morris to William Darlington, 16 April 1844, WDP; Asa Gray to Elizabeth Carrington Morris, 4 November 1843, Box 1, Folder 3, AGP. For more on the gift economy, see Strasser, "Collecting Nature: Practices, Styles, and Narratives," 313–314.

49. For more on the gift economy, social hierarchy, and the market for specimens, see Anne Secord, "Corresponding Interests: Artisans and Gentlemen in Nineteenth-Century Natural History," *British Journal for the History of Science* 27.4

(December 1994): 383–408; Anne Secord, "Science in the Pub: Artisan Botanists in Early Nineteenth-Century Lancashire," *History of Science* 32 (1994): 269–315; Mark V. Barrow Jr., "The Specimen Dealer: Entrepreneurial Natural History in America's Gilded Age," *Journal of the History of Biology* 33.3 (Winter 2000): 493–534; Endersby, "'From Having No Herbarium,'" 343–358; Kuang-Chi Hung, "Finding Patterns in Nature: Asa Gray's Plant Geography and Collecting Networks (1830s–1860s)" (PhD diss., Harvard University, 2013), 22–23; Vetter, "Lay Participation in the History of Scientific Observation," 127–141; Staffan Muller-Wille, "Nature as Marketplace: The Political Economy of Linnaean Botany," *History of Political Economy* 35 (2003): 154–172; Mark Osteen, ed., *The Question of the Gift: Essays Across Disciplines* (Routledge, 2002), 19-20; Jim Endersby, *Imperial Nature: Joseph Hooker and the Practices of Victorian Science* (University of Chicago Press, 2008), 84–111; James Delbourgo, *Collecting the World: Hans Sloane and the Origins of the British Museum* (Belknap Press, 2017), 202–257; Strasser, "Collecting Nature," 313–314.

50. Asa Gray to William Darlington, 27 May 1844, WDP; Asa Gray to Elizabeth Carrington Morris, 11 August 1849, Box 1, Folder 4, AGP; Elizabeth C. Morris to William Darlington, 3 November 1845, WDP; Elizabeth C. Morris to William Darlington, 27 February 1851, WDP.

51. The Morris sisters, through their cousin's wife, Harriet Hare, were closely connected to Dorothea Dix and her larger network of friends in Newport, Rhode Island, and Boston, Massachusetts. Harriet C. Hare to Dorothea Dix, 1848–1857, Box 7, Folder 305, DDP; [Dorothea L. Dix], *Garland of Flora* (B. G. Goodrich, 1829); Miss D. L. Dix, "ART IX—Notice of the Aranea aculeata, the Phaloena antiqua and some species of the Papilio," *AJSA* (January 1831): 61–63; Margaretta H. Morris to Dr. Darlington, 27 May 1844, WDP; Dorothea Lynde Dix to Susan Littell, 4 September 1847, Box 26, Folder 47, LFP; Dorothea L. Dix to Professor Silliman, 3 June 1830, Box 17, Folder 742, DDP; John Torrey to Dorothea Dix, 1832–1871, Box 14, Folder 625, DDP; Thomas J. Brown, *Dorothea Dix: New England Reformer* (Harvard University Press, 1998), 45–46.

52. Elizabeth C. Morris to William Darlington, 23 December 1844, WDP.

53. Morris to Darlington, 23 December 1844; Asa Gray to Elizabeth Carrington Morris, 9 June 1845, Box 1, Folder 3, AGP; Asa Gray to Elizabeth C. Morris, 31 August 1845, Box 1, Folder 3, AGP; Asa Gray to Elizabeth Carrington Morris, 26 November 1845, Box 1, Folder 3, AGP; Asa Gray to Elizabeth Carrington Morris, 8 June 1846, Box 1, Folder 3, AGP.

54. Asa Gray to Elizabeth Carrington Morris, 12 August 1846, Box 1, Folder 3, AGP; Asa Gray to Elizabeth Carrington Morris, 21 July 1847, Box 1, Folder 4, AGP; Elizabeth C. Morris to William Darlington, 14 October 1842, WDP; Asa Gray to Elizabeth Carrington Morris, 18 July 1848, Box 1, Folder 4, AGP; Asa Gray to Elizabeth Carrington Morris, 25 August 1848, Box 1, Folder 4, AGP.

55. Asa Gray to Elizabeth Carrington Morris, 9 June 1845, Box 1, Folder 3, AGP; Hung, "Finding Patterns in Nature: Asa Gray's Plant Geography and Collecting Networks (1830s–1860s)," 165–204.

56. Asa Gray to Elizabeth Carrington Morris, 21 July 1847, Box 1, Folder 4, AGP; Asa Gray to Elizabeth Carrington Morris, 18 July 1848, Box 1, Folder 4, AGP; Asa Gray to Elizabeth Carrington Morris, 25 August 1848, Box 1, Folder 4, AGP; Asa Gray, *United States Exploring Expedition Botany. During the years 1838, 1839, 1840, 1841, 1842. Under the command of Charles Wilkes, U.S.N. Vol. XV. Botany. Phanerogamia, Part 1: With a Folio Atlas of One Hundred Plates* (Printed by C. Sherman, 1854). For more on collecting and territorial expansion, see Strasser, "Collecting Nature," 312–316; Londa Schiebinger, *Plants and Empire: Colonial Bioprospecting in the Atlantic World* (Harvard University Press, 2004); Philip Pauly, *Biologists and the Promise of American Life: From Meriwether Lewis to Alfred Kinsey* (Princeton University Press, 2000), 15–43.

57. Asa Gray to Elizabeth Carrington Morris, 29 June 1846, Box 1, Folder 3, AGP; Asa Gray to Elizabeth Carrington Morris, 27 August 1846, Box 1, Folder 3, AGP; Asa Gray to Elizabeth Carrington Morris, 3 October 1846, Box 1, Folder 3, AGP; Asa Gray to Elizabeth Carrington Morris, 3 October 1848, Box 1, Folder 4, AGP; Elizabeth C. Morris to William Darlington, 16 May 1848, WDP.

58. Asa Gray to Elizabeth Carrington Morris, 24 June 1847, Box 1, Folder 4, AGP.

59. Asa Gray to Elizabeth Carrington Morris, 24 June 1847, Box 1, Folder 4, AGP; Asa Gray to Elizabeth Carrington Morris, 21 July 1847, Box 1, Folder 4, AGP; Asa Gray to Elizabeth Carrington Morris, 2 May 1848, Box 1, Folder 4, AGP; Asa Gray to Elizabeth Carrington Morris, 25 August 1848, Box 1, Folder 4, AGP; Lisa Ann DeCesare, "Jane Lathrop Loring Gray (1821–1909) and the Archives of the Gray Herbarium," *Harvard Papers in Botany* 15.2 (December 2010): 221–230.

60. Asa Gray to Elizabeth Carrington Morris, 18 July 1848, Box 1, Folder 4, AGP; Asa Gray to Elizabeth Carrington Morris, 25 August 1848, Box 1, Folder 4, AGP; Asa Gray to Elizabeth Carrington Morris, 3 October 1848, Box 1, Folder 4, AGP.

61. Elizabeth C. Morris to William Darlington, 13 April 1849, WDP; Elizabeth C. Morris to William Darlington, 11 June 1849, WDP.

62. Asa Gray to Elizabeth Carrington Morris, 18 June 1849, Box 1, Folder 4, AGP; Asa Gray to Elizabeth Carrington Morris, 11 August 1849, Box 1, Folder 4, AGP.

63. Elizabeth C. Morris to William Darlington, 29 August 1843, WDP.

64. "Notes from My Scrap-Book," *GM* 4.11 (November 1862): 329–331; Asa Gray to Elizabeth Carrington Morris, 25 August 1848, Box 1, Folder 4, AGP.

CHAPTER FIVE: ANONYMOUSLY FIERCE

1. "Home Pictures," *Pennsylvania Farm Journal* 4.4 (April 1854): 116; E. C. Morris to Dr. Darlington, 25 April 1854, William Darlington Collection, New-York Historical Society.

2. "Home Pictures," 116; James Parton, *Eminent Women of the Age* (S. M. Betts, 1869), 382–387.

3. "Death of One of Our Contributors," *GM* 7.3 (March 1865): 84; Elizabeth Morris, Contributions to the *American Agriculturist*, Box 1, Folder 18, LFP; Margaretta Hare Morris, draft of article on Apple Moth, *Tortrix Carpocapsa pomonara*, [n.d.], Box 1, Folder 12, AGP.

4. Virginia Woolf, *A Room of One's Own* (John Wiley/Blackwell, 2015), 36–37; Mary Kelley, *Private Woman, Public Stage: Literary Domesticity in Nineteenth-Century America* (Oxford University Press, 1984), 126–137; Robert Gunn, "'How I Look': Fanny Fern and the Strategy of Pseudonymity," *Legacy* 27.1 (2010): 23–42; Nicole Tonkovich, *Domesticity with a Difference: The Nonfiction of Catharine Beecher, Sarah J. Hale, Fanny Fern, and Margaret Fuller* (University of Mississippi Press, 1997), 47–71.

5. E. C. Morris to Dr. Darlington, 25 April 1854, WDP.

6. Albert Lowther Demaree, *The American Agricultural Press, 1819–1860* (Porcupine Press, 1974), 348–351.

7. Ann B. Shteir, "Elegant Recreations? Configuring Science Writing for Women," in *Victorian Science in Context*, ed. Bernard Lightman (University of Chicago Press, 1997), 236–255; Barbara T. Gates, "Ordering Nature: Revisioning Science Culture," in *Victorian Science in Context*, ed. Bernard Lightman (University of Chicago Press, 1997), 179–211; Margaret Welch, *The Book of Nature: Natural History in the United States, 1825–1875* (Northeastern University Press, 1998), 141–147.

8. Robert V. Bruce, *The Launching of Modern American Science, 1846–1876* (Alfred A. Knopf, 1976), 115–119; George H. Daniels, "The Process of Professionalization in American Science: The Emergent Period, 1820–1860," *Isis* 58.2 (Summer 1967): 150–166; Craig James Hazen, *The Village Enlightenment in America: Popular Religion and Science in the Nineteenth Century* (University of Illinois Press, 2000), 8–14; [Mary Townsend], *Life in the Insect World, or, a Conversation upon Insects Between an Aunt and Her Nieces* (Lindsay & Blackiston, 1844); Ann B. Shteir, *Cultivating Women, Cultivating Science* (Johns Hopkins University Press, 1996), 60–103; Bernard Lightman, "'The Voices of Nature': Popularizing Victorian Science," in *Victorian Science in Context*, ed. Bernard Lightman (University of Chicago Press, 1997), 187–211.

9. G. P., "The Midge," *AA* 4.3 (March 1845): 86.

10. "To Our Subscribers," *AA* 3.7 (December 1844): 353; Albert Lowther Demaree, *American Agricultural Press*, 159–179, 351; Stephen Mandravelis, "The *AA*: Art and Agriculture in the United States' First Illustrated Farming Journal, 1842–78," *Nineteenth-Century Art Worldwide* 20.3 (2021). For more on how journalists wrote

for women in newspapers during the same period, see Julia Guarneri, *Newsprint Metropolis: City Papers and the Making of Modern Americans* (University of Chicago Press, 2018), 58–62.

11. "For Farmers' Daughters," *AA* 4.4 (April 1845): 47.

12. "For Farmers' Daughters," 47; G. P., "What Woman May Do," *AA* 4.7 (July 1845): 222–223; Nina Baym, *American Women of Letters and the Nineteenth-Century Sciences* (Rutgers University Press, 2002); Tina Gianquitto, *"Good Observers of Nature": American Women and the Scientific Study of the Natural World* (University of Georgia Press, 2007), 15–56; Mark Stoll, *Inherit the Holy Mountain: Religion and the Rise of American Environmentalism* (Oxford University Press, 2015): 10–53; Peter J. Bowler and Iwan Rhys Morus, *Making Modern Science*, 2nd ed. (University of Chicago Press, 2020), 137–140.

13. G. P., "What Woman May Do."

14. G. P., "What Woman May Do"; Miss Catharine E. Beecher, *A Treatise on Domestic Economy*, 3rd ed. (Harper & Brothers, 1848), 322–324, 331, 341, 350–351; Kathryn Kish Sklar, *Catharine Beecher: A Study in American Domesticity* (Yale University Press, 1973), 151–155; Barbara A. White, *The Beecher Sisters* (Yale University Press, 2003), 43–44.

15. G. P., "What Woman May Do."

16. G. P., "What Woman May Do." Elaine Leong shows that there is a long history of domestic spaces being used for medical and culinary recipe experimentation, extending back at least to early modern England. Elaine Leong, *Recipes and Everyday Knowledge: Medicine, Science, and the Household in Early Modern England* (University of Chicago Press, 2018). For more on the home economics movement at the start of the twentieth century, see Helen Z. Veit, *Modern Food, Moral Food: Self-Control, Science, and the Rise of Modern American Eating in the Early Twentieth Century* (University of North Carolina Press, 2013): 77–100; Sarah A. Leavitt, *From Catharine Beecher to Martha Stewart: A Cultural History of Domestic Advice* (University of North Carolina Press, 2002), 40–72.

17. E. S. "Making Butter," *AA* 4.10 (October 1845): 320; "A Delicious Apple-Pudding," *AA* 4.11 (November 1845): 355; E. S., "Sundry Items," *AA* 7.1 (January 1847): 34; E. S., "How to Make Apple Butter," *AA* 6.9 (September 1847): 290; E. S., "Smoke-Houses," *AA* 6.10 (October 1847): 322; E. S., "How to Make Ginger Syrup," *AA* 7.7 (July 1848): 226; E. S., "Butter Making," *AA* 7.7 (July 1848): 226; E. S., "Potato Starch," *AA* 8.3 (March 1849): 96; E. S., "How to Make Mushroom Catchup," *AA* 7.9 (September 1848): 288; "Appendix No. 24: DAIRY," *Public Documents Printed by Order of the Senate of the United States, First Session of the Twenty-Ninth Congress, Begun and Held at the City of Washington, December 1, 1845* (Ritchie & Hess, 1846), 985–986; Reviewer, "Review of January No. of the Agriculturist," *AA* 6.6 (June 1847): 188.

18. E. S., "Effects of Cosmetics on the Skin," *AA* 7.3 (March 1848): 98; C., "The Effects of Cosmetics on the Skin," *AA* 7.4 (April 1848): 129–130; E. S., "Almond

Paste for the Toilet," *AA* 7.4 (April 1848): 130; "To Make Cold Cream," *AA* 7.6 (June 1848): 192; "To Make a Pleasant Cosmetic Soap," *AA* 7.6 (June 1848): 192.

19. E. S., "Early Rising," *AA* 4.8 (August 1845): 256–257.

20. E. S., "Early Rising"; Thomas M. Allen, *A Republic in Time: Temporality and Social Imagination in Nineteenth-Century America* (University of North Carolina Press, 2008), 114–145.

21. E. S., "Early Rising."

22. Solus, "Knitting," *AA* 5.5 (May 1846): 162; E. L., "Ladies' Department: Knitting," *AA* 5.8 (August 1846): 257.

23. E. L., "Ladies' Department: Knitting."

24. Reviewer, "Review of the August No. of the Agriculturist," *AA* 5.11 (November 1846): 347.

25. E. S., "Ladies' Department: Reply to Reviewer on Knitting, &c.," *AA* 6.2 (February 1847): 65; Reviewer, "Review of the May Number of the Agriculturist," *AA* 6.10 (October 1847): 310.

26. Sally McMillen, *Seneca Falls and the Origins of the Women's Rights Movement* (Oxford University Press, 2009); Lisa Tetrault, *The Myth of Seneca Falls: Memory and the Women's Suffrage Movement, 1848–1898* (University of North Carolina Press, 2014).

27. Almira Lincoln Phelps, "Woman's Rights," *New York Times*, 27 February 1871; Jeanne Boydston, Mary Kelley, and Anne Margolis, *The Limits of Sisterhood: The Beecher Sisters on Women's Rights and Woman's Sphere* (University of North Carolina Press, 1988), 231–232; White, *The Beecher Sisters*, 74–75; Thomas J. Brown, *Dorothea Dix: New England Reformer* (Harvard University Press, 1998), 167–172; Martha S. Jones, *Vanguard: How Black Women Broke Barriers, Won the Vote, and Insisted on Equality for All* (Basic Books, 2020), 43–68; Janice Nimura, *The Doctors Blackwell: How Two Pioneering Sisters Brought Medicine to Women and Women to Medicine* (W. W. Norton, 2021), 67–69.

28. Evidence of these connections is in the many letters in the William Darlington Papers, the Asa Gray Correspondence Files, and the James Family Papers.

29. "Proceedings of the Franklin Institute of the State of Pennsylvania, for the Promotion of the Mechanic Arts, Relative to the Establishment of a School of Design for Women," Moore College of Art and Design records, Archives of American Art, Smithsonian Institution, Reel 3654; Laura R. Prieto, *At Home in the Studio: The Professionalization of Women Artists in America* (Harvard University Press, 2001), 26–28; Margaret Rives King, *Memoirs of the Life of Mrs. Sarah Peter* (R. Clarke, 1889), 1: 69–71; Nina de Angeli Walls, "Educating Women for Art and Commerce: The Philadelphia School of Design, 1848–1932," *History of Education Quarterly* 34.3 (1994): 329–355; *First Annual Report of the Philadelphia School of Design for Women, Incorporated September 24th, 1853* (Crissy & Markley, 1854); Thomas O' Connor, *Civil War Boston: Home Front and Battlefield* (Northeastern University Press, 1997), 19.

30. Lucretia Mott, *Discourse on Woman* (T. B. Peterson, 1850), 4, 12.

31. Frances W. Kaye, "The Ladies' Department of the 'Ohio Cultivator,' 1845–1865: A Feminist Forum," *Agricultural History* 50.2 (April 1976): 414–423.

32. Elizabeth Morris, Contributions to the *American Agriculturist*, Box 1, Folder 18, LFP.

33. "Insects.—No. 1," *AA* 5.2 (February 1846): 65–66.

34. "Insects.—No. 1"; Margaretta Hare Morris, draft of article on Apple Moth, Tortrix Carpocapsa pomonara, [n.d.], Box 1, Folder 12, AGP.

35. "Insects.—No. 1"; "To the Girls," *AA* 5.5 (May 1846): 161–162.

36. "Book-Farming—The Shrew-Mole and Cut-Worm," *AA* 6.7 (July 1847): 223–224; Donald Worster, *Nature's Economy: A History of Ecological Ideas*, 2nd ed. (Cambridge University Press, 1994, 1997), 58–111; Laura Dassow Walls, "Textbooks and Texts from the Brooks: Inventing Scientific Authority in America," *American Quarterly* 49.1 (1997): 17–19; Laura Dassow Walls, *Seeing New Worlds: Henry David Thoreau and Nineteenth-Century Natural Science* (University of Wisconsin Press, 1995); Richard Judd, *The Untilled Garden: Natural History and the Spirit of Conservation in America, 1740–1840* (Cambridge University Press, 2009), 156–182.

37. "Book-Farming—The Shrew-Mole and Cut-Worm"; "Ladies' Department: Natural history of the Chinche," *AA* 6.1 (January 1847): 33.

38. "Insects.—No. 2," *AA* 5.3 (March 1846): 97–98, 345; "Natural history of the Chinche"; M., "The Cotton Moth," *AA* 6.1 (January 1847): 22; M., "The Army Worm," *AA* 6.2 (February 1847): 50; Morris, draft of article on Apple Moth; Victor de Motschulsky, *Études Entomologiques* 5 (Helsingfors: Société de Litérature Finnoise, 1856), 17–18; Samuel Stehman Haldeman to Margaretta Hare Morris, 21 June 1854, Box 1, Folder 8, AGP; Walter Horn and Sigkund Schenkling, *Index Litteraturae Entomologicae* 1.3 (Berlin-Dahlem, 1925): 839–840.

39. E. S., "For Farmers' Daughters," *AA* 4.4 (April 1845): 47; E. S., "Directions for Gathering Garden Seeds," *AA* 7.8 (August 1848): 257; E. S., "Transplanting," *AA* 4.10 (October 1845): 321; E. S., "Sundry Items," *AA* 7.1 (January 1847): 34; "A Chapter on Grasses—No. 1," *AA* 5.1 (January 1846): 35; Reviewer, "Review of the September No. of the Agriculturist," *AA* 5.12 (December 1846): 374; E. S., "A Chapter on Grasses—No. 3," *AA* 6.3 (March 1847): 97–98; E. S., "A Chapter on Grasses—No. 4," *AA* 6.4 (April 1847): 130.

40. E. S., "Ladies' Department: Reply to Reviewer on Knitting, &c.," *AA* 6.2 (February 1847): 65; E. S., "The Garden," *AA* 5.3 (March 1846): 98; Elizabeth C. Morris to William Darlington, 14 October 1842, WDP.

41. William Kirby and William Spence, *An Introduction to Entomology* (Longman, Rees, Orme, Borwn, and Green, 1826), 525; E. S., "The Garden," *AA* 5.4 (April 1846): 127.

42. E. S., "The Garden"; "Insects.—No. 2"; "Hints to Housekeepers," *AA* 5.11 (November 1846): 353.

43. E. S., "The Garden"; Edith Denner, "Dress Reform: Science and Long Skirts," *Water-Cure Journal* 20.1 (July 1855): 7; "Women's Dresses," *Water-Cure*

Journal 8.6 (December 1849): 186; Gayle V. Fischer, *Pantaloons & Power: A Nine-teenth-Century Dress Reform in the United States* (Kent State University Press, 2001), 79–109.

44. E. S., "Farmers' Wives," *AA* 4.12 (December 1845): 378; S. H. R., "Country Schools," *AA* 5.2 (February 1846): 66; E. S., "Country Schools," *AA* 5.4 (April 1846): 128. For more on the Albany Normal School, see *A Historical Sketch of the State Normal School at Albany, N.Y.* (Press of Brandow & Barton, 1884); *Annual Register and Circular of the State Normal School, Albany, N.Y.* (C. Van Benthuysen & Co., 1846).

45. S. H. R., "Rearing of Silk Worms," *AA* 6.7 (July 1847): 224; E. S., "On the Culture of Silk," *AA* 7.5 (May 1848): 160–161.

46. Laurel Thatcher Ulrich, *The Age of Homespun: Objects and Stories in the Cre-ation of an American Myth* (Vintage, 2001), 377–381; Emily Pawley, *The Nature of the Future: Agriculture, Science, and Capitalism in the Antebellum North* (University of Chicago Press, 2020), 103–129.

47. E. S., "On the Culture of Silk."

48. E. S., "On the Culture of Silk."

49. E. S., "Country Life," *AA* 5.1 (January 1846): 33–34; E. S., "Hints to Coun-try Housekeepers," *AA* 5.6 (June 1846): 194–195; E. S., "Rural Pastimes by Social Labor.—No. 1," *AA* 6.8 (August 1847): 257; E. S., "Rural Pastimes by Social Labor—No. 2," *AA* 6.11 (November 1847): 353; E. S., "Rural Pastimes by Social Labor—No. 3," *AA* 6.12 (December 1847): 377–378; E. S., "Rural Pastimes by Social Labor—No. 4," *AA* 7.9 (September 1848): 286; "Harvest Home," *AA* 7.10 (October 1848): 321; Harry M. Tinkcom, Margaret B. Tinkcom, and Grant Miles Simon, *Historic Germantown: From the Founding to the Early Part of the Nineteenth Century* (American Philosophical Society, 1955); Elizabeth M. Geffen, "Industrial Development and Social Crisis, 1841–1854," in *Philadelphia: A 300-Year History*, ed. Russell F. Weigley (W. W. Norton, 1982), 307–362.

50. Margaretta H. Morris to Dr. Darlington, 27 May 1844, WDP; M. H. Mor-ris to Dr. Darlington, 18 June 1844, WDP.

51. E. C. Morris to Dr. Darlington, 25 April 1854, WDP.

CHAPTER SIX: HIDDEN AT THE ROOT

1. "American Scientific Association," *Litchfield Enquirer*, 22 August 1850; *PAAAS, Fourth Meeting, Held at New Haven, Conn., August 1850* (S. F. Baird, 1851); "Scientific Convention at New-Haven," *New-York Tribune*, 22 August 1850; Sally Gregory Kohlstedt, *The Formation of the American Scientific Community: The Amer-ican Association for the Advancement of Science, 1848–1860* (University of Illinois Press, 1976), 41–46, 79.

2. *PAAAS, Fourth Meeting*, 49–50, 346; "Scientific Convention at New-Haven"; Kohlstedt, *Formation of the American Scientific Community*, 78–99.

3. "Scientific Convention at New-Haven"; "Science in America," *New York Daily Herald*, 23 August 1850; *PAAAS, Fourth Meeting*, 49; Margaretta Hare Morris, manuscript on cicada, [n.d.], Box 1, Folder 9, LFP; Kohlstedt, *Formation of the American Scientific Community*, 94.

4. "Stated Meeting, December 15, 1846," *PANSP* 3.6 (November–December 1846): 131–134; M. H. Morris, "Apple and Pear Trees Destroyed by the Locust," *AA* 6.3 (March 1847): 86–87; M. H. M., "Destruction of Fruit Trees by the Seventeen-Year Locust," *AA* 7.9 (September 1848): 279; Harry Bluff, "Letter from Philadelphia," *GT*, 18 March 1846; "The Late Rains," *GT*, 18 March 1846; "The Weather–The Crops," *GT*, 8 April 1846.

5. M. H. Morris, "Apple and Pear Trees Destroyed by the Locust"; Thaddeus William Harris, *A Report on the Insects of Massachusetts, Injurious to Vegetation* (Folsom, Wells and Thurston, 1841), 166–167.

6. Morris, "Apple and Pear Trees Destroyed by the Locust."

7. "A Delicious Apple-Pudding," *AA* 4.11 (November 1845): 355; E. S., "How to Make Apple Butter," *AA* 6.9 (September 1847): 290; "A Day to Myself," *AA* 7.11 (November 1848): 352–353; E. S., "Sagacious Horse," *AA* 7.12 (December 1848): 378; E. S., "Apple-Paring Bee," *AA* 8.9 (September 1849): 288; Andrew Jackson Downing, *The Fruits and Fruit Trees of America* (Wiley & Putnam, 1845), 551–561; Andrew Jackson Downing, *Rural Essays* (George P. Putnam, 1852), 440; Philip Pauly, *Fruits and Plains: The Horticultural Transformation of America* (Harvard University Press, 2007), 51–79; Emily Pawley, *The Nature of the Future: Agriculture, Science, and Capitalism in the Antebellum North* (University of Chicago Press, 2020), 166–171.

8. M. H. M., "Destruction of Fruit Trees by the Seventeen-Year Locust"; "Stated Meeting, December 15, 1846."

9. M. H. M., "Destruction of Fruit Trees by the Seventeen-Year Locust."

10. "Stated Meeting, December 15, 1846"; Morris, "Apple and Pear Trees Destroyed by the Locust"; Morris, manuscript on cicada.

11. "Stated Meeting, December 15, 1846"; J. B. W., "How to Renovate an 'Outcast,'" *Horticulturist and Journal of Rural Art and Rural Taste* 1.5 (November 1846): 225–227.

12. "Stated Meeting, December 15, 1846"; J. B. W., "How to Renovate an 'Outcast.'"

13. William Travis, *A History of The Germantown Academy* (Press of S. H. Burbank, 1910), 1:158–164; DNLD [transcriptions], 1815–1839, Barbara Jones, transcriber, 1 April 1825, VIII: 9.

14. "Stated Meeting, December 15, 1846."

15. "Stated Meeting, December 15, 1846"; Patricia Tyson Stroud, *Thomas Say: New World Naturalist* (University of Pennsylvania Press, 1992), 270–271.

16. An Amateur, "Cultivation of Fruit Trees," *AA* 6.1 (January 1847): 30.

17. Morris, "Apple and Pear Trees Destroyed by the Locust."

18. Morris, "Apple and Pear Trees Destroyed by the Locust"; Virginia Woolf, *Three Guineas* (Hogarth Press, 1938).

19. Morris to Johnson, 5 March 1847, Coll. 567, Folder 276, ANSP; "Stated Meeting, March 16th, 1847," *PANSP* 3.8 (March–April 1847): 189–192; "Meteorological Observations," *GT*, 3 March 1847; *GT*, 17 March 1847; "Meteorological Observations," *GT*, 14 April 1847.

20. Morris to Johnson, 5 March 1847; "Stated Meeting, March 16th, 1847."

21. Morris to Johnson, 5 March 1847; "Stated Meeting, March 16th, 1847."

22. Morris to Johnson, 5 March 1847; *McElroy's Philadelphia City Directory* (A. McElroy & Co., 1847), 175.

23. Louis Agassiz, *Twelve Lectures on Comparative Embryology Delivered Before The Lowell Institute, in Boston, December and January, 1848–1849* (Redding, 1849); "Professor Agassiz," *GT*, 14 March 1849; Asa Gray to John Torrey, 4 January 1847, John Torrey Papers (PP), Archives, The New York Botanical Garden; Jules Marcou, *Life, Letters, and Works of Louis Agassiz*, 2 vols. (Macmillan, 1896), 2:27–30; Pennsylvania Horticultural Society Minute Book, vol. 5, August 1848–December 1850, 268 (2010.03.16); Pennsylvania Horticultural Society McLean Library; "The Seventeen Year Locust," *The Friend: A Religious and Literary Journal* 24.25 (8 March 1851): 193–194; Louis Agassiz, *An Introduction to the Study of Natural History, in a Series of Lectures Delivered in the Hall of the College of Physicians and Surgeons, New York* (Greeley and McElrath, 1847), 4; Christoph Irmscher, *Louis Agassiz: Creator of American Science* (Houghton Mifflin Harcourt, 2013), 104; Ann Fabian, *The Skull Collectors: Race, Science, and America's Unburied Dead* (University of Chicago Press, 2010), 112–115; Edward Lurie, *Louis Agassiz: A Life in Science* (Johns Hopkins University Press, 1960), 127; Ralph Waldo Emerson, quoted in A. Hunter Dupree, *Asa Gray: American Botanist, Friend of Darwin* (Johns Hopkins University Press, 1959, 1988), 225.

24. Asa Gray to Elizabeth Carrington Morris, 18 December 1846, Box 1, Folder 3, AGP; Asa Gray to Elizabeth Carrington Morris, 25 July 1847, Box 1, Folder 4, AGP; Irmscher, *Louis Agassiz*, 88.

25. One of Agassiz's earliest essays on the origin of human races was published a year later in 1850: Louis Agassiz, "The Diversity of Origin of the Human Races," *Christian Examiner and Religious Miscellany* (4th Series) 14.1 (July 1850): 110–145; Marcou, *Life, Letters, and Works of Louis Agassiz*, 2:28–29; Lurie, *Louis Agassiz*, 124, 257–259; Fabian, *Skull Collectors*, 80–87, 112–115; Richard Conniff, *The Species Seekers: Heroes, Fools, and the Mad Pursuit of Life on Earth* (W. W. Norton, 2011), 179–192; Britt Rusert, *Fugitive Science: Empiricism and Freedom in Early African American Culture* (New York University Press, 2017), 126–131. For Frederick Douglass's repudiation of Morton's and Agassiz's claims, see Frederick Douglass, *The Claims of the Negro, Ethnologically Considered, An Address* (Printed by Lee, Mann, & Company, 1854).

26. "Donations to Museum, in March and April, 1849," *PANSP* 4 (1848–1849): 175.

27. "Stated Meeting, March 16th, 1847," *PANSP*. For more on how women scientists shaped their images, see Sally Gregory Kohlstedt and Donald L. Opitz, "Re-imag(in)ing Women in Science: Projecting Identity and Negotiating Gender in Science," in *The Changing Image of the Sciences*, Ida H. Stamhuis, Teun Koetsier, Cornelis de Pater, and Albert van Helden, eds. (Kluwer Academic Publishers, 2002), 105–140.

28. Seventh Census of the United States, 1850; Records of the Bureau of the Census, Record Group 29; National Archives, Washington, DC; Elizabeth C. Morris to William Darlington, 23 December 1844, WDP; E. C. Morris to Dr. Darlington, 12 July 1853, WDP; Michael Horn to Lady Morris, 1 July 1853, WDP; Edwin C. Jellett, *Germantown Gardens and Gardeners* (Horace F. McCann, 1914), 57–60.

29. Pennsylvania Horticultural Society Minute Book, vol. 5, August 1848–December 1850, 268 (2010.03.16); Pennsylvania Horticultural Society McLean Library; "The Seventeen Year Locust," 193–194. A rare glimpse into one of the lives of their later gardeners, Michael Horn, is possible in a set of letters preserved in an exchange with William Darlington where Elizabeth asks William to help translate the German: E. C. Morris to Dr. Darlington, 12 July 1853, WDP; Horn to Lady Morris; E. C. Morris to Dr. Darlington, 9 November 1853, WDP.

30. Kohlstedt, *Formation of the American Scientific Community*.

31. *PAAAS, Second Meeting, Held at Cambridge, August 1849* (Henry Flanders, 1850), 2; Photographic Views of Harvard Hall, 1764, 1841–1968, HUV 30.2, Harvard University Archives, Cambridge, MA; Thaddeus William Harris to Margaretta Hare Morris, August 15, 1849, TWHP.

32. Harris to Morris, 15 August 1849; Lurie, *Louis Agassiz*, 162.

33. Harris to Morris, 15 August 1849.

34. Harris to Morris, 15 August 1849.

35. Harris to Morris, 15 August 1849; Thaddeus William Harris, *A Treatise on Some of the Insects of New England Which Are Injurious to Vegetation*, 2nd ed. (White & Potter, 1852): 184–185; Margaretta Hare Morris to Thaddeus William Harris, 12 September 1850, TWHP.

36. Samuel Stehman Haldeman to Margaretta Hare Morris, 21 May 1849, Box 1, Folder 8, AGP; Samuel Stehman Haldeman to Margaretta Hare Morris, 10 July 1849, Box 1, Folder 8, AGP; Margaretta Hare Morris to W. R. Johnson, 21 July 1849, ANSP Correspondence, Coll. 567, Folder 276, ANSP; "August 21st," *PANSP* IV (1849): 194.

37. E. C. Morris to William Darlington, 9 August [1849], William Darlington Papers, New-York Historical Society, New York; M. H. Morris, "The Potato Curculio," *Valley Farmer* 2.7 (August 1850): 241; Margaretta Hare Morris to Thaddeus William Harris, 28 August [1849], TWHP.

38. Margaretta Hare Morris to Thaddeus William Harris, 12 September 1850, TWHP; E. C. Morris to William Darlington, 18 June 1850, WDP.

39. "The Seventeen Year Locust," 193–194.

40. Pauly, *Fruits and Plains*, 51–79; Pennsylvania Horticultural Society Minute Book, vol. 5, August 1848–December 1850, 231, 263–272.

41. Pennsylvania Horticultural Society Minute Book, vol. 5; "The Seventeen Year Locust," 193–194.

42. E. Nichols, "In the first . . . ," *Ohio Cultivator* 3.17 (1 September 1847): 123; "More about the Seventeen Year Locust," *The Friend: A Religious and Literary Journal* 24.28 (29 March 1851): 221–222.

43. "The Seventeen Year Locust"; M. H. M., "Destruction of Fruit Trees by the Seventeen-Year Locust," *AA* 7.9 (September 1848): 279.

44. "Tree Fruit Insect Pest—Periodical Cicada," *Penn State Extension*, https://extension.psu.edu/tree-fruit-insect-pest-periodical-cicada, updated 25 October 2017 [accessed 16 April 2021]; JoAnn White and Charles E. Strehl, "Xylem Feeding by Periodical Cicada Nymphs," *Ecological Entomology* 3 (1978): 323–327; Richard Karban, "Periodical Cicada Nymphs Impose Periodical Oak Tree Wood Accumulation," *Nature* 287.25 (September 1980): 326–327; Chris T. Maier, "A Mole's-Eye View of Seventeen-Year Periodical Cicada Nymphs, *Magiciada septendecim* (*Hemiptera: Homoptera Cicadida*)," *Annals of the Entomological Society of America* 73.2 (1980): 147–152; R. Karban, "Experimental Removal of 17-Year Cicadas and Growth of Host Apple Trees," *Journal of New York Entomological Society* 90 (1982): 74–81; R. Karban, "Induced Responses of Cherry Trees to Periodical Cicada Oviposition," *Oecologia* 59 (1983): 226–231; R. Karban, "Addition of Periodical Cicada Nymphs to an Oak Forest: Effects on Cicada Density, Acorn Production, and Rootlet Density," *Journal of Kansas Entomological Society* 58 (1985): 269–272; Sergio Rasmann and Anurag A. Agrawal, "In Defense of Roots: A Research Agenda for Studying Plant Resistance to Belowground Herbivory," *Plant Physiology* 146.3 (March 2008): 878–880; S. Flory and W. Mattingly, "Response of Host Plants to Periodical Cicada Oviposition Damage," *Oecologia* 156 (2008): 649–656; James H. Speer, Keith Clay, Graham Bishop, and Michelle Creech, "The Effect of Periodical Cicadas on Growth of Five Tree Species in Midwestern Deciduous Forests," *American Midland Naturalist* 164.2 (October 2010): 173–186; Louie H. Yang and Richard Karban, "The Effects of Pulsed Fertilization and Chronic Herbivory by Periodical Cicadas on Tree Growth," *Ecology* 100.6 (27 March 2019).

45. *PAAAS, Fourth Meeting*, v, ix, xiii, xv; Kohlstedt, *Formation of the American Scientific Community*, 102–103; Margaretta Hare Morris, manuscript on cicada, [n.d.], Box 1, Folder 9, LFP.

46. Glaucus, "American Scientific Association: The Springfield Session," *New York Times*, 10 August 1859; Kohlstedt, *Formation of the American Scientific Community*, 95, 102–103.

47. Renee Bergland, *Maria Mitchell and the Sexing of Science: An Astronomer Among the American Romantics* (Beacon Press, 2008).

48. Margaret W. Rossiter, *Women Scientists in America: Struggles and Strategies to 1940* (Johns Hopkins University Press, 1982), 1:76–77.

49. *PAAAS, Fourth Meeting*, 354; Glaucus, "American Scientific Association"; "The Seventeen Year Locust"; Louis Agassiz, *The Natural History of the United States of America* (Little, Brown, 1857): xxx–xxxi, 88–89.

50. Asa Fitch, "Insects Injurious to Vegetation: The Hessian Fly (Continued)," *American Journal of Agriculture and Science* 5.9 (January 1847): 11–12; M. H. Morris, "Controversy Respecting the Hessian Fly," *American Journal of Agriculture and Science* 5.12 (April 1847): 206–208.

51. Asa Fitch, "Wheat Insects—Joint-Worm," *Cultivator* (October 1851): 8, 10.

52. *PAAAS, Fourth Meeting*, xxv. For more on the AAAS, see Kohlstedt, *Formation of the American Scientific Community*.

53. Fitch, "Wheat Insects—Joint-Worm."

54. E. C. Morris to William Darlington, 1 April 1851, WDP; Louis Agassiz to Margaretta Hare Morris, 9 December [1853], Box 1, Folder 7, AGP; E. S., "The Garden," *AA* 5.4 (April 1846): 127; Mary P. Winsor, *Reading the Shape of Nature: Comparative Zoology at the Agassiz Museum* (University of Chicago Press, 1991), 1–42.

55. Morris, partial manuscript on cicada.

56. Morris, partial manuscript on cicada; "The Philadelphia *North American* announces . . . ," *Trenton State Gazette*, 6 June 1851; "The Sting of a Locust," *The Sun*, 6 June 1851; "The Locust Has No Sting," *The Sun*, 12 June 1851; "The Sting of the Locust," *The Sun*, 13 June 1851; "Do the Locusts Sting," *The Sun*, 18 June 1851.

57. "The Song of the Locusts," *The Sun*, 30 June 1851.

58. Morris, partial manuscript on cicada; "Stated Meeting, December 15, 1846"; Morris, "Apple and Pear Trees Destroyed by the Locust"; Margaretta Hare Morris to Thaddeus William Harris, 12 September 1850, TWHP.

59. "May 27th, 1851," "August 12th, 1851," "August 19th, 1851," "September 30th, 1851," *PANSP* (Matthew & Thompson, 1852), 5:209, 237, 272–274; Membership Card: Margaretta Hare Morris, Coll. 723 Academy Membership 1812–1924, ANSP.

60. "October 1, 1851," *Proceedings of the Boston Society of Natural History* 4 (1851–1854): 110–111.

61. Thaddeus William Harris, *A Treatise on Some of the Insects of New England Which are Injurious to Vegetation* (White & Potter, 1852), 187; Asa Fitch, "Report on the Noxious, Beneficial, and Other Insects of the State of New York," *Transactions of the New York State Agricultural Society for 1854* 14 (1855): 747–748.

CHAPTER SEVEN: LITTLE TIME TO CALL MY OWN

1. Margaretta Hare Morris to Thaddeus William Harris, 12 September 1850, TWHP; Catharine Beecher, "Articles and Conveniences for the Sick," *A Treatise on Domestic Economy*, 3rd ed. (Harper & Brothers, 1846), 209–216.

2. Elizabeth C. Morris to William Darlington, 27 February 1851, WDP; E. C. Morris to William Darlington, 1 April 1851, WDP; E. C. Morris to William Darlington, 5 June 1851, WDP.

3. For more on unmarried women and the social and cultural pressures at home, see Lee Virginia Chambers-Schiller, *Liberty, a Better Husband: Single Women in America: The Generations of 1780-1840* (Yale University Press, 1984), 107–126.

4. M. A. Donaldson to Ann Haines, 14 December 1836, Series 2, Box 27, Folder 409, WAP.

5. Miss Morris, "The Yellows, Caused by an Insect," *Horticulturist and Journal of Rural Art and Rural Taste* 4.11 (May 1850): 502–503; "Discovery of the Cause of the Yellows in the Peach Tree," *AA* 9.5 (May 1850): 144–145.

6. "Reports on Peach Trees," *The Plough, the Loom, and the Anvil* 5.1 (January 1853): 37–40; Erwin F. Smith, *Department of Agriculture Botanical Division Bulletin No. 9: Peach Yellows: A Preliminary Report* (US Government Printing Office, 1888); Morris, "The Yellows, Caused by an Insect"; "Discovery of the Cause of the Yellows in the Peach Tree," *AA* 9.5 (May 1850): 144–145; "The Yellows, Caused by an Insect," *Southern Planter* 10.5 (June 1850): 178.

7. Samuel Stehman Haldeman to Margaretta Hare Morris, 10 July 1849, Box 1, Folder 8, AGP; S. S. Haldeman, "Report on the Progress of Entomology in the United States during the year 1849," *PANSP* 5 (1850–1851): 5–8; Agricola, "A Chapter on Various Subjects," *AA* 9.7 (July 1850): 219; "Review of the May Number of the Agriculturist," *AA* 9.7 (July 1850): 221; Samuel Stehman Haldeman to Margaretta Hare Morris, 21 June 1854, Box 1, Folder 8, AGP; A. J. Downing, *The Fruits and Fruit Trees of America* (Wiley & Putnam, 1845), 460–470. For more on plant pathology and the Peach Yellows, see C. Lee Campbell, Paul D. Peterson, and Clay S. Griffith, *The Formative Years of Plant Pathology in the United States* (American Philosophical Society Press, 1999), 24–27, 175–180, 220–221.

8. M., "Remedy for Diseased Peach-Trees," *AA* 6.5 (May 1847): 154; Morris, "The Yellows, Caused by an Insect."

9. Campbell, Peterson, and Griffith, *The Formative Years of Plant Pathology in the United States*, 24–27, 175–180, 220–221; Morris, "The Yellows, Caused by an Insect"; Downing, *The Fruits and Fruit Trees of America*.

10. John Reader, *Potato: A History of the Propitious Esculent* (Yale University Press, 2009), 153–168, 194; Austin Bourke, "Emergence of the Potato Blight, 1843-1846," *Nature* 203.4947 (1964): 805–808; Amanda C. Saville, Michael D. Martin, and Jean B. Ristaino, "Historic Late Blight Outbreaks Caused by a

Widespread Dominant Lineage of *Phytophthora infestans* (Mont.) de Bary," *PLoS One* 11.2 (December 2016).

11. *Catalogue of the Mount Airy Agricultural Institute* (T. K. and P. G. Collins, Printers, 1849); John Wilkinson, "Cultivation of Potatoes at Mount Airy College, Penn.," in New York State Legislature, *Documents of the Assembly of the State of New York* 7.175 (1850): 735–736; M. H. Morris, "The Potato Curculio," *AA* 9.4 (April 1850): 113–114. For more on experiments in antebellum farms, see Timothy K. Minella, "A Pattern for Improvement: Pattern Farms and Scientific Authority in Nineteenth-Century America," *Agricultural History* 90.4 (Fall 2016): 434–458; Emily Pawley, *The Nature of the Future: Agriculture, Science, and Capitalism in the Antebellum North* (University of Chicago Press, 2020), 63–80.

12. Reader, *Potato*, 196–206; "The Potato Rot," *AA* 4.12 (December 1845): 362; Margaretta Hare Morris to Thaddeus William Harris, 28 August [1849], TWHP; M. H. Morris to S. S. Haldeman, 28 August 1849, Collection 73, Box 54, Folder 108, ANSP.

13. Morris, "The Potato Curculio"; Morris to Harris, 28 August [1849]; M. H. Morris to S. S. Haldeman, 28 August 1849, Collection 73, Box 54, Folder 108, ANSP; S. S. Haldeman, "Report on the Progress of Entomology."

14. Morris, "The Potato Curculio"; M. H. Morris, "The Potato Curculio," *Valley Farmer* 2.7 (August 1850): 241.

15. "Review of the April Number of the Agriculturist," *AA* 9.6 (June 1850): 190–191; "Review of the July Number of the Agriculturist," *AA* 9.9 (September 1850): 286.

16. "The Potato Rot," *New England Farmer* 2.3 (1850): 53; T. W. Harris, "Insects on Potatoes," *New England Farmer* 2.13 (1850): 204.

17. Margaretta Hare Morris to Thaddeus William Harris, 12 September 1850, TWHP; Thaddeus William Harris, *A Treatise on Some of the Insects of New England Which Are Injurious to Vegetation*, 2nd ed. (White & Potter, 1852), 71–72.

18. Morris to Harris, 12 September 1850; Morris to Harris, 6 November 1850.

19. Thaddeus William Harris to Margaretta Hare Morris, December 1852, TWHP.

20. Reader, *Potato*, 194–208; Campbell, Peterson, and Griffith, *The Formative Years of Plant Pathology in the United States*, 29–49.

21. The *Baridius trinotatus* is now considered a synonym for *Trichobaris trinotatus*. F. H. Chittenden, "The Potato Stalk Weevil," in *Some Insects Injurious to Vegetable Crops*, US Department of Agriculture, Division of Entomology—Bulletin No. 33, New Series (US Government Printing Office, 1902), 9–18; W. S. Blatchley and Charles W. Leng, *Rhynchophora or Weevils of North Eastern America* (Nature Publishing Company, 1915), 375; Jens Prena, "Apropos *Trichobaris* Leconte (*Coleoptera Curculiniodae*)—An Inquiry by the Late Horace R. Burke," *Coleopterists Bulletin* 73.4 (2019): 901.

22. Thaddeus William Harris to Margaretta Hare Morris, 15 August 1849, TWHP; "Harris to Miss Morris," 23 October 1849, in *Entomological Correspondence of Thaddeus William Harris*, ed. Samuel Hubbard Scudder (Boston Society of Natural History, 1869), 242; Margaretta Hare Morris to Thaddeus William Harris, 12 September 1850, TWHP.

23. "Harris to Miss Morris," 25 September 1850, in *Entomological Correspondence of Thaddeus William Harris*, 242–244; Thaddeus William Harris to Margaretta Hare Morris, 25 September 1850, Box 1, Folder 9, AGP; "Harris to Miss Morris," 29 October 1850, in *Entomological Correspondence of Thaddeus William Harris*, 245–247; Robert Spear, *The Great Gypsy Moth War: The History of the First Campaign in Massachusetts to Eradicate the Gypsy Moth, 1890-1901* (University of Massachusetts Press, 2005); *Entomological Society of America*, "Spongy Moth Adopted as New Common Name for *Lymantia dispar*," press release, 2 March 2022.

24. Morris to Harris, 12 September 1850; Thomas J. Brown, *Dorothea Dix: New England Reformer* (Harvard University Press, 1998), 32, 124; Clark A. Elliott, *Thaddeus William Harris (1795–1856): Nature, Science, and Society in the Life of an American Naturalist* (Lehigh University Press, 2008), 172.

25. E. C. Morris to William Darlington, 10 November 1850, WDP; E. C. Morris to William Darlington, 20 January 1851, WDP; Elizabeth C. Morris to Dr. Darlington, 2 January 1854, WDP; John Jay Smith, *Recollections of John Jay Smith* (Press of J. B. Lippincott, 1892), 301–302; G. P., "What Woman May Do," *AA* 4.7 (July 1845): 222–223; E. S., "Ladies' Department: Early Rising," *AA* 4.8 (August 1845): 256–257.

26. Sally Gregory Kohlstedt, "Parlors, Primers, and Public Schooling: Education for Science in Nineteenth-Century America," *Isis* 81.3 (1990): 428–434; Londa Schiebinger, *The Mind Has No Sex? Women in the Origins of Modern Science* (Harvard University Press, 1989), 30–32.

27. Elizabeth C. Morris to William Darlington, 7 April 1852, WDP; John W. Jordan, ed., *Colonial and Revolutionary Families of Pennsylvania* (Clearfield, 1911), 1:72.

28. Elizabeth C. Morris to Dr. Darlington, 8 October 1852, WDP; John S. Littell, "To the Memory of Thomas Willing Morris," Margaretta Hare Morris, album, collection of S. C. Doak, Hood River, Oregon; S[usan] Littell, "My Brother's Grave," 27 November 1852, Margaretta Hare Morris, album, collection of S. C. Doak, Hood River, Oregon.

29. Morris to Dr. Darlington, 8 October 1852; E. C. Morris to Dr. Darlington, 30 November 1852, WDP; E. C. Morris to Dr. Darlington, 1 January 1853, WDP; E. C. Morris to Dr. Darlington, 6 January 1853, WDP.

30. Morris to Dr. Darlington, 8 October 1852; M. H. Morris to Dr. Darlington, 7 January 1853, WDP.

31. "Friend Sorrow," 21 May 1853, Margaretta Hare Morris, album, personal collection of S. C. Doak, Hood River, Oregon.

32. E. C. Morris to Dr. Darlington, 30 November 1852, WDP; E. C. Morris to Dr. Darlington, 16 December 1852, WDP; M. H. Morris to Dorothea Lynde Dix, 13 February 1858, DDP. For more on the relationships of nineteenth-century women, see Carroll Smith-Rosenberg, "The Female World of Love and Ritual: Relations Between Women in Nineteenth-Century America," *Signs* 1.1 (Autumn 1975): 1–29; small, wood-beaded cross from Dorothea Dix, Box 21, Item 44, LFP.

33. Morris to Dr. Darlington, 1 January 1853; E. C. Morris to Dr. Darlington, 6 January 1853, WDP.

34. Margaretta H. Morris to Dr. Darlington, 3 February 1853, WDP; "At Germantown . . . ," *Public Ledger*, 12 January 1853.

35. Margaretta H. Morris to Dr. Darlington, 3 February 1853, WDP; E. C. Morris to Dr. Darlington, 3 February 1853, WDP.

36. "Will of Ann Willing Morris," in *The Morris Family of Philadelphia: Descendants of Anthony Morris*, ed. Robert C. Moon (R. C. Moon, 1908): 4: 190–191.

37. Asa Gray to Elizabeth Carrington Morris, 25 April 1853, Box 1, Folder 4, AGP; E. S., "Ladies' Department: Reply to Reviewer on Knitting, &c.," *AA* 6.2 (February 1847): 65; E. S., "The Garden," *AA* 5.3 (March 1846): 98.

38. Margaretta H. Morris to Harriet Clark Hare, 17 February 1853, Robert Hare Papers, 1764–1858 (Mss. B.H22), Box 3, APS; Margaretta H. Morris to Dr. Darlington, 3 February 1853, WDP.

39. E. C. Morris to Asa Gray, 18 April 1853, AGCF; E. C. Morris to Dr. Darlington, 19 November 1851, WDP; E. C. Morris to Dr. Darlington, 21 December 1851, WDP; E. C. Morris to Dr. Darlington, 26 June 1853, WDP; E. C. Morris to Dr. Darlington, 12 July 1853, WDP; Asa Gray to Elizabeth Carrington Morris, 19 September 1853, Box 1, Folder 4, AGP; E. C. Morris to Dr. Darlington, 9 November 1853, WDP; E. C. Morris to Isabella James, 21 November 1853, Box 1, Folder 27, JFP.

40. R. Robinson Scott, "Curiosities of Vegetation," *Philadelphia Florist and Horticultural Journal* 1.12 (1853): 355–356.

41. Louis Agassiz to Margaretta Hare Morris, 15 August 1853, Box 1, Folder 7, AGP; Margaretta H. Morris to Dr. Darlington, 11 November 1853, WDP; Louis Agassiz to Margaretta Hare Morris, 9 December [1853], Box 1, Folder 7, AGP; E. C. Morris to Dr. Darlington, 22 June 1853, WDP.

42. Robert Hare, *Lecture on Spiritualism Delivered Before an Audience of Three Thousand* (Partridge & Brittan, 1855), 13; Robert Hare, *Experimental Investigation of the Spirit Manifestations* (Partridge & Brittan, 1855); Craig James Hazen, *The Village Enlightenment in America: Popular Religion and Science in the Nineteenth Century* (University of Illinois Press, 2000), 65–112; Bret E. Carroll, *Spiritualism in Antebellum America* (Indiana University Press, 1997), 69–70; Timothy W. Kneeland, "Robert Hare: Politics, Science, and Spiritualism in the Early Republic," *Pennsylvania Magazine of History and Biography* 132.3 (2008): 245–260.

43. Craig James Hazen, *The Village Enlightenment in America: Popular Religion and Science in the Nineteenth Century* (University of Illinois Press, 2000), 73–76;

Benjamin Silliman to Miss Morris, 29 September 1859, Box 1, Folder 6, AGP; Edgar Fahs Smith, *The Life of Robert Hare* (J. B. Lippincott Co., 1917), 502; "ART XVI—The Late Dr. Robert Hare," *AJSA* Ser. 2, 26 (November 1858): 100–105; "Death of Dr. Robert Hare," *New York Times*, 18 May 1858.

44. Robert Hare, *Experimental Investigation of the Spirit Manifestations* (Partridge & Brittan, 1855), 168.

45. Elizabeth C. Morris to Dr. Darlington, 2 January 1854, WDP.

46. J. C. Sidney, *Map of the Township of Germantown* (R. P. Smith, c. 1848).

47. Andrew Heath, *In Union There Is Strength: Philadelphia in the Age of Urban Consolidation* (University of Pennsylvania Press, 2019).

48. Russell F. Weigley, "The Border City in Civil War, 1854–1865," *Philadelphia: A 300-Year History*, ed. Russell F. Weigley (W. W. Norton, 1982), 363–372.

49. "Another Fire in Germantown," *GT*, 5 April 1854; "Fires," *GT*, 15 February 1854; "Another Relic Gone!," *GT*, 8 March 1854; "Incendiarism," *GT*, 29 March 1854.

50. "Another Fire in Germantown." For more on firefighting in 1850s Philadelphia and other regional cities, see John F. Watson and Willis P. Hazard, *Annals of Philadelphia, and Pennsylvania in the Olden Time* (E. S. Stuart, 1884), 413–431; Bruce Laurie, *Working People of Philadelphia, 1800–1850* (Temple University Press, 1980), 58–62; Russell F. Weigley, ed., *Philadelphia: A 300-Year History* (W. W. Norton, 1982), 275–276; Marc Tebeau, *Eating Smoke: Fire in Urban America, 1800-1950* (Johns Hopkins University Press, 2005), 26; Amy S. Greenberg, *Cause for Alarm: The Volunteer Fire Department in the Nineteenth-Century City* (Princeton University Press, 2014).

51. "More Fires and the Arrest of the Incendiary," *GT*, 19 April 1854.

52. E. C. Morris to Dr. Darlington, 25 April 1854, WDP; "More Fires and the Arrest of the Incendiary"; John F. Watson and Willis P. Hazard, *Annals of Philadelphia, and Pennsylvania in the Olden Time* (E. S. Stuart, 1884), 426-427; Andrew Heath, *In Union There Is Strength* (University of Pennsylvania Press, 2019), 9–11, 22; Bruce Laurie, *Working People of Philadelphia, 1800–1850* (Temple University Press, 1980), 58–62; Amy S. Greenberg, *Cause for Alarm: The Volunteer Fire Department in the Nineteenth-Century City* (Princeton University Press, 2014), 84–86.

53. Morris to Dr. Darlington, 25 April 1854; "Another Fire in Germantown."

54. Policy Number Survey 20692 made 13 June 1854, for Margaretta H. Morris, Franklin Fire Insurance Company of Philadelphia Surveys, HSP.

55. Policy Number Survey 20342 made 7 April 1854, for Margaretta H. Morris, Franklin Fire Insurance Company of Philadelphia Surveys, HSP.

CHAPTER EIGHT: A LIFE OF EXPERIENCE

1. Margaretta Hare Morris to Richard Chandler Alexander, 17 June 1855, "Letter no. 1701," DCP; Samantha Evans, *Darwin and Women: A Selection of Letters* (Cambridge University Press, 2017), 110–112.

2. In 1849, the same year that Richard Alexander visited Margaretta Morris, he inherited property in Britain under the requirement that he assume his mother's maiden name, Prior. Some of his historical records, therefore, are filed under Richard Chandler Alexander Prior, though he still signed letters with his former last name in the years following. G. H. Brown, *Lives of the Fellows of the Royal College of Physicians of London, 1826–1925* (Published by the College, 1955) 4:24; Evans, *Darwin and Women*, 110–112; Janet Browne, *Charles Darwin: Voyaging* (Princeton University Press, 1995), 511–543.

3. Brown, *Lives of the Fellows of the Royal College of Physicians of London*, 4:24; Asa Gray to Elizabeth Carrington Morris, 11 August 1849, Box 1, Folder 4, AGP; Morris to Alexander, 17 June 1855.

4. Terri L. Premo, "'Like a Being Who Does Not Belong': The Old Age of Deborah Norris Logan," *Pennsylvania Magazine of History and Biography* 107.1 (January 1983): 94; Mary Donaldson to Ann Haines, 16 August [1830], Series 2, Box 26, Folder 404, Wyck Association Papers, APS; Elizabeth Carrington Morris, "Offerings of Friendship," Box 1, Folder 12, LFP; Margaretta Hare Morris to Susan Morris Littell, 17 March [c. 1860], Box 2, Folder 35, LFP.

5. Morris to Alexander, 17 June 1855; M., "The Army Worm," *AA* 6.2 (February 1847): 50; William Kirby and William Spence, *An Introduction to Entomology* (Lea and Blanchard, 1846), 473–476.

6. The Woodbourne estate that Margaretta visited is now part of the Woodbourne Forest and Wildlife Preserve, managed by the Nature Conservancy in Montrose, Pennsylvania. Morris to Alexander, 17 June 1855; Louis Compton Miall, *The Natural History of Aquatic Insects* (Macmillan, 1895), 39–61.

7. Charles Darwin to J. D. Hooker, 7 April 1855, "Letter no. 1661," DCP.

8. Janet Browne, *Charles Darwin: The Power of Place* (Alfred A. Knopf, 2002), 10–14.

9. Darwin to Hooker, 7 April 1855; Charles Darwin, *On the Origin of Species: A Facsimile of the First Edition* (1859; Harvard University Press, 1964), 2–3.

10. Darwin, *On the Origin of Species*, 346–410; James T. Costa, *Darwin's Backyard: How Small Experiments Led to a Big Theory* (W. W. Norton, 2017), 60, 86–87.

11. Elizabeth C. Morris to William Darlington, 3 February 1845, WDP; Browne, *Charles Darwin: Voyaging*, 457–472.

12. Browne, *Charles Darwin: Voyaging*; James A. Secord, *Victorian Sensation: The Extraordinary Publication, Reception, and Secret Authorship of Vestiges of the Natural History of Creation* (University of Chicago Press, 2000).

13. Darwin, *On the Origin of Species*, chapters 11 and 12; Charles Darwin to C. J. F. Bunbury, 21 April [1856], "Letter no. 1856," DCP; Charles Darwin to J. D. Hooker, [10 February 1845], "Letter no. 826," DCP; Browne, *Charles Darwin: Voyaging*, 517; Costa, *Darwin's Backyard*, 149–181.

14. Charles Darwin to Joseph Dalton Hooker, 13 April 1855, "Letter no. 1667," DCP; on Darwin and beetles, see Browne, *Charles Darwin: Voyaging*, 99–104.

15. Joseph Dalton Hooker to Charles Darwin, 2 [March] 1846, DCP.

16. Richard C. Alexander [Prior] to Asa Gray, 6 January 1855, AGCF.

17. Charles Darwin to Asa Gray, 25 April 1855, "Letter no. 1674," DCP; Asa Gray to Charles Darwin, 30 June 1855, "Letter no. 1707," DCP; A. Hunter Dupree, *Asa Gray: American Botanist, Friend of Darwin* (Johns Hopkins University Press, 1959, 1988), 192.

18. Morris to Alexander, 17 June 1855.

19. Darwin, *On the Origin of Species*, 2.

20. Robert V. Bruce, *The Launching of Modern American Science, 1846–1876* (Alfred A. Knopf, 1976), 7–13.

21. Charles Lyell to Charles Darwin, 1–2 May 1856, "Letter no. 1862," DCP; Leonard G. Wilson, ed., *Sir Charles Lyell's Scientific Journals on the Species Question* (Yale University Press, 1970), 52–53, 83; Browne, *Charles Darwin: Voyaging*, 186–190, 324; Browne, *Charles Darwin: The Power of Place*, 90–91.

22. Darwin, *On the Origin of Species*, 385–386.

23. Morris to Alexander, 17 June 1855.

24. Daniel Pauly, *Darwin's Fishes: An Encyclopedia of Ichthyology, Ecology, and Evolution* (Cambridge University Press, 2004), 140–141; Sharon McCormick and Gary Polis, "Arthropods That Prey on Vertebrates," *Biological Reviews* 57.1 (1982): 29–58; Miall, *The Natural History of Aquatic Insects*, 39–61; Philip Emanuel Hirsch, Anouk N'Guyen, Roxane Muller, Irene Adrian-Kalchhauser, and Patricia Burkhardt-Holm, "Colonizing Islands of Water on Dry Land—On the Passive Dispersal of Fish Eggs by Birds," *Fish and Fisheries* 19 (2018): 502–510; Jose W. Valdez, "Arthropods as Vertebrate Predators: A Review of Global Patterns," *Global Ecology and Biogeography* 29 (2020): 1691–1703.

25. Evans, *Darwin and Women*, xix–xxv, 79–88, 105–121; Mary Treat to Charles Darwin, 20 December 1871, DCP; Charles Darwin to Asa Gray, 8 January 1873, DCP.

26. Charles Darwin, *The Descent of Man and Selection in Relation to Sex* (1871; Cambridge University Press, 2009), 2:327; Browne, *Charles Darwin: The Power of Place*, 76–77; Evelleen Richards, "Redrawing the Boundaries: Darwinian Science and Victorian Women Intellectuals," in *Victorian Science in Context*, ed. Bernard Lightman (University of Chicago Press, 1997), 119–142.

27. E. Otis Kendall to [John L.] LeConte, 5 February 1855, John L. LeConte Papers, Mss. B.L493, Series I, APS; Cynthia Eagle Russett, *Sexual Science: The Victorian Construction of Womanhood* (Harvard University Press, 1989).

28. [Elizabeth Blackwell], "The Position of Women," *Philadelphia Press*, 25 August 1857; Janice Nimura, *The Doctors Blackwell: How Two Pioneering Sisters Brought Medicine to Women and Women to Medicine* (W. W. Norton, 2021), 208.

29. Pnina G. Abir-Am and Dorinda Outram, ed., *Uneasy Careers and Intimate Lives: Women in Science, 1789–1979* (Rutgers University Press, 1987); Vera Norwood, *Made from This Earth: American Women in Nature* (University of North Carolina Press, 1993); Suzanne Le-May Sheffield, *Revealing New Worlds: Three*

Victorian Women Scientists (Routledge, 2001); Robert L. Herbert and Daria D'Arienzo, *Orra White Hitchcock: An Amherst Woman of Art and Science* (University Press of New England, 2011); Renee Bergland, *Maria Mitchell and the Sexing of Science: An Astronomer Among the American Romantics* (Beacon Press, 2008); J. F. M. Clark, *Bugs and the Victorians* (Yale University Press, 2009), 154–186; Naomi Oreskes, "Objectivity or Heroism? On the Invisibility of Women in Science," *Osiris* 11 (1996): 87–113.

30. Asa Gray to Elizabeth Carrington Morris, 27 August 1856, Box 1, Folder 4, AGP.

31. P. Barry, "To the Readers of the Horticulturist," *Horticulturist and Journal of Rural Art and Rural Taste* 10.7 (July 1855): 297–298; J. Jay Smith, "The Editor to the Reader," *Horticulturist and Journal of Rural Art and Rural Taste* 10.7 (July 1855): 298–299; "Pears and Locusts," *Horticulturist and Journal of Rural Art and Rural Taste* 10.11 (November 1855): 479.

32. Miss Margaretta Morris, "The Pear Slug-Worm," *Horticulturist and Journal of Rural Art and Rural Taste* 6 (September 1856): 410–411; Margaretta H. Morris, "Original Observations on Insects Injurious to Our Fruits—The Curculio," *Horticulturist and Journal of Rural Art and Rural Taste* 14 (November 1859): 506–509; Miss M. H. Morris, "Notes on the Peach," *GM* 2.5 (May 1860): 130; Margaretta H. Morris, "The Peach-Tree and Its Enemies," *Horticulturist* 15 (March 1860): 118–120; "Insects," *GM* 2.11 (November 1860): 341.

33. Morris, "The Pear Slug-Worm"; Morris, "Original Observations on Insects Injurious to Our Fruits"; Morris, "Notes on the Peach"; Morris, "The Peach-Tree and Its Enemies."

34. "Yellows in the Peach Tree," *GM* 2.4 (April 1860): 117; F. Dana, "Yellows in the Peach," *GM* 2.5 (May 1860): 137; Morris, "Notes on the Peach"; "The Curculio and Black Knot in Plum and Cherry Trees," *Valley Farmer* 12.5 (May 1860): 149; "The Curculio Not the Cause of the Black Knot," *Valley Farmer* 13.2 (February 1861): 49.

35. Morris, "Notes on the Peach."

CHAPTER NINE: PLANTING AND PRESERVING

1. Elizabeth C. Morris to William Darlington, 22 January 1860, WDP.

2. Morris to Darlington, 22 January 1860; Elizabeth M. Geffen, "Industrial Development and Social Crisis, 1841–1854," in *Philadelphia: A 300-Year History*, ed. Russell F. Weigley (W. W. Norton, 1992), 307–362.

3. Philip Pauly, "Summer Resort and Scientific Discipline: Woods Hole and the Structure of American Biology, 1882–1925," in *American Development of Biology*, ed. Ronald Rainger, Keith R. Benson, and Jane Maienschein (University of Pennsylvania Press, 1988), 121–150; Lynn Nyhart, "Natural History and the 'New' Biology," in *Cultures of Natural History*, ed. N. Jardine, J. A. Secord, and E. C. Spary (Cambridge University Press, 1996), 426–443; Bruno Strasser, *Collecting*

Experiments: Making Big Data Biology (University of Chicago Press, 2019), 1–28; Peter J. Bowler and Iwan Rhys Morus, *Making Modern Science: A Historical Survey* (University of Chicago Press, 2005), 172–196; Robert E. Kohler, *Landscapes and Labscapes: Exploring the Lab-Field Border in Biology* (University of Chicago Press, 2002); Robert E. Kohler, *All Creatures: Naturalists, Collectors, and Biodiversity* (Princeton University Press, 2006); Robert E. Kohler and Jeremy Vetter, "The Field," in *A Companion to the History of Science*, ed. Bernard Lightman (John Wiley & Sons, 2016), 282–295.

4. Morris to Darlington, 22 January 1860; E. C. Morris to William Darlington, 26 December 1855, WDP.

5. E. C. Morris to Dr. Darlington, 11 April 1855, WDP; E. C. Morris to Dr. Darlington, 19 March 1855, WDP; Elizabeth C. Morris to Daniel Cady Eaton, 13 March 1861, Series 1, Box 7, Folder 373, DCEP; "Forest Mosses," 17 February 1855, in Elizabeth Morris, copybook, 1832–1856, Box 1, Folder 14, LFP.

6. Luke Keogh, *The Wardian Case: How a Simple Box Moved Plants and Changed the World* (University of Chicago Press, 2020).

7. Elizabeth C. Morris to William Darlington, 29 March 1860, WDP; Elizabeth C. Morris to Daniel Cady Eaton, 30 April 1860, Series 1, Box 7, Folder 373, DCEP; Elizabeth C. Morris to Daniel Cady Eaton, 17 June 1860, Series 1, Box 7, Folder 373, DCEP; N. B. Ward, *On the Growth of Plants in Closely Glazed Cases*, 2nd ed. (J. Van Voorst, 1852); Asa Gray, "N. B. Ward, F. R. S., &c., On the Growth of Plants in Closely Glazed Cases," *AJSA* 16.46 (July 1853): 133–134; Sarah Whittingham, *The Victorian Fern Craze* (Shire Publications, 2009); David Elliston Allen, *The Victorian Fern Craze: A History of Pteridomania* (Hutchison of London, 1969); Lynn Barber, *The Heyday of Natural History* (Doubleday, 1980), 111–124; David Allen, "Tastes and Crazes," in *Cultures of Natural History*, ed. N. Jardine, J. A. Secord, and E. C. Spary (Cambridge University Press, 1996), 394–407; Lindsay Wells, "Close Encounters of the Wardian Kind: Terrariums and Pollution in the Victorian Parlor," *Victorian Studies* 60.2 (2018): 158–170; Luke Keogh, "The Wardian Case: Environmental Histories of a Box for Moving Plants," *Environment and History* 25.2 (2019): 219–244; Keogh, *Wardian Case: How a Simple Box Moved Plants and Changed the World*; Jen Maylack, "How a Glass Terrarium Changed the World," *Atlantic*, 12 November 2017.

8. William M. Canby, ed., "Autobiography and Some Reminiscences of the Late August Fendler," *Botanical Gazette* 10.6 (June 1885): 285–290; Elizabeth A. Shaw, "Augustus Fendler's Collection List: New Mexico, 1846–1847," *Contributions from the Gray Herbarium of Harvard University* 212 (1982): 1–70; Elizabeth A. Shaw, "Changing Botany in North America: 1835–1860, The Role of George Engelmann," *Annals of the Missouri Botanical Garden* 73.3 (1986): 508–519; Jane Camerini, "Remains of the Day: Early Victorians in the Field," in *Victorian Science in Context*, ed. Bernard Lightman (University of Chicago Press, 1997): 354–377;

Anne Larsen, "Equipment for the Field," in *Cultures of Natural History*, ed. N. Jardine, J. A. Secord, and E. C. Spary (Cambridge University Press, 1997), 358–377; Kuang-Chi Hung, "Finding Patterns in Nature: Asa Gray's Plant Geography and Collecting Networks (1830s–1860s)" (PhD diss., Harvard University, 2013), 197–201.

9. Asa Gray to Elizabeth Carrington Morris, 17 January 1849, Box 1, Folder 4, AGP; Augustus Fendler to Asa Gray, 11 August 1853, AGCF; Jane Loring Gray to Elizabeth Carrington Morris, 19 March 1858, Box 1, Folder 5, AGP.

10. Hung, "Finding Patterns in Nature," 27–38, 149–150, 260–263; Kuang-Chi Hung, "Subscribing to Specimens, Cataloging Subscribed Specimens, and Assembling the First Phytogeographical Survey in the United States," *Journal of the History of Biology* 52 (2019): 391–431.

11. Augustus Fendler to Asa Gray, 16 December 1854, AGCF; Augustus Fendler to Asa Gray, 25 November 1855, AGCF.

12. Fendler to Gray, 25 November 1855; Elizabeth C. Morris to William Darlington, 16 March 1857, WDP.

13. Augustus Fendler to Asa Gray, 25 April 1856, AGCF; Jane Loring Gray to Elizabeth Carrington Morris, 8 April 1856, Box 1, Folder 5, AGP; Elizabeth C. Morris to William Darlington, 26 April 1856, WDP.

14. E. C. Morris to William Darlington, 26 March 1856, WDP; Asa Gray to Elizabeth Carrington Morris, 27 August 1856, Box 1, Folder 4, AGP; Carol A. Todzia, "Augustus Fendler's Venezuelan Plant Collection," *Annals of the Missouri Botanical Garden* 76.1 (1989): 310–329; Asa Gray to Elizabeth Carrington Morris, 21 February 1859, Box 1, Folder 4, AGP.

15. Elizabeth C. Morris to William Darlington, 26 February 1857, WDP; Elizabeth C. Morris to William Darlington, 16 March 1857, WDP; Elizabeth C. Morris to William Darlington, 12 March 1858, WDP.

16. Elizabeth C. Morris to William Darlington, 14 October 1842, WDP.

17. Elizabeth C. Morris to William Darlington, 22 January 1860, WDP; Judith Wellman, *The Road to Seneca Falls: Elizabeth Cady Stanton and the First Woman's Rights Convention* (University of Illinois Press, 2004), 194; Elizabeth C. Morris to William Darlington, 29 March 1860, WDP; Asa Gray to Elizabeth Carrington Morris, 19 July 1859, Box 1, Folder 4, AGP; A. Hunter Dupree, *Asa Gray: American Botanist, Friend of Darwin* (Johns Hopkins University Press, 1959, 1988), 200.

18. Elizabeth C. Morris to Daniel Cady Eaton, 30 April 1860, Series 1, Box 7, Folder 373, DCEP; Elizabeth C. Morris to Daniel Cady Eaton, 21 May 1860, Series 1, Box 7, Folder 373, DCEP.

19. Elizabeth C. Morris to Daniel Cady Eaton, [undated], Series 1, Box 7, Folder 373, DCEP; Morris to Darlington, 29 March 1860.

20. Morris to Eaton, 30 April 1860; Morris to Eaton, 17 June 1860; Morris to Eaton, 13 March 1861.

21. Edward Tatnall, *Catalogue of the Phænogamous and Filicoid Plants of Newcastle County, Delaware* (Wilmington Institute, 1860); "Books, Catalogues, &c.," *GM* 2.11 (1860): 343–344.

22. Tatnall, *Catalogue of the Phænogamous and Filicoid Plants*, 44.

23. "Books, Catalogues, &c."

24. M., "The Rose of Jericho," *GM* 1.6 (June 1859): 83-84; Miss E., "House Culture of Camellias," *GM* 3.4 (April 1861): 116; Miss E., "Derivation of the Word 'Nosegay,'" *GM* 2.1 (January 1860): 6; Miss E., "Ye Lazie Fevre," *GM* 2.4 (April 1860): 108–109; "We have received from E. S. the following," *GM* 4.2 (February 1862): 53; E., "Note on Wistaria Sinensis," *GM* 4.7 (July 1862): 195; E., "Notes from My Scrap-Book," *GM* 4.11 (November 1862): 329–331; M., "Dwarf Pears in Grass," *GM* 5.2 (February 1863): 47; M., "A Pretty Native Shrub," *GM* 5.3 (March 1863): 100; M., "Flower Markets," *GM* 5.5 (May 1863): 166.

25. E., "Address to a Land Tortoise," *GM* 5.9 (September 1863): 166; E. C. M., "Note on the Land Tortoise," *GM* 5.10 (October 1863): 302.

26. Elizabeth C. Morris to William Darlington, 16 July 1858, WDP; Asa Gray, "Terminations of Botanical Names," *GM* 4.4 (April 1862): 109; "Books, Catalogues, &c.," 90–91; Thomas Meehan to Asa Gray, 7 July 1859, AGCF.

27. Morris to Eaton, 13 March 1861.

28. Gray to Morris, 19 July 1859.

29. Louis Agassiz, *Lake Superior: Its Physical Character, Vegetation, and Animals* (Gould, Kendall, and Lincoln, 1850), 152; Hung, "Finding Patterns in Nature," 374–442; Dupree, *Asa Gray*, 136–139.

30. Asa Gray, Review: "Explanations: A Sequel to the Vestiges of the Natural History of Creation," *North American Review* 62.131 (April 1846): 468; Elizabeth C. Morris to William Darlington, 3 February 1845, WDP; Elizabeth C. Morris to William Darlington, 12 March 1845, WDP.

31. Asa Gray to Charles Darwin, 30 June 1855, "Letter no. 1707," DCP.

32. Asa Gray, "Analogy Between the Flora of Japan and that of the United States," *AJSA* 2.4 (1846): 135–136; Asa Gray, "Diagnostic Characters of New Species of Phaenogamous Plants, Collected in Japan by Charles Wright, Botanist of the U.S. North Pacific Exploring Expedition with Observations upon the Relations of the Japanese Flora to That of North America, and of Other Parts of the Northern Temperate Zone," *Memoirs of the American Academy of Arts and Sciences* 6.2 (1859): 377–452; Hung, "Finding Patterns in Nature," 1–9, 617–659; Christoph Irmscher, *Louis Agassiz: Creator of American Science* (Houghton Mifflin Harcourt, 2013), 121–167.

33. Elizabeth C. Morris to William Darlington, 14 September 1859, WDP.

34. Asa Gray to Charles Darwin, 7 July 1857, "Letter no. 2120," DCP; Charles Darwin to Asa Gray, 20 July [1857], "Letter no. 2125," DCP; Asa Gray to Charles Darwin, [August 1857], "Letter no. 2129," DCP; Charles Darwin to Asa Gray, 5 September [1857], "Letter no. 2136," DCP; Janet Browne, "Asa Gray and Charles Darwin: Corresponding Naturalists," *Harvard Papers in Botany* 15.2

(2010): 209–220; Hung, "Finding Patterns in Nature," 535–543; Janet Browne, *Charles Darwin: The Power of Place* (Alfred A. Knopf, 2002), 3–42.

35. Asa Gray, "Review of Darwin's Theory on the Origin of Species by Means of Natural Selection," *AJSA* 29.86 (March 1860): 152–184; Browne, *Charles Darwin: The Power of Place*, 132–135.

36. Asa Gray, "Darwin and His Reviewers," *Atlantic Monthly* 6.36 (October 1860): 406–425; Browne, "Asa Gray and Charles Darwin"; Browne, *Charles Darwin: The Power of Place*, 134–135; Dupree, *Asa Gray*, 233–306; A. Hunter Dupree, "The First Darwinian Debate in America: Gray versus Agassiz," *Daedalus* 88.3 (Summer 1959): 560–569; Irmscher, *Louis Agassiz*.

37. "Of the Progressive scale or chain of Beings in the Universe," from Smellie's *Philosophy of Natural History*, in Elizabeth C. Morris's copybook 1829–1831, Box 1, Folder 13, LFP.

38. Dupree, *Asa Gray*, 275–278.

CHAPTER TEN: SHE IS EVERYTHING NOW, TO ME

1. Elizabeth C. Morris to William Darlington, 1 January 1857, WDP; Elizabeth C. Morris to William Darlington, 9 February 1857, WDP; "Cold Weather," *Weekly Patriot and Union*, 31 December 1856; "The Recent Cold Weather," *GT*, 31 December 1856; Lock from the Morris-Littell house, S. E. Corner Main and High Streets, Torn Down 1915, Germantown Historical Society.

2. Elizabeth C. Morris to William Darlington, 1 January 1857, WDP; Morris to Darlington, 9 February 1857.

3. Linda Nash, *Inescapable Ecologies: A History of Environment, Disease, and Knowledge* (University of California Press, 2006), 42, 210; Conevery Bolton Valencius, *The Health of the Country: How American Settlers Understood Themselves and Their Land* (Basic Books, 2002), 53–84.

4. E. C. Morris to William Darlington, 15 July 1855, WDP; Patricia Cline Cohen, "Safety and Danger: Women on American Public Transport, 1750–1850," in *Gendered Domains: Rethinking Public and Private in Women's History*, ed. Dorothy O. Helly and Susan M. Reverby (Cornell University Press, 1992), 109–122; Patricia Cline Cohen, "Women at Large: Travel in Antebellum America," *History Today* 44.12 (December 1994): 44–50.

5. Elizabeth C. Morris to William Darlington, 31 August 1855, WDP; E. C. Morris to William Darlington, 15 October 1855, WDP.

6. Elizabeth C. Morris to William Darlington, 16 March 1857, WDP.

7. Elizabeth C. Morris to William Darlington, 11 May 1857, WDP; Jane Loring Gray to Elizabeth Carrington Morris, [22] July 1857, Box 1, Folder 5, AGP; "Harvard's College Commencement," *New York Times*, 16 July 1857; Asa Gray to Elizabeth Carrington Morris, 17 July 1857, Box 1, Folder 4, AGP; Thomas J. Brown, *Dorothea Dix: New England Reformer* (Harvard University Press, 1998), 251.

8. Elizabeth C. Morris to William Darlington, 27 July 1857, WDP.

9. Morris to Darlington, 27 July 1857.

10. M. H. Morris to Dr. Darlington, 7 January 1853, WDP; Elizabeth C. Morris to William Darlington, 16 December 1857, WDP; Elizabeth C. Morris to William Darlington, 16 July 1858, WDP.

11. Morris to Darlington, 16 December 1857; E. C. Morris to Dorothea Lynde Dix, 13 February 1858, DDP; "The Executive Committee of the Relief Association of Germantown . . . ," *GT*, 21 April 1858; Russell F. Weigley, "The Border City in Civil War, 1854–1865," in Russell F. Weigley, ed., *Philadelphia: A 300-Year History* (W. W. Norton, 1982), 381.

12. "Woman's Help for Farmers' Family," *GT*, 31 December 1856; Morris to Darlington, 1 January 1857; Margaretta Hare Morris to Richard Chandler Alexander, 17 June 1855, DCP.

13. Elizabeth C. Morris to William Darlington, 17 December 1858, WDP; Elizabeth C. Morris to William Darlington, 24 October 1859, WDP; Miss E., "Ye Lazie Fevre," *GM* 2.4 (April 1860): 108–109; Miss E., "House Culture of Camellias," *GM* 3.4 (April 1861): 116; "We have received from E. S. the following," *GM* 4.2 (February 1862): 53; E., "Note on Wistaria Sinensis," *GM* 4.7 (July 1862): 195; E., "Notes from My Scrap-Book," *GM* 4.11 (November 1862): 329–331; M., "Dwarf Pears in Grass," *GM* 5.2 (February 1863): 47; M., "A Pretty Native Shrub," *GM* 5.3 (March 1863): 100; M., "Flower Markets," *GM* 5.5 (May 1863): 166; E., "Address to a Land Tortoise," *GM* 5.9 (September 1863): 166; E. C. M., "Note on the Land Tortoise," *GM* 5.10 (October 1863): 302; M., "Scraps," *GM* 6.6 (June 1864): 171; M. "Old American Horticulturists," *GM* 6.9 (September 1864): 260–261.

14. Elizabeth C. Morris to William Darlington, 17 November 1858, WDP; William Henry Harvey, *The Sea-Side Book: Being an Introduction to the Natural History of the British Coasts* (J. Van Voorst, 1857), 9; E. C. Morris to Dr. Darlington, 12 July 1853, WDP; E. C. Morris to Dr. Darlington, 13 February 1859, WDP.

15. Asa Gray to Elizabeth Carrington Morris, 8 June 1846, Box 1, Folder 3, AGP; Asa Gray to Elizabeth Carrington Morris, 29 June 1846, Box 1, Folder 3, AGP; William Henry Harvey, *Nereis Boreali-Americana, or, Contributions to the History of the Marine Algae of North America* (Smithsonian Institution, 1852–1858), 1:48; 2:38, 48, 107, 121, 170–171, 217, 224–225, 238; 3:41, Pl. XLV; Jacques S. Zaneveld, "The Benthic Marine Algae of Delaware, U.S.A.," *Chesapeake Science* 13.2 (June 1972): 120–138; Asa Gray to Elizabeth Carrington Morris, 21 February 1859, Box 1, Folder 4, AGP.

16. Margaretta Hare Morris Membership Card, Collection 723, Academy Membership, ANSP.

17. Rough Minutes, 27 September 1859, Collection 502, vol. 7, ANSP; "Meeting for Business, 27 September 1859," and "Ordinary Meeting, 21 October 1859," Minutes of the Academy of Natural Sciences, Collection 502, vol. 11,

ANSP; Articles of Association, Academy of Natural Sciences, 1814, Collection 104, Folder 1, ANSP.

18. Proposals for Resident Memberships, American Entomological Society, Collection 218, Box 1, Folder 3, ANSP; *Proceedings of the Entomological Society of Philadelphia*, vol. 1 (Printed by the Society, 1863); Jessica C. Linker, "The Fruits of Their Labor: Women's Scientific Practice in Early America, 1750–1860" (PhD diss., University of Connecticut, 2017), 112–113.

19. W. Conner Sorensen, *Brethren of the Net: American Entomology, 1840–1880* (University of Alabama Press, 1995), 15–59.

20. Margaretta Hare Morris to Susan Morris Littell, [c. November 1859], Series 3, Box 2, Folder 35, LFP.

21. At the time of the raid, Harpers Ferry was in the Commonwealth of Virginia, though today it is within the boundaries of West Virginia. For more on the raid, see Tony Horwitz, *Midnight Rising: John Brown and the Raid That Sparked the Civil War* (Henry Holt and Company, 2011); John Stauffer and Zoe Trodd, eds., *The Tribunal: Responses to John Brown and the Harpers Ferry Raid* (Belknap Press, 2012); David W. Blight, *Frederick Douglass: Prophet of Freedom* (Simon & Schuster, 2018), 280–309; Kellie Carter Jackson, *Force and Freedom: Black Abolitionists and the Politics of Violence* (University of Pennsylvania Press, 2019), 106–134.

22. Horwitz, *Midnight Rising*; Blight, *Frederick Douglass*; Carter Jackson, *Force and Freedom*.

23. Margaretta Hare Morris to Susan Morris Littell, [30 October 1859], Series 3, Box 2, Folder 35, LFP; Brown, *Dorothea Dix*, 262–263; Douglass quoted in Blight, *Frederick Douglass*, 308–309; Stauffer and Trodd, *The Tribunal*, xxxv, 105–109.

24. Margaretta Hare Morris to Thomas Gardiner Littell, [30 October 1859], Series 3, Box 2, Folder 35, LFP; Osborne P. Anderson, *A Voice from Harper's Ferry* (Printed for the Author, 1861), 59–63.

25. Margaretta Hare Morris to Thomas Gardiner Littell, [30 October 1859], Series 3, Box 2, Folder 35, LFP; Stauffer and Trodd, *The Tribunal*, xxi; Oswald Garrison Villard, *John Brown, 1800-1859* (Houghton Mifflin, 1910), 563; Gary B. Nash, *First City: Philadelphia and the Forging of Historical Memory* (University of Pennsylvania Press, 2002), 223–260.

26. Elizabeth C. Morris to William Darlington, 3 February 1861, WDP; Elizabeth C. Morris to Daniel Cady Eaton, 13 March 1861, MS 581, Series 1, Box 7, Folder 373, DCEP; Elizabeth C. Morris to William Darlington, 6 June 1861, WDP.

27. Elizabeth C. Morris to William Darlington, 4 July 1861, WDP; Nash, *First City*, 231–233; Daniel Kilbride, *An American Aristocracy: Southern Planters in Antebellum Philadelphia* (University of South Carolina Press, 2006).

28. Charles Darwin, *On the Origin of Species* (New York: D. Appleton, 1860); "The Origin of Species," *New-York Times*, 28 March 1860; Randall Fuller, *The Book That Changed America* (Viking, 2017); Janet Browne, *Charles Darwin: The Power of Place* (Alfred A. Knopf, 2002), 132–135.

29. Weigley, "The Border City in Civil War, 1854–1865," 363–416; Elizabeth C. Morris to William Darlington, 4 July 1861, 26 August 1861, WDP; William Dusinberre, *Civil War Issues in Philadelphia, 1856–1865* (University of Pennsylvania Press, 1965), 127–150; "The Apparent Want of Loyalty," *GT*, 10 April 1861; "The Southern Rebellion, the War Begun!" *GT*, 17 April 1861.

30. Elizabeth C. Morris to William Darlington, 26 August 1861, WDP; Brown, *Dorothea Dix*, 267–283; Judith Ann Giesberg, *Civil War Sisterhood: The U.S. Sanitary Commission and Women's Politics in Transition* (Northeastern University Press, 2000).

31. Dorothea L. Dix to Margaretta Hare Morris [n.d.], Box 26, Folder 47, LFP.

32. Robert V. Bruce, *The Launching of Modern American Science, 1846–1876* (Alfred A. Knopf, 1976), 271–312; Asa Gray to Charles Darwin, 31 March [1862], DCP; "Horticulture Just Now," *GM* 6.9 (September 1864): 170; Dorothy Godfrey Wayman, *Edward Sylvester Morse* (Harvard University Press, 1943), 169.

33. Elizabeth C. Morris to William Darlington, 14 September 1859, WDP; Elizabeth C. Morris to William Darlington, 20 January 1862, WDP; Elizabeth C. Morris to William Darlington, 26 April 1862, WDP; "Death of Dr. Darlington," *GM* 5.5 (May 1863): 157; "The Late Dr. William Darlington," *GM* 5.6 (June 1863): 182–183.

34. "The Late Dr. William Darlington"; M., "Old American Horticulturists," *GM* 6.9 (September 1864): 260–261.

35. "Victory!! Waterloo Eclipsed!! The Desperate Battle Near Gettysburg!" *Philadelphia Inquirer*, 6 July 1863; "The War for the Union! The State Invasion!" *GT*, 8 July 1863; "Scenes in and around Gettysburg," *GT*, 29 July 1863; "Great Central Fair," [newspaper clipping], 4 June 1864, in Scrapbook, Box 2, Folder 48, LFP; Drew Gilpin Faust, *This Republic of Suffering: Death and American Civil War* (Vintage Books, 2008), 137–170.

36. Scrapbook, c. 1830–1876, Box 3, Folder 48, LFP; "Washington Elm," Box 19, Item 7-2, LFP; Judith Giesberg, ed., *Emilie Davis's Civil War: The Diaries of a Free Black Woman in Philadelphia, 1863–1865* (Pennsylvania State University Press, 2014), 137; Giesberg, *Civil War Sisterhood*, 105–107; Nash, *First City*, 246–248; Weigley, "The Border City in Civil War, 1854–1865," 412; Thomas Campanella, *Republic of Shade: New England and the American Elm* (Yale University Press, 2003), 45–68; J. G. Jack, "The Cambridge Washington Elm," *Arnold Arboretum Harvard University Bulletin of Popular Information*, Series 3, 5.18 (December 1931): 69–73; J. Matthew Gallman, *Mastering Wartime: A Social History of Philadelphia During the Civil War* (Cambridge University Press, 1990), 146–169; Elizabeth Milroy, "Avenue of Dreams: Patriotism and the Spectator at Philadelphia's Great Central Sanitary Fair," in *Making and Remaking Pennsylvania's Civil War*, ed. William Blair and William Pencak (Pennsylvania State University Press, 2001), 23–57.

37. Elizabeth C. Morris to William Darlington, 29 March 1860, WDP; Death Certificate: Elizabeth C. Morris, "Pennsylvania, Philadelphia City Death Certificates, 1803–1915," *FamilySearch*, Philadelphia City Archives and Historical Society

of Pennsylvania, Philadelphia; Giesberg, *Emilie Davis's Civil War*, 145; "The Snow Storm of the Winter," *GT*, 15 February 1865.

38. Gary Laderman, *The Sacred Remains: American Attitudes Toward Death, 1799-1883* (Yale University Press, 1996), 29; Miss E., "House Culture of Camellias," *GM* 3.4 (April 1861): 116.

39. Elizabeth C. Morris, Philadelphia, Pennsylvania, United States, Death Certificates Index, 1803–1915; Laderman, *The Sacred Remains*, 27–50, 164–175; Drew Faust, *This Republic of Suffering*.

40. "Death of One of Our Contributors," *GM* 7.3 (March 1865): 84.

41. Will of Elizabeth Carrington Morris, Philadelphia, Wills, Probate Records [microform] Book 55–56, 1865; Will of M. H. Morris, Philadelphia, Wills, Probate Records [microform] Book 60, 1866–1867, HSP, Record 262.

42. "Great Freshet in the Schuylkill," *Philadelphia Inquirer*, 18 July 1865; "The Storm on Sunday," *New York Times*, 18 July 1865; "The Summer Freshet, From the Philadelphia Ledger," *New York Times*, 20 July 1865.

43. "Evening Hour," in Margaretta Hare Morris, album, personal collection of S. C. Doak, Hood River, Oregon.

44. Margaretta Hare Morris, album, collection of S. C. Doak, Hood River, Oregon.

45. Death Certificate: Margaretta H. Morris, "Pennsylvania, Philadelphia City Death Certificates, 1803–1915," *FamilySearch*, Philadelphia City Archives and HSP, Philadelphia.

Chapter Eleven: Forgetting

1. In 2010, the Wissahickon Valley Park separated from the Fairmount Park System. "The Water Question," *Philadelphia Inquirer*, 15 October 1867; John Cresson, Surveyor, "Map of the Survey of Wissahickon Creek, 1868" (adopted by the Commissioners of Fairmount Park, May 15, 1869); "Map of Fairmount Park and Vicinity" (O. S. Senter, 1872); Russell F. Weigley, ed., *Philadelphia: A 300-Year History* (W. W. Norton, 1982), 376–379, 413–414, 426–427; Elizabeth Milroy, *The Grid and the River: Philadelphia's Green Places, 1682–1876* (Pennsylvania State University Press, 2016), 286–287, 302–303; Nate Gabriel, "Visualizing Urban Nature in Fairmount Park: Economic Diversity, History, and Photography in Nineteenth-Century Philadelphia," in *A Green Country Towne: Philadelphia's Ecology in the Cultural Imagination*, ed. Alan C. Braddock and Laura Turner Igoe (Pennsylvania State University Press, 2016), 65–80.

2. See, for instance, Phebe Mitchell Kendall, comp., *Maria Mitchell: Life, Letters, and Journals* (Lee and Shepard Publishers, 1896); Helen Wright, *Sweeper in the Sky: The Life of Maria Mitchell* (Macmillan, 1949); Grace Hathaway Melin, *Maria Mitchell: Girl Astronomer* (Bobbs-Merrill, 1954); Katharine Elliott Wilkie and Paul E. Kennedy, *Maria Mitchell, Stargazer* (Garrard Publishing, 1966); Elizabeth Fraser Torjesen, *Comet over Nantucket: Maria Mitchell and Her Island: The*

Story of America's First Woman Astronomer (Friends United Press, 1984); Beatrice Gormley, *Maria Mitchell: The Soul of an Astronomer* (Eerdmans, 1995); Deborah Hopkinson, *Maria's Comet* (Atheneum Books for Young Readers, 1999); Renee Bergland, *Maria Mitchell and the Sexing of Science: An Astronomer Among the American Romantics* (Beacon Press, 2008); Jeannine Atkins, *Finding Wonders: Three Girls Who Changed Science* (Atheneum Books for Young Readers, 2016); Hayley Barrett, *What Miss Mitchell Saw* (Beach Lane Books, 2019).

3. Samuel H. Scudder, ed., *Entomological Correspondence of Thaddeus William Harris, M.D.* (Printed for the Boston Society of Natural History, 1869); *Robert Hare, M.D.* ([n.p.], 1858); "The Late Dr. Hare," *AJSA* 26 (1858): 100–105; "Death of Dr. Robert Hare," *New York Times*, 18 May 1858; Washington Townsend, *Memorial of William Darlington, M.D.* (E. F. James, 1863); Thomas Potts James, *An Obituary Notice of Dr. William Darlington, Read before the American Philosophical Society*, February 19, 1864 (1864); "The Late Dr. Darlington," *Lancaster Examiner*, 29 April 1863; *Memorial Biographies of New England Historic Genealogical Society* (New England Historic Genealogical Society, 1894) 5:202–220.

4. *PANSP* 19 (1867): 82, 238.

5. W. Conner Sorensen, *Brethren of the Net: American Entomology, 1840–1880* (University of Alabama Press, 1995), 188–193; Jeffrey I. Barnes, "Insects in the New Nation: A Cultural Context for the Emergence of American Entomology," *Bulletin of the Entomological Society of America* (Spring 1985): 21–30.

6. For more on this, see Chapter 3. Thaddeus William Harris, *A Report on Insects of Massachusetts Injurious to Vegetation* (Folsom, Wells, and Thurston, 1841): 429–430; Asa Fitch, *The Hessian Fly: Its History, Character, Transformations, and Habits* (C. Van Benthuysen, 1847), 2–4, 23, 42–43; Margaretta Hare Morris to Thaddeus William Harris, 12 September 1847, TWHP; M. H. Morris, "Controversy Respecting the Hessian Fly," *American Journal of Agriculture and Science* 5.12 (April 1847): 206–208.

7. Asa Fitch, "Wheat Insects—Joint-Worm," *Cultivator* 8.10 (October 1851): 321; Thaddeus William Harris, *A Treatise on Some of the Insects of New England which are Injurious to Vegetation* (White & Potter, 1852), 465–466; *PANSP* 4 (1849): 194; S. S. Haldeman, "Report on the Progress of Entomology in the United States during the year 1849," *PANSP* 5 (1850–1851): 5–8; William Draper Brinckle, *Remarks on Entomology: Chiefly in Reference to an Agricultural Benefit* (W. B. Wiley, 1852), 5; E. Emmons, *Agriculture of New-York* (Printed by C. Van Benthuysen, 1854), 5:179–180; Benedict Jaeger and Henry C. Preston, *The Life of North American Insects* (Harper & Brothers, 1859), 306; Entomologischer Verein in Stettin, *Linnaea Entomologica: Zeitschrift Herausgegeben von dem Entomologischen Vereine in Stettin* (Ernst Siegfried Mittler, 1860) 14:263; Entomologischer Verein in Stettin, *Entomologische Zeitung* 22 (1861): 420; United States Bureau of Entomology, *Bibliography of the More Important Contributions to American Economic Entomology* (US Government Printing Office, 1889), 66–68; Lawrence Bruner, *The Insect Enemies of Small Grains* (Nebraska State Board of Agriculture, 1893), 361, 367;

Lawrence Bruner, *A Preliminary Introduction to the Study of Entomology* (J. North, 1894), 205.

8. A. S. Packard, *The Hessian Fly: Its Ravages, Habits, Enemies, and Means of Preventing Its Increase* (US Government Printing Office, 1880); Herbert Osborn, *The Hessian Fly in the United States* (US Government Printing Office, 1898); Paul Marchal, "Les Cécidomyies des Céréales et Leurs Parasites," *Annales de la Société Entomologique de France* 66 (Au Siege de Société, 1897): 64; J. J. Kieffer, *Monographie des Cécidomyides d'Europe et d'Algérie* (Société Entomologique de France, 1900), 413–414; E. P. Felt, "Hosts and Galls of American Gall Midges," *Journal of Economic Entomology* 4, no. 10 (October 1911): 473; H. F. Barnes, *Gall Midges of Economic Importance*, vol. 7: *Gall Midges of Cereal Crops* (Crosby Lockwood & Son, 1956), 84; Marion Harris, email to author, 4 August 2017; Raymond J. Gagné, *The Plant-Feeding Gall Midges of North America* (Cornell University Press, 1989), 30–32; Christopher Kemp, *The Lost Species: Great Expeditions in the Collections of Natural History Museums* (University of Chicago Press, 2017).

9. Like many of his peers, Erwin F. Smith was not consistently sexist during his career. He would go on to hire a groundbreaking number of women botanists at his lab in the USDA's Bureau of Plant Industries, further reflecting the complex views of these scientific men. Miss Morris, "The Yellows, Caused by an Insect," *Horticulturist and Journal of Rural Art and Rural Taste* 4.11 (May 1850): 502–503; "Discovery of the Cause of the Yellows in the Peach Tree," *AA* 9.5 (May 1850): 144–145; Erwin F. Smith, *Peach Yellows: A Preliminary Report, Prepared Under the Direction of the Commissioner of Agriculture* (US Government Printing Office, 1888); Erwin F. Smith, *Legal Enactments for the Restriction of Plant Diseases* (US Government Printing Office, 1896); Geoffrey Clough Ainsworth, *Introduction to the History of Plant Pathology* (Cambridge University Press, 1981), 70, 81, 182; Margaret W. Rossiter, *Women Scientists in America: Struggles and Strategies to 1940* (Johns Hopkins University Press, 1982), 1:60–63; L. O. Kunkel, "Insect Transmission of Peach Yellows," *Contributions Boyce Thompson Institute* 5 (1933): 19–28; Wendell M. Stanley, *Louis Otto Kunkel, 1884-1960: A Biographical Memoir* (National Academy of Sciences, 1965).

10. H. F. Wilson, "The Peach-Tree Barkbeetle," in U.S. Department of Agriculture Bureau of Entomology Bulletin No. 68 (US Government Printing Office, 1909), 91–108.

11. *Umbilicaria mammulata* (Ach.) Tuck. (Morris, 1846, Philadelphia Co., Pennsylvania), 1058999, NYBG; *Cladonia furcate* (Huds.) Schrad. (Morris, 1846, Philadelphia Co., Pennsylvania), 1219376, NYBG; *Teloschistes chrysophthalmus* (L.) Th.Fr. (Morris, 1846, Philadelphia Co., Pennsylvania), 1704140, NYBG; Barbara M. Thiers, *Herbarium: The Quest to Preserve & Classify the World's Plants* (Timber Press, 2020), 122–124.

12. J. Mason Heberling, "Herbaria as Big Data Sources of Plant Traits," *International Journal of Plant Sciences* 183.2 (February 2022); Barbara M. Thiers, *Herbarium: The Quest to Preserve and Classify the World's Plants* (Timber Press, 2020), 219–242;

Courtney Fullilove, *The Profit of the Earth: The Global Seeds of American Agriculture* (University of Chicago Press, 2017), 201–203.

13. William Coleman, *Biology in the Nineteenth Century: Problems of Form, Function, and Transformation* (John Wiley & Sons, 1971); Philip Pauly, "Summer Resort and Scientific Discipline: Woods Hole and the Structure of American Biology, 1882–1925," in *American Development of Biology*, ed. Ronald Rainger, Keith R. Benson, and Jane Maienschein (University of Pennsylvania Press, 1988), 121–150; Lynn K. Nyhart, "Natural History and the 'New' Biology," in *Cultures of Natural History*, ed. N. Jardine, J. A. Secord, and E. C. Spary (Cambridge University Press, 1996), 426–443; Garland E. Allen, "The Changing Image of Biology in the Twentieth Century," in *The Changing Image of the Sciences*, ed. Ida H. Stamhuis, Teun Koetsier, Cornelis de Pater, and Albert van Helden (Kluwer Academic Publishers, 2002), 43–83; Robert E. Kohler, *Landscapes and Labscapes: Exploring the Lab-Field Border in Biology* (University of Chicago Press, 2022); Bruno J. Strasser, "Collecting Nature: Practices, Styles, and Narratives," *Osiris* 27.1 (2012): 303–340.

14. Margaret Rossiter, "The Organization of Agricultural Improvement in the United States, 1785–1865," in *The Pursuit of Knowledge in the Early American Republic: American Scientific and Learned Societies from Colonial Times to the Civil War*, ed. Alexandra Oleson and Sanborn C. Brown (Johns Hopkins University Press, 1976), 292–296; Sorensen, *Brethren of the Net*, 60–91.

15. F. M. Webster, "One Hundred Years of American Entomology," *Thirteenth Annual Report of the Entomological Society of Ontario, 1899* (L. K. Cameron, 1900), 32–41; L. O. Howard, "The Rise of Applied Entomology in the United States," *Agricultural History* 3.3 (July 1929): 131–139; H. B. W., "Early Feminine Entomologists," *Journal of the New York Entomological Society* 60 (1947): 280.

16. Bergland, *Maria Mitchell and the Sexing of Science*, 53–71, 239–249.

17. Elizabeth Morris to Asa Gray, 18 April 1853, AGCF; Asa Gray to Elizabeth Carrington Morris, 1 January 1848, Box 1, Folder 4, AGP; Elizabeth Morris to Asa Gray, 16 July 1859, AGCF.

18. AGP; Maurita Baldock, Library of Congress, email to author, 6 August 2018; Obituary: John Johnson, *Rutland Herald*, 7 July 2003; Michel-Rolph Trouillot, *Silencing the Past: Power and the Production of History* (Beacon Press, 1995, 2015), 47–53.

19. Harriet Hare Littell, 30 October 1885, Case 1164, Pennsylvania, U.S., Wills and Probate Records, 1683–1993; "Want High School in Germantown," *Philadelphia Inquirer*, 3 April 1912; "To Take Butler Home for New High School," *Philadelphia Inquirer*, 10 May 1913; "School District Acquires Tract," *Philadelphia Inquirer*, 2 October 1913; Atlas of Philadelphia, 1895 (G. W. Bromley & Co., 1895); Edwin C. Jellett, *Germantown Gardens and Gardeners* (Horace McCann, 1914).

20. Minute Book, 7 October 1907, Germantown and Chestnut Hill Improvement Association, GHS; "Germantown Has a Chance," *GT*, 25 February 1910;

"High School Mass Meetings," *GT*, 22 March 1912; "High School Sites Under Inquiry," February 1913, Schools—Public, Germantown High School, Box 3, GHS; "To Build High School on Old Butler Estate," 15 May 1913, JCS, 23; "Butler Site Bought for High Schools," 16 May 1913, JCS, 23; "School District Acquires Tracts," *Philadelphia Inquirer*, 2 October 1913; "Old Mansion to Make Way for High School," Schools—Public, Box 2, GHS; "How High School Got Its Site," 1919, Schools—Public, Germantown High School, Box 3, GHS; Erika M. Kitzmiller, "The Roots of Educational Inequality: Germantown High School, 1907–2011" (PhD diss., University of Pennsylvania, 2012), 17–79; Erika M. Kitzmiller, *The Roots of Educational Inequality: Philadelphia's Germantown High School, 1907–2014* (University of Pennsylvania Press, 2022), 13–27.

21. "Historic Houses of Germantown: XXIII—The Morris-Littell House, 13 February 1914, Historic Buildings, Box 4, GHS; "To Save Trees on High School Site," 13 June 1913, JCS, 23:62; "Want Old Trees Saved," 19 June 1913, JCS, 23: 57; "Would Save Landmark," 20 June 1913, JCS, 23:62; "Want Giant Oak Tree to Remain," 20 November 1913, JCS, 24:113; "Fine Old Oak Tree Doomed," 21 November 1913, JCS, 24:119.

22. The article was reprinted more than two dozen times throughout 1913 and 1914 in states ranging from Mississippi to New York, though most printings occurred in midwestern states like Michigan and Wisconsin. "To Preserve Famous House," *Crete Democrat*, 23 July 1913; *Germantown History, Consisting of Papers Read Before the Site and Relic Society of Germantown* (Site and Relic Society, 1915, v, 3–7; "Would Save Landmark," 20 June 1913.

23. James M. Lindgren, "'A Constant Incentive to Patriotic Citizenship': Historic Preservation in Progressive-Era Massachusetts," *New England Quarterly* 64.4 (December 1991): 594–608; Jennifer B. Goodman and Gail Dubrow, *Restoring Women's History Through Historic Preservation* (Johns Hopkins University Press, 2003); Max Page and Randall Mason, eds., *Giving Preservation a History: Histories of Historic Preservation in the United States* (Routledge, 2004).

24. "Brumbaugh Tells of Germantown's Early Scholars," *Philadelphia Inquirer*, 27 September 1914; "Cornerstone for New High Schools," 4 September 1914, JCS, 27:19; "Cornerstone Laying at Germantown High School," *Evening Public Ledger*, 26 September 1914.

25. JCS, 23:5, 6, 51; 24:58, 127; 25:43; 26:7; 27:3; 28:40, 109, 110, 153, 154, 183; Ellen Carol Dubois, *Suffrage: Women's Long Battle for the Vote* (Simon & Schuster, 2020), 166–204; Martha S. Jones, *Vanguard: How Black Women Broke Barriers, Won the Vote, and Insisted on Equality for All* (Basic Books, 2020), 149–174.

26. Dr. Naaman H. Keyser, C. Henry Kain, John Palmer Garber, Horace F. McCann, *History of Old Germantown* (Horace F. McCann, 1907); "Laying of Corner-stone," 19 September 1914, JCS, 27:19; "Cornerstone Laying at Germantown High School," *Evening Public Ledger*, 26 September 1914; "Cornerstone for New High Schools," 4 September 1914, JCS, 27:19; "To Lay Schools' Cornerstone," 18

September 1914, JCS, 27:19; "History Packed in Cornerstone," October 1914, JCS 27:54; "Clash Over Old House," 14 May 1915, JCS, 28b:159–160; "The Morris-Littell House," 15 May 1915, JCS, 28b:160.

27. "A Botanic Landmark," May 1915, JCS 28b; Morris-Littell House Plaque, GHS.

28. "Clash Over Old House"; "The Morris-Littell House"; 15 May 1915, JCS, 28b:160; Naaman Keyser, Notes for Volume II, *History of Old Germantown*, GHS; Goodman and Dubrow, *Restoring Women's History Through Historic Preservation*.

29. "Clash Over Old House"; "The Morris-Littell House"; "Communicated," May 1915, JCS, 28b:160; "Fail to Save Landmark," 23 July 1915, JCS, 28b.

30. "To Build High School on Old Butler Estate," 15 May 1913, JCS, 23; "Clash Over Old House."

31. Morris-Littell House Plaque, GHS; Board Minutes, 1908–1919, GHS Administration, Box 1; John McA. Harris Jr. to Mary Morris Littell, 30 December 1967; Morris-Littell House, Historic Buildings, Box 4, GHS.

32. Margaret W. Rossiter, "The ~~Matthew~~ Matilda Effect in Science," *Social Studies of Science* 23.2 (May 1993): 325–341; Virginia Scharff, "Introduction," in *Seeing Nature Through Gender*, ed. Virginia Scharff (University Press of Kansas, 2003), xiii–xxii; Katie Holmes and Ruth Morgan, "Placing Gender: Gender and Environmental History," *Environment and History* 27.2 (2021): 187–191.

33. Robert N. Proctor and Londa Schiebinger, eds., *Agnotology: The Making and Unmaking of Ignorance* (Stanford University Press, 2008).

34. Jeff Gammage, "Germantown High School's Emotional Farewell," *Philadelphia Inquirer*, 20 June 2013; Kitzmiller, *Roots of Educational Inequality*, 212–229; Philadelphia Historical Commission, "The Minutes of the 691st Stated Meeting of the Philadelphia Historical Commission," 13 March 2020, 28–36.

35. Asa Gray to Elizabeth C. Morris, 31 August 31, 1845, Box 1, Folder 3, AGP; Elizabeth C. Morris to William Darlington, 15 May 1859, WDP; Edward Tatnall, *Catalogue of the Phænogamous and Filicoid Plants of Newcastle County, Delaware* (Wilmington Institute, 1860), 44; "Books, Catalogues, &c.," *GM* 2.11 (1860): 343–344; Asa Gray, *A Manual of the Botany of the Northern United States* (J. Munroe, 1848), 250–251.

36. In 1992, the geneticist Elizabeth Wagner Reed, herself a marginalized scientist, did extensive research on women who worked in science before the Civil War, logging many of Margaretta's publications, but her self-published book languished in obscurity. Rachel May, "Overlooked No More: Elizabeth Wagner Reed, Who Resurrected Legacies of Women in Science," *New York Times*, 22 April 2023; Sally Gregory Kohlstedt, *The Formation of the American Scientific Community: The American Association for the Advancement of Science, 1848–1860* (University of Illinois Press, 1976), 103; Rossiter, *Women Scientists in America*, 1:76; Sally Gregory Kohlstedt, "In from the Periphery: American Women in Science, 1830–1880," *Signs* 4.1 (Autumn 1978): 85–86; Londa Schiebinger, *The Mind Has No Sex? Women in the Origins of Modern Science* (Harvard University Press, 1989); Cynthia Eagle Russett,

Sexual Science: The Victorian Construction of Womanhood (Harvard University Press, 1989); Sally Gregory Kohlstedt, "Women in the History of Science: An Ambiguous Place," *Osiris* 10 (1995): 39–58; Erika Lorraine Milam and Robert A. Nye, eds., "Scientific Masculinities," *Osiris* 30 (2015); Jenna Tonn, "Extralaboratory Life: Gender Politics and Experimental Biology at Radcliffe College, 1894–1910," *Gender and History* 29.2 (2017): 329–358.

37. L. Rebecca Johnson Melvin, email to author, 28 November 2022.

38. Margot Lee Shetterly, *Hidden Figures: The American Dream and the Untold Story of the Black Women Mathematicians Who Helped Win the Space Race* (William Morrow, 2016).

39. Catherine McNeur, "The Woman Who Solved a Cicada Mystery—but Got No Recognition," *Scientific American*, 9 May 2021; John R. Cooley, email to author, 9 May 2021; John R. Cooley, email to author, 25 April 2022; Harriet Ritvo, "Zoological Nomenclature and the Empire of Victorian Science," in *Victorian Science in Context*, ed. Bernard Lightman (University of Chicago Press, 1997), 334–353; Stephen B. Heard, *Darwin's Barnacle and David Bowie's Spider: How Scientific Names Celebrate Adventurers, Heroes, and Even a Few Scoundrels* (Yale University Press, 2020).

Index

407

pomology. *See* orchards
Pontine Marshes, 111, 137, 273
popular science, 5–6, 10, 20, 151–152,
    155–157, 166–170, 176, 180, 190,
    192–193, 244, 249–250, 269, 305.
    See also lectures, science writing.
potato blight, 216–222
potato curculio (*Baridius trinotatus*),
    217–222, 385n21
potato stalk weevil. *See* potato curculio
Powel, Elizabeth Willing, 32, 35, 38,
    39–41, 43, 65–66, 76
Powel, Samuel, 38
Prior, Richard Alexander. *See* Richard
    Chandler Alexander
Proctor, Robert, 332
professionalization, 11–17
    comparisons with Europe, 16–17,
        26, 51, 55, 112, 125, 142, 244, 296
    and conferences, 146, 179–181,
        195–197, 202–205, 223, 235, 244,
        259–260
    and employment, 12–14, 70, 112,
        94–96, 105, 109–110, 121–122,
        279–280
    and institutions, 12–13, 26–27,
        53–54, 188–189, 296–297
    and marginalization of women,
        11–12, 14–16, 54, 76, 118, 132,
        152, 188–189, 221–222, 257–260,
        262–263, 280, 314, 316, 337
    and memoirs, 8, 315–316
    and publications, 14–15, 54, 56–57,
        87–89, 96, 106, 205
    and terminology, 10–11, 13, 15–16,
        26, 54–55, 93–94, 109–110
    and trust, 8, 93, 109, 113, 119–120,
        134, 146, 206, 221, 248, 251,
        252–255, 261, 281, 332
    See also Academy of Natural
        Sciences, amateurs, American
        Association for the Advancement
        of Science, American

Philosophical Society, masculinity,
    popular science, science writing,
    Smithsonian Institution
Proliferous Rose, 115–116
Pteridomania. *See* ferns

Quakers. *See* Society of Friends

Rafinesque, Constantine Samuel,
    128
railroads, 117, 175, 183, 195, 217, 236,
    258, 266, 291
Reed, Elizabeth Wagner, 404n36
Relief Association of Germantown,
    291–292
religion. *See* Catholicism,
    Episcopalianism, evolution,
    natural theology, Society of
    Friends, Spiritualism
"Reviewer," 161–162, 165, 214, 219
Rogers, Henry Darwin, 281
Rossiter, Margaret, 334–335
Rush, Benjamin, 34–35

St. Louis, Missouri, 269, 279
St. Luke's Episcopal Church, 42, 65,
    285, 308, 333, 354n34
Say, Lucy Sistare, 14, 73, 189, 259, 295,
    326
Say, Thomas, 15, 52–56, 70, 73–74,
    84–88, 95, 102, 108, 189
Schiebinger, Londa, 332
science writing, 6, 9, 12, 57, 89–90,
    151–158, 165–172, 177, 189–190
scientific correspondence networks,
    96, 105–106, 111–117, 119–125,
    139–141, 146–147, 247–249, 266,
    279, 283, 322
scientific equipment and clothing, 3,
    5, 19, 25, 34, 45, 79, 81, 104, 107,
    113, 127, 153, 166, 170–172,
    184, 199, 217, 223, 240, 261, 266.
    See also Wardian cases

CATHERINE McNEUR is an associate professor of history at Portland State University and the author of *Taming Manhattan*. She is the recipient of several awards, including the American Society for Environmental History's George Perkins Marsh Prize. She lives in Portland, Oregon.